$$-\frac{31521}{512}e^3e'\frac{n'^4}{n^4}-\frac{451337}{6144}ee'\frac{n'^6}{n^6}-\frac{17913673}{73728}ee'\frac{n'^7}{n^7}-\frac{435}{128}e^3e'\frac{n'^4}{n^4}+\frac{27}{64}e^3e'\frac{n'^4}{n^4}+\frac{1875}{1024}e^3e'\frac{n'}{n^4}$$

[36 · · · 98] [37 · · 110] [38 · · 20] [39 · · · 108]

$$e'\frac{n'^4}{n^4}+\frac{11463}{256}e^3e'\frac{n'}{n^4}$$

24] [33 · · · 23]

Snowbird, Utah
August 5-7, 1981

$$\frac{3858053}{576}ee'\frac{n'^7}{n^7}-\frac{6}{5}$$

12] [32

$$e^3e'\frac{n'^4}{n^4}+\frac{5569}{1024}ee'\frac{}{n''}$$

$$\frac{473245063}{32768}$$

$$\frac{n'^6}{n^6}-\frac{71501675673}{16777216}e$$

$$\frac{801191314}{12582912}$$

$$\frac{48657}{1024}e^3e'\frac{n'^4}{n^4}-\frac{10}{8192}$$

128] [51

$$\frac{594845}{2048}$$

$$\frac{n'^7}{n^7}+\frac{3561859}{24576}e^3e'\frac{n'}{n^4}$$

[75 · · · 16]

$$\frac{077183}{096}ee'\frac{n'^6}{n^6}+\frac{403}{1638}$$

$$\frac{n'^6}{n^6}-\frac{510431563}{196608}ee'\frac{}{n}$$

16]

Proceedings of the 1981 ACM Symposium on Symbolic and Algebraic Computation

$$\frac{61401}{1024}e^3e'\frac{n'^4}{n^4}-\frac{130}{4096}$$

SYMSAC '81

$$\frac{87}{}\gamma^2e^2+\frac{23}{16}\gamma^2e'^2-$$

$$-\frac{2054241}{32768}$$

$$\frac{9}{1024}e^3e'\frac{n'^4}{}$$

$$\frac{33}{16}$$

$$\frac{0122205}{26214}$$

$$\frac{583}{9216}ee'$$

$$\frac{2685993}{4608}$$

$$\frac{0959}{128}$$

[59 · ·

$$\frac{357}{24576}ee'$$

$$\frac{62945}{536}ee'$$

$$\frac{49661}{1024}e$$

Paul S. Wang, Editor, Kent State University

$$e'\frac{n'^4}{n^4}-\frac{661}{10}$$

[67 · · ·

$$\frac{7436965}{98304}e^3e'\frac{n'^4}{n^4}-\frac{10430755}{16384}e^3e'\frac{n'^4}{n^4}+\frac{219278541689}{9437184}ee'\frac{n'^6}{n^6}+\frac{240769601998o3}{113246208}ee'\frac{n'^7}{n^7}\frac{9931}{10}$$

[42 · · · 134] [43 · · · 118] [41 ·

$$+\frac{38206425}{16384}e^3e'\frac{n'^4}{n^4}+\frac{40477898939}{1572864}ee'\frac{n'^6}{n^6}+\frac{2803111543447}{18874368}ee'\frac{n'^7}{n^7}-\frac{1197}{256}ee'\frac{n'^7}{n^7}-\frac{1995}{256}ee'\frac{n'^7}{n^7}$$

[57 · · · 7] [58 · · 2] [58 · ·

ACM ORDER No. 505810

The Association for Computing Machinery
1133 Avenue of the Americas
New York, New York 10036

ISBN O-89791-047-8

Additional copies of these Proceedings may be ordered prepaid from:

ACM Order Department
P.O. Box 64145
Baltimore, MD 21264

Prices:
Members $17.00 prepaid
All others $23.00 prepaid

ACM Order No. 505810

DEDICATION

William A. Martin
(1938-1981)

Professor William A. Martin of MIT, died on June 2, 1981, following a long illness. Bill was closely associated with algebraic manipulation and SIGSAM in its early years, primarily during the period 1963-1972. As a graduate student in the Electrical Engineering Department at MIT between 1963 and 1967, he wrote his doctoral thesis entitled "Symbolic Mathematical Laboratory". His "laboratory" still possesses what is probably the best automatic expression display routines. It also had a provocative probabilistic approach to expression simplification. The power of such probabilistic techniques was not fully appreciated until recently.

This thesis and related work by Joel Moses and Carl Engleman led Martin to the concept of a large and powerful symbolic manipulation system, combining many diverse techniques and representations. Martin was in charge of the development of this system, eventually called MACSYMA, between 1968 and 1972. It is clear that without his vision and leadership MACSYMA would not have existed.

After 1972, Bill Martin changed his field of interest to automatic programming and finally to natural language understanding by computer. He was always interested in large, practical systems. In fact, he was among the first to popularize the notion of knowledge-based systems.

Bill was undaunted by the shear magnitude of a problem such as natural language understanding. Though adept at dealing with grand theories of the world, he would not shy away from the details that reality and practicality required. His perseverance would never allow him to take easy or hollow paths. If a problem required the handling of 200 special cases, they would be dealt with thoroughly, not dismissed as being unimportant nor relegated to future research.

This volume is dedicated to the memory of William A. Martin, a friend of SIGSAM and one of the pioneers in the field of algebraic manipulation.

SIGSAM Chairman:
Richard D. Jenks, IBM Research, USA

Conference Chairman:
B. F. Caviness, Rensselaer Polytechnic Institute,
and GE Research and Development Center, USA

Program Committee Chairman:
Arthur C. Norman, University of Cambridge, England

Program Committee Members:
Richard J. Fateman, University of California, Berkeley, USA
Ruediger Loos, Universitaet Karlsruhe, West Germany
Maurice Mignotte, Université Louis Pasteur, France
Edward W. Ng, Jet Propulsion Laboratory, USA
Michael Rothstein, Kent State University, USA
Charles Sims, Rutgers University, USA
Richard Zippel, Massachusetts Institute of Technology, USA

Proceedings Editor:
Paul S. Wang, Kent State University, USA

Publicity:
James H. Griesmer and Richard D. Jenks, IBM Research, USA

Local Arrangements:
Martin L. Griss, University of Utah, USA

Treasurer:
M. A. Hussain, GE Research and Development Center, USA

PREFACE

SYMSAC '81 is the continuation of a well-established series of conferences sponsored or co-sponsored by ACM SIGSAM for the presentation of the latest advances in the field of symbolic and algebraic computation. The program consists of three invited talks, an invited banquet address, the presentation of thirty-nine submitted papers, two one-hour poster sessions, and demonstrations of various computer algebra systems. In addition the Program Committee made a conscious effort to provide ample time and facilities for the informal exchange of ideas among the attendees.

The planning for the meeting was initiated by Dick Jenks, SIGSAM Chairman, who has continued to play a guiding role. Moreover, he has handled many of the details necessary in putting together a meeting of this scope. The Program Committee did a yeoman's job of reading and evaluating many papers in a short period of time. Jim Griesmer, much more than his official role of Publicity Chairman would indicate, has been especially helpful in settling many aspects of the meeting. Paul Wang has worked hard to make the Proceedings both attractive and valuable. Martin Griss, as Local Arrangements Chairman, has attended to numerous details that can only be done by someone on the scene. Moayyed Hussain handled our accounts conscientiously. Helpful support has come also from the staff of both ACM Headquarters and the Snowbird Ski and Summer Resort.

It is a pleasure to acknowledge the contributions of all these persons, the entire organizing committee, and especially those persons who submitted manuscripts for consideration by the Program Committee. Unfortunately, due to the time constraint of three days without parallel sessions, many interesting papers could not be accepted. Finally we wish to acknowledge the generous financial support of the U.S. Army Research Office* which enabled us to provide the outstanding slate of invited speakers.

B. F. Caviness	A. C. Norman
General Chairman	Program Chairman
	June, 1981

* The views, opinions, and/or findings contained in this report are those of the author(s) and should not be construed as an official Department of the Army position, policy, or decision, unless so designated by other documentation.

TABLE OF CONTENTS

WEDNESDAY, AUGUST 5

8:45 - 9:00

THURSDAY, AUGUST 6

9:00 - 10:20

Session 5. Algorithm Implementations
Chairman: E. Goto, U. of Tokyo

10:20 - 10:50 **Coffee**

10:50 - 12:30

Session 6. Performance Issues
Chairman: S. J. Harrington, SUNY Brockport

12:30 - 2:00 **LUNCH**

2:00 - 4:00

Session 7. Linear Algebra Algorithms
Chairman: A. C. Hearn, RAND Corporation

4:00 - 6:00 **POSTER SESSIONS 1 AND 2, and SYSTEM DEMONSTRATIONS**

7:00 **SYMPOSIUM BANQUET**

FRIDAY, AUGUST 7

9:00 - 10:20

Session 8. Groups, Rings and Algebras
Chairman: H. J. Zassenhaus, Ohio State U.

10:20 - 10:50 **Coffee**

10:50 - 12:30

Session 9. Polynomials and Rational Functions
Chairman: M. Rothstein, Kent State U.

12:30 - 2:00 **LUNCH**

2:00 - 4:00

Session 10. Semi-Symbolic Computation
Chairman: H. Khalil, U. of Delaware

The Basis of a Computer System for Modern Algebra

John J. Cannon
Department of Pure Mathematics
University of Sydney

§1. Introduction

So-called general purpose systems for algebraic computation such as ALTRAN, MACSYMA, SAC, SCRATCHPAD and REDUCE are almost exclusively concerned with what is usually known as "classical algebra", that is, rings of real or complex polynomials and rings of real or complex functions. These systems have been designed to compute with elements in a fixed algebraic structure (usually the ring of real functions). Typical of the facilities provided are: the arithmetic operations of the ring, the calculation of polynomial gcd's, the location of the zeros of a polynomial; and some operations from calculus: differentiation, integration, the calculation of limits, and the analytic solution of certain classes of differential equations. For brevity, we shall refer to these systems as CA systems.

Up until about 1900, the various algebraic objects that had been studied were based on the real or complex numbers. However, the work of such people as Cayley in group theory, Weber in fields, Wedderburn in linear associative algebras, and Noether in ring and ideal theory, established the notion of abstract groups, fields, division algebras, rings and ideals. This type of algebra has come to be known as modern algebra. With this in mind I define a modern algebra system (or MA system) to be an algebra system which enables a user to compute in a wide range of different structures. Within reasonable bounds, an MA system should permit a user to define the structure in which he wishes to compute. To illustrate the distinction that I am attempting to draw between CA systems and MA systems, consider the problem of solving the system of linear equations

$$\Sigma_{j=1}^{n} a_{ij}x_j = b_i, \quad i = 1,\ldots,m .$$

where the a_{ij} and b_j are elements of a field K. In a CA system the user has to be satisfied with the one or two choices for K provided by the system (probably the rational field). In a true MA system the user would have a great deal of choice in specifying K. Thus, for example, he could take K to be a Galois field or a finite algebraic extension of the rationals.

A further distinction that can be made between CA systems and MA systems is that, whereas in CA systems the interest is in performing calculations with individual elements, in MA systems it is desirable to have the capability to compute some global properties of an algebraic structure. Here, I have in mind such things as the centre of a group, the radical of a ring, or a system of fundamental units for an algebraic number field.

To summarise then, a CA system works at the element level (locally) of an enormously rich but fixed algebraic structure, while an MA system allows the user to define the structure in which he wishes to work (subject to limitations imposed by the implementation). Further, an MA system will often provide tools which can be used to investigate a structure globally.

§2. Issues in the Design of MA Systems.

While there is overlap in the problems faced by the designers of the two types of algebraic system, the designer of an MA system is faced with some novel problems. Chief among these must be the types of algebraic object (group, ring, field etc) permitted under the system, and, within a particular type of algebraic structure, the possible modes of specification (permutation group, matrix group etc). It is in this area that efficiency/generality trade-offs must be made. While it is possible, in principle, to provide the user with machinery sufficiently powerful to enable him to define and calculate in almost any algebraic structure, this would be achieved at the cost of unacceptable inefficiency. An alternative approach is to look at the ways in which algebraists describe structures of a particular type, and, from among these, choose the ones that appear to have the greatest versatility.

It is also necessary for an MA system to permit the coexistence of several distinct structures. For example, in group theory it is a standard technique when investigating a group G which is too complex to be investigated directly, to examine various homomorphic images onto less complex groups. Structure-preserving mappings play such an important role in abstract algebra that an MA system without some abilities to work with mappings would be seriously deficient.

I summarise below the more important issues in the design of an MA system:

(i) What kinds of algebraic structure will be included?

(ii) How will these various types of structure be represented?

(iii) How should simultaneous computation in distinct structures be organized?

(iv) How are sets of elements of these structures to be represented?

(v) What kind of mappings are to be permitted and how are they to be represented?

It should be noted that there are two parts to each of the above representation problems: What is the appropriate syntax to describe the object (in the user language of the system); and how is the object to be represented internally in the computer?

§3. The MA System Cayley

Over the past eight years I have been engaged in the development of what I believe to be the first serious attempt to build an MA system (Cannon [1,2,3]). This system is a package primarily intended for computing with discrete groups. Although it contains some facilities for working with infinite groups, the system is heavily biased towards finite groups. The user can specify a group either as a permutation group, as a group of matrices (over a finite field or the integers), or by a finite presentation. Apart from providing for element calculations, the system enables the user to compute such global information as conjugacy classes, normal subgroups, derived series, Sylow subgroups, centralizers and normalizers.

Since the study of matrix groups naturally involves the study of their actions on vector spaces and modules, Cayley allows the user to define and compute in various types of vector space and module.

As a system, Cayley comprises a language (see Cannon [2] for an early draft of the language), an interpreter, and a large library containing implementations of group theoretic and other algebraic algorithms. Apart from code written in Sydney, the library contains implementations of group theory algorithms produced by Neubüser's team in Aachen and Newman's team in Canberra.

Although Cayley was conceived as a group theory package, the fact that it allows computation in certain types of ring, field, vector space and module as well as in groups, means that the lessons learnt with Cayley have general applicability to the design of MA systems for areas other than group theory. In the remainder of this paper we discuss the principle features of the Cayley design.

§4. The Specification of Algebraic Structures

4.1. Groups

A basic decision taken early in the Cayley development was that the system should be powerful enough to compute with groups sufficiently large and complex to be relevant to contemporary research in finite group theory. For example, the classification of finite simple groups has thrown up computational problems involving permutation groups having degrees exceeding 1000, and, occasionally, very much larger degrees (Sims [5,6]).

In the most general sense a group is specified by defining a set, and a rule of composition for the elements of the set. The implementation of this general concept presents serious difficulties. Further, element multiplication would be extremely slow. For these reasons it was decided to restrict groups to those types which are widely employed and for which reasonable computation techniques are known. At present Cayley allows groups to be specified in the following ways:

(i) As a subgroup of the symmetric group Σ_n, for some integer n. In practice, n is restricted to around 10,000.

(ii) As a subgroup of the general linear group GL(n,K), where n is the dimension and K is an integral domain. Currently, Cayley allows K to be a finite field or the ring of integers. It is not difficult to extend the possibilities for K.

(iii) As a finite presentation by generators and relations.

4.2 Syntax for Groups

The definition of a permutation or matrix group is done in two distinct steps. The first step defines the *type* of permutation or matrix group with which we wish to work. The second step defines the actual permutation or matrix group by listing a set of generators for the group. It should be noted that for certain problems, the first step alone suffices. For example, it is sufficient to specify the group type if it is merely wished to form some products of particular elements.

The following two group specifications will serve to illustrate the flavour of the Cayley syntax.

```
(i)  G : PERM(8);
     G.GENERATORS: A = (1,2,3,4,6,7,8),
                   B = (2,3,5,7,8,6,4);
```

```
(ii) G : MATRIX(3,GF(2));
     G.GENERATORS: X = (1,0,0, 1,1,0, 0,1,1),
                   Y = (1,1,0, 0,1,1, 0,1,0);
```

Both examples define G to be the simple group PSL(2,7): the first defines it as the subgroup of Σ_8 generated by the permutations (1,2,3,4,6,7,8) and (2,3,5,7,8,6,4), while the second example defines it as the subgroup of GL(3,K), where K is the field of two elements, generated by the pair of matrices

$$\begin{pmatrix} 1 & 0 & 0 \\ 1 & 1 & 0 \\ 0 & 1 & 1 \end{pmatrix}, \quad \begin{pmatrix} 1 & 1 & 0 \\ 0 & 1 & 1 \\ 0 & 1 & 0 \end{pmatrix}.$$

As with permutation and matrix groups, the specification of a finitely presented group involves two steps. The first step lists the r generator symbols and the second gives the actual defining relations. In effect the first step defines the free group of rank r. We give the Cayley statements needed to define PSL(2,7) as the finitely presented group

$$\{R,S \mid R^3 = S^3 = (RS)^4 = (R^{-1}S)^4 = E\}$$

```
G : FREE(R,S);
G.RELATIONS : R↑3 = S↑3 = (R*S)↑4 = (R↑-1*S)↑4 = 1;
```

4.3 Fields

I shall confine myself to finite algebraic extensions of a field K. Let K(α): K be a simple algebraic extension of the field K, where α has minimum polynomial m(x) over K. Then any element of K(α) has a unique representation in the form p(α), where p is a polynomial over K and the degree of p(x) is less than the degree of m(x).

Further, the prime subfield of an arbitrary field K is isomorphic either to the field Q of rationals or the field Z_p of integers modulo a prime number p. Thus, calculation in finite algebraic extensions of either of these fields reduces to calculating with polynomials. Here it is desirable to have access to the polynomial facilities provided in most CA systems.

Finite fields present few difficulties. If the required finite field is small (i.e. having at most a few thousand elements) Cayley computes the table of Zech logarithms for the field.

At the time of writing no machinery has been provided for working with extensions of Q in Cayley. However, in the near future it is planned to implement simple extensions of the rationals. I am loath to implement the general case in view of the fact that field calculations would often be extremely expensive.

The extension field K(a) of K is defined in Cayley using a statement of the form

```
L : FIELD(K,a,x,m(x))
```

where m(x) is the minimum polynomial of a over K. (It is assumed that the field K has been defined previously.)

4.4 Vector Spaces

The definition and representation of vector spaces over those fields definable in Cayley present no particular problems. Since all vector spaces of dimension n over a field K are isomorphic, it is convenient to take the space V(n,K) of n dimensional row vectors over K as the standard vector space. The type of vector space can then be specified by stating n and K. If desired, a basis can be supplied either for V(n,K) or for a subspace of V(n,K).

The following Cayley statements define the subspace of V(3,GF(9)) spanned by the vectors $(1,w,w^2)$ and (1,0,w), (where w is a primitive element of the field):

```
K : FIELD(Z(3),w,x,x↑3 + 2*x↑2 + 1);
V : VECTOR SPACE(3,K);
V.BASIS : X = (1,w,w↑2), Y = (1,0,w);
```

Vector spaces are an example of a structure which can be acted on by other structures. Thus, in Cayley, subgroups of GL(n,K) are permitted to act on V(n,K).

§5. Representation of Algebraic Structures

In the previous section the various types of algebraic structures that are specifiable in Cayley were outlined. In most cases the machine representation of their elements should be clear to the reader. How are the structures themselves to be represented in a computer? It might seem that having laid down rules as to the kind of structures permitted (e.g. permutation groups), the problem has already been settled. The problem, however, is more complex than this. I shall discuss the problem in terms of permutation groups. For certain computations with a permutation group, such as evaluating words in the generators, or testing whether the group is abelian, knowing the generators alone suffices. In order to answer deeper questions about the group, such as, for example, determining its conjugacy classes of elements, one needs to have available the set of its elements in some form. At this stage one might simply conclude that it is necessary to create and store the complete list of elements of the group. However, if one analyzes how the set is actually used, it becomes clear that a large amount of group theoretic computation can be carried out provided that the following operations are available:

(i) Determine the order of G;

(ii) Determine if an element of the symmetric group Σ_n is in G;

(iii) Run through the elements of G without repetition.

In [4], Sims introduced the notion of *base* and *strong generating set* and used them to give a very elegant solution to this problem for permutation groups.

The great majority of permutation group algorithms have been designed to utilize the strong generating set concept. Two points should be made here: firstly, the cost of obtaining a strong generating set is often high compared to the cost of subsequent computation (e.g. computing a centralizer); and secondly, in a typical session computing with a permutation group, one tends to deploy a series of algorithms that assume the availability of a strong generating set for the group. It is therefore clear that once a strong generating set has been calculated for G, then it should be retained throughout the computation.

In Cayley I have adopted the philosophy of saving not only such critical information as strong generators, but also other information that may be useful in subsequent computation. Thus, for example, if the classes are computed for any reason they will be automatically saved. Any subsequent calculation requiring the conjugacy classes will check to see if they are already known before attempting to construct them.

It is necessary then to keep track of the information that has accumulated for an algebraic structure. This is done by means of a *structure table*, one of which is associated with each algebraic structure stored in the machine. This table stores such information as:

(i) Type of structure;

(ii) Definition of structure;

(iii) Method of representing individual elements;

(iv) Representation of the set of elements (e.g. strong generating set plus associated information in the case of a permutation group).

(v) The properties and structural information already computed which could be useful in subsequent calculations.

A Cayley computation, then, can be viewed, from the point of view of the machine, as the construction of a rather complex data structure. As new facts are discovered about the object, they are added to this data structure. As the computation progresses it is expected that at least some of the users' requests can be met by doing little more than extracting information from this data structure. The construction of this data structure is invisible to the user.

It is necessary to distinguish between information that clearly belongs to a single structure (e.g. classes of a group) from information whose definition involves two or more structures (e.g. coset table of a subgroup). In the course of computing with an algebraic structure A, it is usual to generate substructures of A and homomorphic images of A. The relationship between A and a substructure or homomorphic image of A must be noted, and there must be provision for the storage of information common to a pair of structures related in such a manner. In Cayley this is achieved by means of a *structure relationship table*.

Finally, given a request from a user, the system must have suitable machinery which matches the desired information with what is currently known and then produces a "micro program" to compute the answer as economically as possible.

I conclude this section by summarizing the issues raised:

(i) Very compact methods for representing the set of elements of certain types of group are known.

(ii) Having computed certain information about an algebraic structure, it is often desirable to retain it so as to avoid subsequent recalculation.

(iii) An algebraic structure is stored in the machine by means of a structure table. This table stores not only the definition of the structure but also what is known about the structure and its whereabouts.

(iv) Information common to a pair of related structures is stored in a structure relationship table.

(v) Sophisticated machinery is necessary to properly utilize the data structure which is constructed during a computation session.

Acknowledgement

This work was supported by a grant from the Australian Research Grants Committee.

References

1. J.J. Cannon: A general purpose group theory program. Procs. 2nd Internal. Conf. Theory of Groups. Lecture Notes in Math., Vol. 372, Springer, Berlin, 1974, 204-217.

2. J.J. Cannon: A draft description of the group theory language Cayley. SYMSAC'76. Proceedings of the 1976 ACM Symposium on Symbolic and Algebraic Computation. Association for Computing Machinery, New York, 1976, 66-84.

3. J.J. Cannon: Software tools for group theory. Proceedings of a Symposium in Pure Mathematics, 37, American Mathematical Society, Providence, R.I. 1981.

4. C.C. Sims: Computational methods in the study of permutation groups. Computational Problems in Abstract Algebra (Proc. Conf., Oxford, 1967). Edited by J. Leech, Pergamon, Oxford, 1970, 169-183.

5. C.C. Sims: Some group-theoretic algorithms. Topics in Algebra. Lecture Notes in Math., Vol. 697, Springer, Berlin, 1978, 108-124.

6. C.C. Sims: A method for constructing a group from a subgroup. Topics in Algebra. Lecture Notes in Math., Vol. 697, Springer, Berlin, 1978, 125-136.

A Language for Computational Algebra

by
Richard D. Jenks
Barry M. Trager
Mathematical Sciences Department
IBM Research Center
Yorktown Heights, New York

ABSTRACT. **This paper reports ongoing research at the IBM Research Center on the development of a language with extensible parameterized types and generic operators for computational algebra. The language provides an abstract data type mechanism for defining algorithms which work in as general a setting as possible. The language is based on the notions of *domains* and *categories*. Domains represent algebraic structures. Categories designate collections of domains having common operations with stated mathematical properties. Domains and categories are computed objects which may be dynamically assigned to variables, passed as arguments, and returned by functions. Although the language has been carefully tailored for the application of algebraic computation, it actually provides a very general abstract data type mechanism. Our notion of a category to group domains with common properties appears novel among programming languages (cf. image functor of RUSSELL) and leads to a very powerful notion of abstract algorithms missing from other work on data types known to the authors.**

1. Introduction.

This paper describes a language with parameterized types and generic operators particularly suited to computational algebra. A flexible framework is given for building algebraic structures and defining algorithms which work in as general a setting as possible. This section will be an overview of our main concepts: "domains" and "categories". In section 2, we give more precise definitions and some examples. Section 3 contains our conclusions.

©1981 ACM 0-89791-047-8/81-0800-0006 $00.75

A language for computational algebra should be able to express algorithms for dealing with algebraic objects at their most natural level of abstraction. We can illustrate this concept with two simple algorithms. First, we wish to write a function *max* which computes the maximum of two elements of any set on which an ordering predicate is defined. One approach to this problem is to explicitly pass the ordering predicate as an additional argument to max. Thus max might be defined by:

max(x,y,lessThan)==**if** lessThan(x,y) **then** y **else** x

where lessThan is the ordering predicate. In more complicated algorithms, the number of additional arguments required gets out of hand. Our approach is instead to require the arguments x and y of max to be elements of some specific algebraic structure which has a "less than" operation '<' implemented by some function. We will call such algebraic structures **domains**. Thus '<' is a "generic" operation which has different function definitions for different domains. Our definition of max, with suitable declarations, becomes:

max(x,y) == **if** x < y **then** y **else** x

The requirement that a generic operation have a particular name does not characterize its algebraic properties. In the above definition of max, it is implicitly assumed that '<' provides a total ordering on the elements of its domain. To this end, our domains will also have a set of *attributes* which permit a description of the algebraic properties of its operations (e.g., so as to distinquish between totally-ordered and partially-ordered sets).

As a second example, we examine the classical algorithm for computing the *gcd*, the greatest common divisor, of two integers:

gcd(x,y) == **if** x=0 **then** y **else** gcd(y,remainder(x,y))

Although this algorithm was originally intended to be used only on integers, a cursory examination shows that only a few properties of the integers are actually required. In fact, the same algorithm can be used on gaussian integers, polynomials over fields, or any other domain which has an appropriate remainder function. We wish to specify the minimum requirements of an algebraic structure for the gcd algorithm to be applicable. To do this, we introduce a grouping of domains called a **category**, in this case, "the category of Euclidean domains". Any domain of this category will be an integral domain with a generic function *remainder* satisfying two requirements. The first is that remainder(x,y) = x−q×y for some q in the domain (this implies gcd(x,y) = gcd(y,remainder(x,y))). The remainder function must also have the property that the remainder sequence generated by any two elements of the domain always reaches 0 in a finite number of steps. These two requirements are sufficient to guarantee the correctness of our gcd algorithm.

Categories provide a set of required generic operations together with a set of attributes describing the required algebraic properties of these operations. Domains provide specific functions which implement the operations and satisfy the attributes. Thus we may speak of "the category of totally-ordered sets" as the class of all domains which have the above '<' operation with specific algebraic properties, and the "the domain of the integers" as an example of a member of that category since it has a function 'integer<' which provides that operation and satisfies those properties.

Once an abstract algorithm has been written, its author specifies the category of domains to which it applies, either by explicitly listing the operations and attributes it requires or by referencing a predefined category. For example, having defined the category EuclideanDomain ("the category of Euclidean domains"), a complete definition of the gcd function could be written:

gcd(x:R,y:R):R where R:EuclideanDomain ==
if x=0 **then** y **else** gcd(y,remainder(x,y))

Expressions of the form "A:B" are called *declarations*. The arguments x and y of gcd are *declared* to be elements of some domain R which, in turn, is declared to be a member of EuclideanDomain. Declaring "A:B" means "A is a member of B" in the following sense. Declaring a domain to be a member of a category indicates that all the operations of the category are implemented in that domain as functions which satisfy the attributes of the category. Similarly declaring an object to be a member of a domain means that all the functions provided in that domain are applicable to the object.

Domains and categories are both computed objects that can be assigned to a variable, passed as arguments, and returned from functions. A category may be produced by explicitly listing operations and attributes, or by invoking a function which returns a category. Categories may be augmented, diminished, or "joined" with other categories to produce a new category containing all of the operations and attributes of the individual categories.

A domain is always created by a function which we call a **functor**. Some simple domains are "the integers" and "the booleans" which are produced by functors of no arguments. Other domains such as "the integers mod 7" are produced by functors which take arguments (such as the modulus 7). Most algebraic domains are built up from other domains which, along with other parameters, are passed as arguments to the functors that construct them. For example, the domain "polynomials in X over the integers" is created by a polynomial functor which takes a variable (e.g. "X") and an underlying domain (e.g. "the integers") as arguments. With the few exceptions noted in section 2, functors and categories are definable in the language and may be freely modified. A user is free to introduce new categories and define new functors in order to make more domains available for computation.

Max and gcd were both defined in terms of generic operations from domains implicitly passed as arguments. As required by some algebraic algorithms, domains are dynamically created and assigned to local variables. Objects of these newly created local domains can be created, manipulated, and converted to objects of other domains.

Both categories and domains may be organized into hierarchies. Figure 1 shows an algebraic hierarchy of categories, listing the operations (but not attributes) introduced by the successive categories. Set is a category with a single operation '='. SemiGroup extends Set by adding an operation '×', etc. More complicated cases will be discussed in Section 2. We also allow one domain to extend another. For example, one can write a "free-module functor" to provide the

Figure 1. Algebraic Categories

category	extends	operations
Set		=
AbelianGroup	Set	0,+,−
OrderedSet	Set	<
QuotientObject(S:Set)	Set	reduce,lift
SemiGroup	Set	×
Finite	Set	size,random
Monoid	SemiGroup	1
Group	Monoid	inv
Ring	(Monoid,AG)	characteristic, recip
Module(R:Ring)	AbelianGroup	scalar×
Algebra(R:Ring)	(Ring,Module(R))	
DifferentialRing	Ring	deriv
IntegralDomain	Ring	isAssociate,//
SkewField	Ring	/
Unique-FactorizationDomain	IntegralDomain	gcd,factor, isPrime
EuclideanDomain	UFD	size<,quo,rem
Field	(ED,SkewField)	
GaloisField	(Field,Finite)	
VectorSpace(S:Field)	Module(S)	

module-theoretic aspects of a polynomial ring (addition and multiplication by scalars). One can then write various polynomial, algebraic-extension, and sparse-matrix functors as extensions of the free-module functor. The polynomial functor, for example, would augment the functions provided by the free-module functor, adding explicit definitions only for the other polynomial functions such as multiplication. Similarly, a "localization functor" can be written to provide computations with quotients of numerators and denominators where denominators may be from a different domain than that of the numerator, such as "the odd integers", "powers of 2", "products of factored polynomials in X". The localization functor is thus a function of two arguments, one for the numerator domain, the other for the denominator domain. A "quotient field functor" can then be written which extends the localization functor for the special case when the two argument domains are the same integral domain. From the quotient field functor, one can produce all of the rational function domains and "the rational numbers" as special cases.

To summarize, our language design provides the useful notions of "domains" and "categories" for the abstract description of algorithms for computational algebra. The facility for categories is unique to our language and its use seems to be invaluable for describing algorithms for computational algebra. Our domains are similar to "modes" in EL1 **[WGB74]** and "types" in RUSSELL **[DMD79]** and other languages. As in EL1, but in contrast to Ada **[ADA80]**, domains are computed values. Our notions of categories and functors are based on concepts in universal algebra **[CHN65]** developed by the ADJ group **[ADJ78]** and Burstall and Goguen **[BRG77]**. In addition, categories extend of the idea of "type constaint" in CLU **[LSK79]** and Alphard **[WLS76]** where functions can require that their arguments have certain specific operations available to them. For related work in computer algebra, see **[ASM79]**, **[GRS76]**, **[JNK74]**, and **[LOS74]**.

2. Concepts.

In this section, we give precise definitions and examples of the concepts of domain, category, and functor.

Domain. By a **domain of computation**, or, simply, **domain**, we mean:

(1) a set of generic **operations**;
(2) a **representation**;
(3) a set of **functions** which implement the operations in terms of the representation;
(4) a set of **attributes**, which designate useful facts such as axioms and mathematical theorems which are true for the operations as implemented by the functions.

The simplest examples of domains are those corresponding to the basic data-types offered by the underlying system, such as Integer ("the integers"), Boolean ("the booleans"), etc. Other examples of domains are RationalNumber ("the rational numbers"), Matrix(Integer) ("rectangular matrices with integer coefficients"), and Polynomial(X,RationalNumber) ("polynomials in X with rational number coefficients")

The generic operations are given by **signatures**, expressions consisting of an *operation name*, a *source*, and a *target*. The domain Integer, for example, has the operation "less than" expressed by the signature: '<': (Integer,Integer) → Boolean with '<' as operation name, (Integer,Integer) as source, and Boolean as target. The source part of the signature is any sequence of domains, and the target part is any domain.

The representation for a domain describes a data structure used to represent the objects of the domain.

The functions component is a set of compiled functions providing a domain-specific implementation for each generic operation. For example, domain Integer has a function "Integer<" which implements "less than". If a domain has an operation signature op:(D1,...,Dn) → D0, then the associated function must take arguments from the representations of D1,...,Dn respectively and return a result in the representation of D0.

The attribute component of a domain is described either by a name, e.g. "finite", or by a form with operator names as parameters, e.g. "distributive('×','+')". The purpose of attributes is not to provide complete axiomatic descriptions of an operation, rather to assert facts which programs can query.

Category. A "category" designates a class of domains with common operations and attributes but with different functions and representations. The categories of interest here will be those of algebraic structures such as Ring ("the class of all rings"), Field ("the class of all fields"), and Set ("the class of all sets").

By a **category** we mean:

(1) a set of generic **operations**;

(2) a set of **attributes**, which designate facts which must be true for any implementation of the operations.

As with domains, the generic operations of categories are given by signatures, consisting of an operation name, a source, and a target. In addition to the domains which may appear in the source and target, a special symbol $ is used to designate an arbitrary member domain of the category. The set of operations and attributes are those which member domains have in common. A simple example of a category is Set, a category which has one operation '=': ($,$) → Boolean and no attributes. Another is SemiGroup, which, besides the '=' operation, has the operation '×': ($,$) → $ and the attribute associative('×').

We say a domain D is "a member of" a category C, equivalently, D is **of** C, if D contains every operation and attribute of C with $ replaced by D. For example, Integer is **of** Set because it contains an operation '=': (Integer,Integer) → Boolean. We say that a category B extends a category A if all of the operations and attributes of A are contained in B. SemiGroup extends Set since all of the operations ('=') and attributes (none) of Set are contained in SemiGroup.

Figure 2. Examples of Category Definitions

Set: Category == **category**
    [*operations*] '=': ($,$) → Boolean

SemiGroup: Category == Set **with**
    [*operations*] '×': ($,$) → $
    [*attributes*] associative('×') [$(x\times y)\times z = x\times(y\times z)$]

Figure 2 illustrates our language for defining categories. Set is defined by explictly listing its operations ('=') and its attributes (none). SemiGroup is defined as an extension of Set. Square-bracketed expressions are comments. The '==' signifies a rewrite-rule definition for the category Set. Evaluation of "Set" causes "Set" to be rewritten by the category value indicated to the right of the '=='. Evaluation of "SemiGroup" similarly causes "SemiGroup" to be rewritten by the corresponding right-hand expression. Further evaluation causes Set to be replaced by its value, a category to which the '×' operation and associative('×') attribute are added by the **with** operation. As implied by this evaluation mechanism, two categories are equivalent iff they have equivalent sets of operations and attributes, irrespective of how they were created.

Figure 3. Examples of Category Definitions

Module(R:Ring): Category==AbelianGroup **with**
    [*operations*] '×': (R,$) → $
    [*attributes*] ...

Algebra(R:Ring):Category==(Ring,Module(R)) **with**
    [*attributes*] ...

Figure 3 gives two examples of parameterized categories, that is, categories that are produced by functions of one or more arguments. The function Module creates the category of all R-modules, that is, modules over a given ring R. For example, the function Module applied to Integer produces the category of all ***Z***-modules, domains D which are abelian groups with the additional operation '×': (Integer,D) → Integer. This category includes domain Integer since Integer is an abelian group and has the operation '×': (Integer,Integer) → Integer. The function Algebra(R) extends the *join* of a Ring and a Module(R), written (Ring,Module(R)). The join designates the category formed by directly combining the operations and attributes of Ring with those of Module(R).

Another way of parameterizing categories is by operator names. For example, the above definition of SemiGroup could be extended to take a binary operation as a parameter:

```
SemiGroup(op): Category == Set with
    [operations] op: ($,$) → $
    [attributes] associative(op) [op(op(x,y),z)=op(x,op(y,z))]
```

after which we may refer to the multiplicative form of SemiGroup in Figure 2 by SemiGroup('×').

Functor. By a **functor**, we mean any function which returns a domain. A functor creates a domain, a member of some category. A category never creates anything; it simply acts as a template for domains, describing which operations and attributes must be present. A functor creates a domain by storing functions into a template given by its target category. Categories never specify representations for objects; functors always do.

Domains can only be produced by functors. Basic domains (e.g. "the integers") are produced through functors bound to identifiers (e.g. Integer). In addition four built-in functors, List, Vector, Struct, and Union, build aggregate domains from other domains passed as arguments. The functor List can be applied to any domain (e.g. Integer) to produce a composite domain (e.g. List(Integer)) with a set of functions which provide operations on lists (e.g. first, rest, cons). The functor Vector takes two arguments, a positive integer n and a domain D, and produces the domain "the set of all vectors of length n with elements from D". Struct produces a domain represented by a set of name-value pairs, e.g. Struct(real:Integer,imag:Integer) describes an appropriate representation for "the Gaussian integers". Union(A,...,B) creates a new domain D from domains A,...,B which is the disjoint union of the domains A,...,B.

The language permits the building of new functors from these basic functors. A simple example is FiniteField in Figure 4. The functor FiniteField applied to p, a prime number, creates a domain "the integers modulo p", a member of its target category GaloisField ("the class of all Galois fields"). The sets of operations and attributes of this domain are given by GaloisField, the representation and set of functions, by the *capsule* part of the definition which appears to the right of the of the '=='. The representation is always defined by the distinquished symbol Rep in terms of a "lower level"

Figure 4. Example of Functor Definition

```
FiniteField(p:PrimeNumber): GaloisField ==
  capsule
    [representation]
      Rep == Integer
    [declarations]
      x,y: $
    [definitions]
      0 == Integer.0
      1 == Integer.1
      x+y == if (w←x Integer.+ y) > p
                 then w−p
                 else w
      ...
```

functor. For FiniteField, Rep is defined to be Integer (meaning that elements of a finite field are represented by integers). In a more complicated example, Rep might be defined in terms of a functor Matrix, whose Rep, in turn, might be defined in terms of the built-in functor Vector.

Figure 5. Example of Functor Definition

```
IntegerMod(m: Integer | m > 1): T == C where
  T == (Ring,Finite) with
           if isPrime m then GaloisField
  C == capsule
      ...
    [definitions]
      if isPrime m then
        x / y == ...
      ...
```

Figure 5 illustrates the use of conditional expressions to make the target category of a functor depend upon the parameters of the functor. Here the FiniteField functor of Figure 4 is generalized to IntegerMod, a functor which produces the domain "the integers modulo m" for any positive integer (modulus) m. The domain produced by the IntegerMod functor will be a Galois field if m is prime, a finite ring, if it is not. Conditional expressions are also used in the capsule part of a functor to conditionally provide functions (e.g. operation '/' will be provided by IntegerMod only if m is prime), or to provide alternate versions of functions (e.g. more efficient implementations for some functions when the modulus is small).

Figure 6. Example of HomomorphicImage

```
QuotientObject(C:Category,D:C) == C with
      [operations]
        lift: $ → D
        reduce: D → $

IntegerMod(m): T == C where
  m: Integer | m > 1
  T == (QuotientObject(Ring,Integer),Finite)
          with if isPrime m then GaloisField
  C == HomomorphicImage(Ring,Integer,reduce,lift)
        where
           reduce(x:Integer):$ ==
              if (x←x mod m) < 0 then x+m else x
           lift(x:$): Integer == x
        add
          [definitions]
           characteristic == m
           x / y == ...
```

Another way to define the IntegerMod functor is by use of another built-in functor, HomomorphicImage. Since the desired domain is a homomorphic image of the integers, most operations can be performed over the integers and then reduced modulo m. The HomomorphicImage functor provides a general mechanism for creating a domain which is defined as a quotient object of an existing domain. This construction requires the existence of a pair of maps called lift and reduce. Reduce is the homomorphism into the quotient object, and lift picks a representative in the initial domain such that lift followed by reduce yields the identity function. The target category of the HomomorphicImage functor is defined by the QuotientObject function in Figure 6. This function takes two arguments, a category C and a domain D which is a member of C, and returns a "quotient object category", the category C augmented by lift and reduce operations. IntegerMod calls HomomorphicImage which extracts the Integer functions whose signatures are given by Ring and returns modified versions which call lift on their arguments and reduce on the results. This single call to HomomorphicImage produces nearly all the functions for IntegerMod. Two exceptions are characteristic and "/" whose definitions must be added separately. The HomomorphicImage functor can also be used for such constructions as producing algebraic extensions from polynomial rings.

Figure 7 illustrates a series of functors for localization which illustrate how domains, like categories, can be extended. Localize takes an R-module M, and a denominator domain D which is a monoid contained within R. It produces an R-module of "fractions". LocalAlgebra augments this with a definition of multiplication for fractions producing the localization of an R-algebra. QuotientField uses LocalAlgebra to produce a "field of fractions" in the special situation where the numerators and denominators both come from the same integral domain. When R has a gcd function, QuotientField redefines the arithmetic operations (supplied by LocalAlgebra) to produce reduced fractions. Similarly if R has a derivation defined for it, QuotientField extends this derivation to the field of fractions.

3. Conclusions.

Our language provides the useful notions of "domains" and "categories" for the abstract description of algorithms for computational algebra. Domains are the algebraic structures on which computation is performed. Categories are groupings of domains with common operations and attributes.

There are several advantages to our design. Algorithms can be written to operate over any group, ring, or field, independently of how that algebraic structure is defined or represented in the computer. The algorithm implementor need not know about which domains have actually been created. Rather, he needs only to specify a category which gives the required operations and essential algebraic properties of his algorithm. Also, as required by many algebraic algorithms, domains and categories are dynamically computed objects.

The language we have presented leads to a computer algebra system which is easily extended by any user. All categories are defined in the language and are available for user modification. All domains are created by functions which, with the exception of a few that are built-in, are also defined in the language and can be changed by the user. New domains and categories can be designed and implemented with minimal effort by extending or combining existing structures.

The language permits considerable code economy. An algorithm is implemented by a single function which is applicable to any domain of a declared category. A matrix functor, for example, will use the same compiled function to compute the product of two matrices,

Figure 7. Definition of Localization Functors

```
Localize(isZeroDivisor,M,D): Module(R) == C where
   R: Ring
   M: Module(R)
   D: Monoid | D ⊆ R
   isZeroDivisor: M → Boolean
   C == capsule ...
             [representation]
                Rep == Struct(num: M,den: D)
             [declarations]
                x,y: $
                n: Integer
                r: R;  d: D
             [definitions]
                0  == (0,1)
                - x  == (-x.num,x.den)
                x = y  == isZeroDivisor(y.den × x.num - x.den × y.num)
                x + y  == (y.den × x.num + x.den × y.num, x.den × y.den)
                n × x  == (n×x.num,x.den)
                r × x  == if r=x.den then (x.num,1) else (r×x.num,x.den)
                x / d  == (x.num,d×x.den)

LocalAlgebra(isZeroDivisor,A,D): T == C where
   R: Ring
   A: Algebra(R)
   isZeroDivisor: A → Boolean
   D: Monoid | D ⊆ R
   T == Algebra(R) with if A has commutative('×') then commutative('×')
   C == Localize(isZeroDivisor,A,D) add ...
             1  == (1,1)
             x × y  == (x.num×y.num,x.den×y.den)
             characteristic == A.characteristic

QuotientField(R: IntegralDomain): T == C where
   T == (Field,Algebra(R)) with if R of DifferentialRing then DifferentialRing
   C == LocalAlgebra($1 = 0,R,R) add ...
             if R has gcd: (R,R) → R then
                x + y ==  ...
                x × y ==  ...
                where cancelGcd(x: $): $ ==  ...
             if R of DifferentialRing then
                if R has gcd: (R,R) → R
                   then deriv(x) == ...
                   else deriv(x) == (deriv(x.num)×x.den-deriv(x.den)×x.num,x.den↑2)

RationalFunction(x: Expression,R: Ring) == QuotientField(Polynomial(x,R))

RationalNumber == QuotientField(Integer)
```

regardless of whether the actual matrix coefficients are integers, polynomials, or other matrices. Parameterized functors help to minimize redundant code by providing a set of pre-compiled functions for all domains they can produce. The universal applicability of such functors as QuotientField and HomomorphicImage provide powerful methods for constructing new algebraic objects.

Our primary goal in presenting a language which deals with algebraic objects was to take advantage of as much of the structure implicit in the problem domain as possible. The natural algebraic notions of domains extending one another, and collecting domains with common properties into categories have been shown to be useful computational devices. By preserving this natural structure, we hope to have eased the task of finding computational models for algebraic structures.

Acknowledgements. This language has evolved from previous designs due to R. D. Jenks and J. H. Davenport (IBM Research and Emmanuel College, Cambridge) **[JNKS79,DAJ80a]**. The original notion of categories and functors is due to Davenport **[DAJ80b]**. Our new design has been strongly influenced by the work on algebraic modes by D. R. Barton (U. of Cal., Berkeley) and by many fruitful interactions with J. W. Thatcher (IBM Research). The authors are also grateful to J. L. Archibald (IBM Research), D. Ehrich (Univ. of Dortmund), D. B. Saunders (RPI), D. Y. Y. Yun (IBM Research), and R. E. Zippel (MIT) for many useful discussions. Earlier drafts of this paper were read by J. L. Archibald, V. S. Miller, and J. W. Thatcher, and we are grateful to them for many helpful criticisms and comments.

Bibliography.

[ADA80] *Reference Manual for the Ada Programming Language,* U.S. Dept. of Defense, July 1980, (reprinted November 1980).

[ADJ78] Thatcher, J. W., Wagner, E. G., and Wright, J. B., "Data type specification: Parameterization and the Power of Specification Techniques", *Proceedings SIGACT 10th Annual Symposium on Theory of Computing,* May, 1978, pp. 119-132.

[BRG77] Burstall, R. M. and Goguen, J. A., "Putting Theories Together to Make Specifications," *Proceedings of the 5th International Joint Conference on Artificial Intelligence,* pp. 1045-1056, August, 1977.

[ASM79] Ausiello, G. and Mascari, G. F., "On the Design of Algebraic Data Structures with the Approach of Abstract Data Types", *Proceedings of EUROSAM 79* (Springer-Verlag Lecture Notes in Computer Science 72).

[CHN65] Cohn, P. M., *Universal Algebra,* Harper and Row, New York, 1965

[DAJ80a] Davenport, J. H. and Jenks, R. D., "SCRATCHPAD/370: Modes and Domains" (privately circulated).

[DAJ80b] Davenport, J. H. and Jenks, R. D., "MODLISP", *Proceedings of LISP '80 Conference,* August, 1980 (also available as IBM Research Report RC 8537, October 29, 1980).

[DMD79] Demers, A., and Donahue, J., *Revised Report on Russell,* TR 79-389 Dept. of Computer Science, Cornell U., September 1979.

[GRS76] Griss, Martin L., "The Definition and Use of Data Structures in REDUCE", *Proceedings of the 1976 Symposium on Symbolic and Algebraic Computation,* Yorktown Heights, New York, August 1976, pp. 53-59.

[JNK74] Jenks, R. D., "The SCRATCHPAD Language," *Proceedings of a Symposium on Very High Level Languages,* SIGPLAN Notices, Vol. 9, No. 4, April 1974 (Reprinted in SIGSAM Bulletin, Vol. 8, No. 2, May 1974).

[JNK79] "MODLISP: An Introduction", *Proceedings of EUROSAM 79* (Springer-Verlag Lecture Notes in Computer Science 72) pp. 466 - 480 (also available is revised form as: "MODLISP: A Preliminary Design", IBM Research Report RC 8073, January 18, 1980).

[LSK79] Liskov, B., Atkinson, R., Bloom, T., Moss, E., Schaffert, C., Scheifler, B., and Snyder, A., *CLU Reference Manual,* TR-225, MIT/LCS, October, 1979.

[LOS74] Loos, Ruediger G. D., "Towards a Formal Implementation of Computer Algebra", *Proceedings of Eurosam '74,* SIGSAM Bulletin, Vol. 8., No. 3, August 1974, pp. 9-16

[WLS76] Wulf, W.A., London, R.L., and Shaw, M., "An introduction to the construction and verification of Alphard programs", *IEEE Trans. on Software Eng. SE-2,4,* pp. 253-265, Dec. 1976.

[WGB74] Wegbreit, B., "The Treatment of Data Types in EL1", *Communications of the ACM,* pp. 251-264, vol 74, no. 5, May 1974.

Characterization of VAX Macsyma

John K. Foderaro[1]
Richard J. Fateman

Computer Science Division
and
Center for Pure and Applied Mathematics
University of California
Berkeley, California

1. Introduction

The algebraic manipulation system Macsyma [Grou77, Fate80] has been running for over a year on Digital Equipment Corp. VAX-11 large-address-space medium-scale computers [Stre78]. In order to run Macsyma in this environment, a Lisp system for the VAX, FRANZ LISP[Fode80], was constructed at Berkeley. The goal of running Macsyma provided direction and motivation and is partially responsible for the rapid development of the Lisp system.

Because Lisp is a high level language there are many decisions to be made about the internal framework of the system. Efforts to increase efficiency require that we be able to characterize the demands of a large, compiled, Lisp system. Fortunately, the VAX/UNIX operating system provides useful tools for determining such characteristics. This paper presents some of our data and related analysis[2].

With a years' experience in running Vaxima (Macsyma on the VAX), we began a study of the characteristics of Vaxima in the FRANZ LISP environment. In Section 2 we discuss the space requirements and storage allocation behavior of Vaxima. We concentrate on the issue of garbage collection, since we believe there are significant benefits to be obtained by a modest effort. Furthermore, no unconventional modifications to a standard machine architecture are required for these benefits. We do not attempt to compare these benefits to those obtained by compaction of data, compression of pointers, hash-links, partial reference counts, and other schemes [Stan80]. For the most part, the proposals of section 2 are consistent with most other recently proposed improvements, such as tagged pointers or byte-code Lisp interpretation or compilation.

[1]Work reported herein was supported in part by the U. S. Department of Energy, Contract DE-AT03-76SF00034, Project Agreement DE-AS03-79ER10358.

[2]All measurements were made on a VAX 11/780 with 2.5 million bytes of main memory running Berkeley's 4BSD paging UNIX system.

In Section 3 we describe our analysis of some computational aspects of Vaxima. These observations are at the architectural level rather than at the level of algebraic manipulation: we will not discuss time/space tradeoffs for alternative algorithms one might find in a symbolic algebra system; rather we will view Vaxima primarily as a black box that feeds on CPU time and pages of memory. Our observations will perhaps be most relevant to the designer of a new architecture: We discuss features that should be considered of prime importance in applications of Lisp such as Vaxima.

2. Storage Allocation Issues

The address space Vaxima uses is divided into three segments. Beginning at address 0 is the 1.8 megabyte (Mby) instruction segment, which is read-only and shared among all users of Vaxima. The instruction segment contains the program portion of the Lisp system as well as many of the compiled Lisp modules which make up Vaxima. Also present in the instruction segment are Lisp data that are to be shared among users.

The 1.4Mby data segment follows the instruction segment in the address space. It contains Lisp data plus storage for the variables used in the Lisp system. In our current configuration it can grow dynamically to 3.3Mby.

The stack segment grows downward from the top of memory. The Lisp system uses this stack primarily to store addresses for function call and return.

Each (512 byte) page of memory can contain only one type of Lisp datum, whose type is described by an entry in a fixed-length array called the typetable. The size of the typetable array determines the maximum virtual memory available for a particular Vaxima system.

2.1. Garbage collection

Perhaps the most important topic for performance efficiency in very large Lisps is the interaction of virtual memory systems with Lisp free-list reclamation or *garbage collection.*

The FRANZ LISP garbage collector is based on a classical mark and sweep algorithm using a stack [Knut68, Stan80] The mark bits for all the data are kept in a bitmap array. The marking phase is easily accomplished by the bit manipulation instructions on the VAX, capable of addressing any bit in memory by a base location plus a bit offset in the range -2^{31} to $2^{31}-1$. Marking a Lisp datum, which includes checking the type of the datum and then setting a bit in the bitmap, costs, on average, 61 microseconds per datum[3].

[3]Marking of a cons cell requires approximately 15 machine instructions: a **calls** to enter the mark routine, a test for address in bounds, type extraction and dispatch on type, marking datum and if it hadn't been marked before, marking car and looping to the top to mark the cdr. Other types may take slightly more or less time to mark

The sweep phase of the garbage collection is responsible for recreating a free list for each data type. Sweeping costs roughly 250 microseconds per page, which, for a page of 64 cons cells, is less than 4 microseconds per datum.

These times do not include the extra penalty Vaxima pays for running in a virtual memory environment. The Berkeley UNIX operating system places all free pages in a single pool from which all programs draw. Therefore the working set size for Vaxima is determined in part by the behavior of all other programs running on the VAX. When a page fault occurs, the operating system may find the page in the free list, in which case only a few tables need be updated and the program can continue. Such a *minor* fault takes only 300 microseconds to service. However, if the operating system must go to the disk to retrieve the page, it is a *major* fault. Major faults commonly take 50,000 microseconds of real time to service.

Because the VAX lacks page reference bits it is expensive for the operating system to determine which of a process's pages are in currently being referenced and which are not [Joy81]. The operating system replaces pages globally with an approximate 'least recently used' policy at a fixed maximum rate of 600 pages/second to reduce overhead. Vaxima references so many pages in such a short time during a garbage collection that it would be better served by a page replacement policy with a higher maximum rate. Therefore when Vaxima starts a garbage collection it tells the operating system that its paging behavior will be abnormal, causing the the operating system to change the page replacement policy for Vaxima's pages to 'first in first out' with an effective page replacement rate of 1200 pages/second. After the garbage collection completes, Vaxima tells the operating system that it will return to normal memory referencing behavior which switches Vaxima back to the 'least recently used' policy and invalidates all of Vaxima's data pages, putting them at the end of the global free page list. The rationale is that after a process leaves abnormal paging mode, its current data working set is not likely to be its normal working set, so it would be best to force the process to rebuild its working set from scratch. To test the effect of telling the operating system when garbage collection was occurring we ran the EIGEN demo[4] under a light load and found that there were 7,392 minor faults and 1,529 major faults. We then ran the same demo under the same load but this time did not tell the operating system when we were garbage collecting. The result was 19,140 minor faults and 2,113 major ones.

2.2. Allocation of cons cells

When a program tries to allocate a Lisp datum and there are none left a garbage collection is automatically started. The type of Lisp datum which runs out most frequently is the cons cell, the basic building block for construction of lists and unbounded precision integers (bignums) in Lisp. Henry Baker notes in Bake77 that the optimal storage allocation scheme would divide the storage among data types in such a way that each data type was equally likely to run out. Our experience shows that after allocating a small amount of storage (ten pages) to the other data types, we can generally allocate the rest of the space to cons cells and the cons cell would still be the first data type to run out. One purpose of cons cells is to act as temporary glue to hold data together. Most cons cells are allocated and then discarded before the next garbage collection, though occasionally one is used for a permanent purpose. For a particular Macsyma demonstration script (the Eigen demo), we calculated that an average of 385 cons cells are allocated and quickly discarded before a permanent one is allocated. Furthermore the rate at which cons cells are allocated in that demo is an average of one cell per 302 microseconds of CPU time. This means that one permanent cons cell is allocated every 302*385= 116 milliseconds. The current Vaxima has room for 336,054 cons cells. To fill that up with permanent cons cells at that rate would take at least 10 CPU hours assuming that garbage collection was free. Of course considering the behavior of the garbage collection algorithm, it would be foolish to try to *completely* fill up that space this way since the dwindling returns of the later garbage collections make the cost-per-cell-reclaimed very high. If we were to expand Vaxima to the currently configured operating system limit, 8Mby, then there would be a potential of 726,000 cons cells and the same calculation reveals that it would take 23 hours to fill it up given no cost for garbage collection.

[4]descriptions of all demonstration scripts are given at the end of the paper

2.3. Allocation of bignums

Another behavior the Vaxima system exhibits occurs when calculations are done with numbers which must be represented by more than one machine word. These numbers, called bignums, are represented internally as a sequence of cons cells with a special cell called an *sdot* at the head. The sdot and cons cells in a bignum each contain 30 bits of immediate (numeric) data and 32 bits of pointer data. FRANZ LISP originally represented bignums as sequences of just sdots but this meant that when a calculation began to use bignums heavily the storage allocator would quickly have to allocate a large number of sdot pages to prevent Vaxima from spending most of its time garbage collecting. Once the bignum calculations were finished, those extra sdot pages would be useless yet would still have to be swept during each gc[5]. Thus we decided that since cons and sdot cells were the same size we would use cons cells to store the majority of the bignum and put an sdot cell on the head of the list to differentiate a list from a bignum. This scheme is also used in MacLisp [Stee77]

In order to measure the characteristics of a bignum-intensive calculation, we ran a Vaxima program which computed of the coefficients of the continued fraction expansion of the Binet function [Char80] The results of our analysis showed a markedly different behavior for this type of calculation than for the the type described in the previous section. There was one permanent cons for every 37 allocated and one cons was allocated every 310 microseconds. This implies that one permanent cons is created every 11.47 milliseconds and it would take slightly more than one hour of CPU time to fill the address space of our current Vaxima (assuming again no cost for garbage collection). With an 8Mby Vaxima, it would take 2.31 hours to fill up memory, but as garbage collection times increased one would be likely to give up before the memory was full. It is this type of calculation, where large results are saved for later use, which provides a challenge to the capabilities of the Vaxima system. If the objective of the user is to repeatedly perform such computations, we believe the correct approach to alleviating the total system performance problems would be to consider alternatives to garbage-collection based environments.

[5] FRANZ LISP does not look for totally empty pages and return them to the free page list. Once a page is allocated to a type, it stays allocated forever.

2.4. Improved GC algorithms

In studying this data, we came up with two possible methods of decreasing the amount of time taken in our mark and sweep garbage collection:

a) not clearing the bitmap array before beginning the marking phase; in other words remembering which data was active at the end of the previous garbage collection. This will do two things: eliminate the time it takes to clear the bitmap array (not very much time), and decrease the amount of time in the marking phase (61 microseconds per mark).

b) not sweeping all of the pages. This saves 250 microseconds for each page not swept.

Before proceeding to evaluate these options, we must introduce some notation. A cell is *alive* (or is *living*) after a garbage collection if it could be referenced by a Lisp program. A cell is *dead* otherwise. If a cell was dead after a gc and is alive after the next gc then it has been *born*. A cell *dies* (or exhibits a *death*) if it goes from alive to dead that is if it has survived one garbage collection but not the next.

In the classical mark and sweep algorithm it is the marker's job to locate all of the living cells and the sweeper's job to collect all the dead cells. The two phases communicate by means of the bitmap array which the marker sets and the sweeper reads. If we choose not to clear the bitmaps before the mark phase, then we would

i) gain time by not marking all the living cells again: when the marker comes across a cell which is already marked it assumes that it has already marked everything that cell points to.

ii) lose the use of storage by not sweeping up all the cells that have really died because the bitmap indicates that they are still alive.

iii) lose time whenever a structure modification function (such as rplacd) is executed: we must insure consistency of the mark bits. A small mark phase may be instituted after a modification function to accomplish this.

In analyzing the behavior of Vaxima running the 'Eigen' demo we determined that on average, only 1.6% of the dead cells following a garbage collection had just died (the other 98.6% had been dead following the previous garbage collection and in the time between the two garbage collections they had been allocated and discarded). Thus if we cleared the bitmap only before, say, every fifth garbage collection we would eliminate a large amount (perhaps 50%) of marking time with the insignificant and temporary loss of those cells we failed to notice were dead. We have not implemented this scheme in our Lisp system because of the complications of part (iii). It is easy enough to identify and handle structure modification in Lisp code but some of the structure modification is implicit in the parts of FRANZ LISP which are written in the language C.

The second scheme identified for saving garbage collection time is the sweeping of only those pages which are likely to have a large number of dead cells. We ran the 'FG' demo which causes twelve garbage collections and then looked at each page of cons cells to see what percentage of the cells were free. The first row of Table 1 shows the numer of pages which have a certain percentage of cells free (where we've broken the percentage free into bins of ten percent).

Table 1

	percent free									
	0	10	20	30	40	50	60	70	80	90
# pages	149	38	24	36	27	19	14	23	26	206
% recov	100	98	96	94	90	86	83	79	74	66

There are 20,135 free cells in the 562 pages of cons cells allocated, so if there were a great tendency for cells to spread out we would expect to see a peak at 55% free per page. This is obviously not the case: it demonstrates that there is very little tendency for living cons cells to spread out and fill the pages allocated. Clark and Green noticed the same behavior in InterLisp programs [Clar77]. This is precisely the behavior which will make partial sweeping work as a technique to reduce garbage collection time without interfering significantly with the effective collection of garbage. The second row of Table 1 shows how much garbage you would recover if you only swept pages whose percentage free was greater than or equal a certain amount. For example, if you only swept pages whose percentage free is greater than or equal to 80% then you would recover 74% of the garbage by only sweeping 226 (40%) of the 562 pages. We implemented partial sweeping in FRANZ LISP by maintaining count of the last known percentage of dead cells on each page and using that number to decide whether to sweep a page. Periodically all pages are swept so that large numbers of sudden deaths on a page would eventually be recognized. The 'FG' demo mentioned above was run in a Vaxima with partial sweeping enabled and disabled and the results are shown in Table 2.

Table 2: partial sweeping costs

		time (secs)			
		total		per gc	
method	gcs	mark	sweep	mark	sweep
full sweep	12	17.75	6.63	1.47	.55
partial	15	23.5	6.65	1.56	.44

The directions to the partial sweeper in this run were to only sweep pages with more than 62% free cells and to do a full sweep every five garbage collections. As the table indicates, Vaxima spent more time garbage collecting with the partial sweeper than without. We conclude that this is because once the living data size gets large enough so that partial sweeping becomes effective at reducing sweep time, the cost of doing a garbage collection is dominated by the mark cost. For this reason we have disabled partial sweeping in Vaxima until we implement partial marking (as mentioned at the beginning of this section).

3. Execution Speed

In evaluating the efficiency of a Lisp system, another natural area to study is the program representation and the distribution of instruction frequencies.

In fact, it is valuable to measure the frequency of instructions used by programs in two contexts: static and dynamic. Static instruction counts refer to the frequency with which instructions appear in the text of a program. If instructions with higher frequencies can be encoded in fewer bits, the program text will be shorter, and consequently the overall system will benefit in terms of decreased memory requirements and accessing activity. Dynamic instruction counts refer to the frequency with which instructions are executed. That is, one instruction or a sequence of instructions may occur infrequently in the program text, but because of the position in an "inner loop" may be very important as a target for re-implementation in a more efficient way.

These measurements provide valuable information for (i) design and evaluation of computer architectures, and (ii) optimization within a given architecture when there are alternative ways of producing the same computation with different time/space trade-offs. It has been the case in the past that elaborate special-purpose instructions were included in an architecture but never used because compilers did not generate them, or alternatively, very elaborate compilers were constructed to use esoteric instructions yet without significant benefit since they were executed rarely [Alex75, Fost71]. One would hope to provide an appropriate mix of useful, compact, and efficient operations.

3.1. Dynamic analysis

In order to obtain an accurate dynamic analysis of Vaxima, we ran Vaxima one instruction at a time and recorded information on each instruction executed (the programs for doing this were written by Robert Henry [Henr80]). When Vaxima completed running, the data which the instruction monitor had accumulated was dumped out for later analysis. The overhead for this single stepping operation is approximately 100 to 1, which limits our use of this technique to only short runs.

We ran Vaxima on the 'Begin' demo, which normally takes around 16.5 seconds of CPU time and causes one garbage collection. Table 3 lists the ten most executed instructions[6].

The analysis of the data from this demo show that:

1) 27% of the 6,772,984 instructions executed were **movl** instructions, which simply move one longword (4 bytes) of information. It is not very surprising that this is the most executed instruction because, by varying the type of the operands, the **movl** instruction can accomplish all of the tasks done by many different instructions on other machines. By looking at the types of the operands of the **movl** instructions which were executed, we were able to determine that 22% of the **movl** instructions were pushing Lisp data onto the internal Lisp data stack. Most of the 'pushes' were for the purpose of stacking arguments prior to a function call. In FRANZ LISP all arguments to functions are pushed onto the internal stack. We have considered passing arguments in registers to save some of the 420485 'pushes' which were executed in this short demo. We have not done this for a number of reasons. The C language, in which many of the low level Lisp functions are written, does not permit the programmer to explicitly control the registers. Thus if a Lisp-coded Lisp function were to put arguments in registers and call a C-coded Lisp function, an intermediate routine would have to be called to put the registers where the C program could get at them. Also, during garbage collection the marker would have to locate all Lisp data stored in saved register sets on the C runtime stack. With our mixture of C, Lisp, Fortran and assembler code all using the same stack it would be difficult to tell which stack objects were saved Lisp data.

2) 2% (186376) of the instructions executed were **calls**'s and 1% (71336) were **jsb**'s. Both of these instructions are used when one function calls another. The **calls** instruction performs register saving and stack bookkeeping operations before jumping to the function called whereas the **jsb** instruction just stores the return address on the stack and jumps. As expected, the **calls** instruction is much slower than the **jsb** instruction. Stacking two values and calling *cons* with a **jsb** takes 14.2 microseconds (10 machine instructions executed).

[6]the functions of the instructions are described in Section 4.

Table 3: Instruction usage

Static				Dynamic	
C coded Lisp functions		Lisp coded Lisp functions		Begin demo	
Instruction	pct	Instruction	pct	Instruction	pct
movl	20	**movl**	43	**movl**	27
pushl	12	**movab**	9	**cmpl**	7
calls	10	**calls**	7	**bnequ**	6
pushal	7	**brb**	4	**beqlu**	5
cmpl	4	**clrl**	4	**ashl**	5
beqlu	4	**jsb**	4	**movab**	4
bnequ	3	**beqlu**	3	**tstl**	4
brb	3	**bnequ**	3	**cvtbl**	3
ret	3	**tstl**	3	**brb**	3
clrl	2	**brw**	2	**calls**	2
other	32	other	18	other	34
100 Unique instr.		32 Unique instr.		109 Unique instr.	

Calling *cons* with a **calls** which saves no registers takes 29.9 microseconds and calling *cons* with a **calls** which saves all registers takes 40 microseconds.

The compiled Lisp code could work very well with just the **jsb** but again in order to maintain compatibility with the C coded portion of the Lisp system, which uses **calls**, the normal means for one function to call another is with a **calls** instruction. The exceptions to this are four functions which we noticed were called very frequently. These functions were written in assembler language and are callable by a **jsb** instruction. The Lisp compiler knows about these special functions and compiles references to them in a special way. As the counts given above indicate, 28% of the function calls in this demonstration file were of the **jsb** type, which indicates the heavy use of those few functions.

In order to get an idea of just what functions are being called, we used the UNIX profiling system to generate a list of the number of times each function was called.

The most significant findings from profiling the Begin demo are that the two most called functions were the **jsb** callable functions *cons* and *newint* (which is an internal function to return a pointer to an integer cell given an integer).

3) The arithmetic instructions on the VAX were used very rarely. Only 1% (131,188) of the instructions were subl2 (two operand subtract), 107,587 were subl3 (three operand subtract), 78,883 were addl2 (two operand add) and 49,996 were divl3 (three operand divide). This indicates that even in a system which does algebraic mathematics, most of the time is spent moving data around and little time is spent doing arithmetic. In fact, by observing the operands for the arithmetic instructions that were executed, we deduced that most of these instructions were used in the various C-coded utility routines that form FRANZ LISP, such as the garbage collector, storage allocator and I/O manager.
FRANZ LISP does not have the fancy handling of small integers (fixnums) in compiled Lisp code which occurs in some other Lisps (for example MacLisp). The results above indicate that there would be a very small payoff relative to Vaxima in modifying FRANZ LISP and the Lisp compiler to handle small integers in a more efficient way.

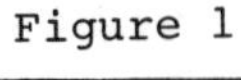

Figure 1

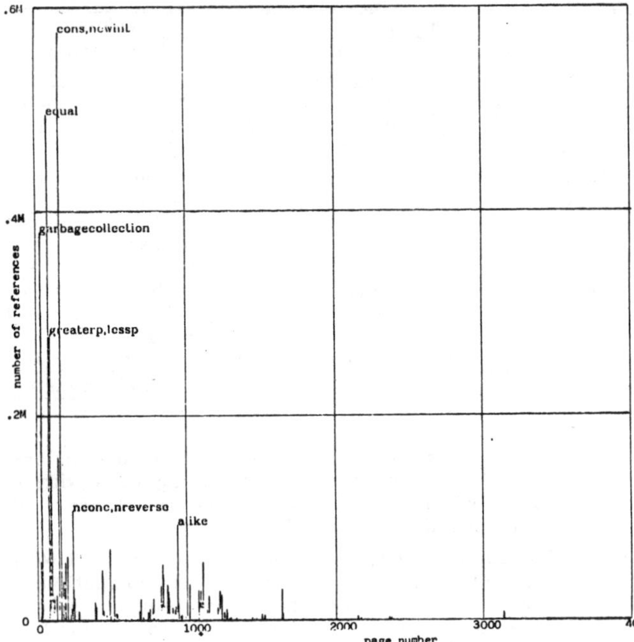

4) Part of the information collected by single stepping is the number of instructions executed on each page. In Figure 1 we have plotted this information with the peaks labeled by the routines on those pages. The text segment of the C coded portion of FRANZ LISP occupies pages 0 to 200, the Lisp coded portion occupies pages 201 to 300 and Vaxima occupies the rest. Approximately 60% of the instructions executed in this demo were executed in the Lisp system and only 40% in the compiled Lisp which makes up Vaxima. This graph emphasizes the importance of the *equal* function to Vaxima. When time permits, the *equal* function will be rewritten in assembler language.

The impact of our choosing a C coded kernel for the Lisp system has been alluded to in the preceding paragraphs but we would like to offer here a more complete discussion of this implementation decision.

In coding a Lisp system, one extreme would be a kernel completely written in the assembler language for the VAX. For a kernel the size of ours, 15,000 lines of C code, this would be a mammoth project, difficult to debug, and would represent a commitment to a specific machine architecture. The opposite extreme would be to write the whole kernel in C without taking advantage of the special characteristics of the VAX. The result would be a very portable system which ran Vaxima too slowly to be interesting. Thus we have stepped back a bit from the totally portable system in those areas where efficiency is critical, such as the garbage collection, storage allocation and multi-precision arithmetic. Where VAX-specific code exists in FRANZ LISP, there are usually comments which describe the portable way of doing the same thing.

There are, however, a variety of sacrifices necessary with our approach. As mentioned above, within C we have insufficient control of the runtime environment to use the registers as much as we could. When a garbage collection occurs, all active Lisp data must be located, and finding Lisp data in the sets of saved registers on the main stack would be very expensive in our system. Thus rather than use registers for passing arguments and storing temporary values within a function, we use a separate stack which contains only valid Lisp data. Although the Lisp compiler does make use of the registers for storing certain temporary values it assumes that none of the temporary registers are preserved through function calls. Another apparent disadvantage of using a C kernel is that the Lisp system has to maintain a function calling sequence compatible with C. While it would be possible to use a **jsb** calling sequence in Lisp and use an intermediate function to convert to the C **calls** calling sequence, this would, we believe, be more costly overall than the current arrangement, as long as many common functions are C-coded. Furthermore, in most cases the difference between **calls** and **jsb** plus the necessary additional stacking operations is swamped by computations internal to the called routine.

Fitch and Norman faced similar problems when they decided to write a Lisp system for an IBM 370 in BCPL [Fitc77]. They were pleased with how quickly they had a reliable and powerful Lisp system which ran only 25% slower than an assembler coded Lisp system on a similar machine. We look forward to the opportunity to compare FRANZ LISP with other Lisp systems of comparable power written with other approaches. This will aid us in judging how much we are paying in execution speed and code size for writing the system in C. Note that even if all **movl**'s **calls**, and **ret**'s were to take no time to execute the 'begin demo' would run only about twice as fast as it does now (estimating a **calls** and **ret** take as much time as ten **movl**'s). In terms of design time we feel that we took the course which resulted in the minimum time to create a working Lisp system.

3.2. Static analysis

Table 3 lists the ten instructions which occur most frequently in the C-coded FRANZ LISP system and nineteen Vaxima files containing Lisp-coded Lisp functions. There is a great deal of similarity between the types of instructions generated by the C and Lisp compilers. The **pushl** instruction, which is used a great deal in compiled C code, has the same effect as a **movl** instruction whose second operand is the main stack. Using **pushl** instead of **movl** saves one byte since the destination operand of **pushl** is implied. The Lisp internal stack is different from the main stack which means that the **movl** instruction is required to push data on the Lisp stack. More detailed analysis of the operands of the **movl** instructions in the files of Lisp coded Lisp functions reveal that 46% of the **movl**'s are pushing data on the Lisp stack.

3.3. Summary

Vaxima spends a great deal of time moving data from one place to another. When this data movement is done in compiled Lisp, much of it is simply stacking arguments prior to calling a function. Since there are a large number of function calls, we would like to optimize the calling sequence, yet maintain compatibility with the C calling sequence. Our compromise was to code in assembler language those small functions which were called most often, enabling us to change the way those functions were called. The results given above demonstrate the efficacy of this approach and suggest that we should recode the *equal* function in assembler as well.

The FRANZ LISP compiler does not do very much to optimize the fixnum arithmetic operations. As our results indicate, it would be a waste of effort to modify the compiler to handle fixnums if all we were interested in was running Vaxima on problems such as those found in the Begin demo. Only a small amount of arithmetic is done in Vaxima and most of that is written in C where it would be unaffected by a compiler change.

4. VAX instructions

For those readers not familiar with the VAX instruction set, we describe below those instructions which appear in Table 3.

movl - move longword
movab - move address of byte (used most often to add a small constant to a register and store the result).
pushl - push longword on main stack
pushab - push address of byte on stack
beqlu, **bnequ** - conditional branch
calls - call subroutine with large overhead
jsb - call subroutine with no overhead
ret - return from subroutine with large overhead
clrl - clear longword
tstl - test longword (set condition code bits)
cmpl - compare two longwords
cvtbl - convert byte to longword

5. Demonstration Scripts

The information provided in this paper was gathered by adding special code to Vaxima to monitor its activity and then running a demonstration script using the 'batch' command. The two scripts we chose have particular characteristics indicated below:

The Begin demo, which is file "demo;begin demo" at MIT-MC, exercises many of the modules of Vaxima yet runs rapidly enough to allow us to single step it. In seventeen CPU seconds, the Begin demo does one integration, calculates two derivatives, does a complicated simplification, a Taylor series expansion, two polynomial factorizations, a matrix multiplication, and solves a sixth degree polynomial equation.

The Eigen demo, file "share;eigen demo" at MIT-MC, starts by loading in the pre-parsed definitions of Vaxima functions. It then proceeds to calculate eigenvalues and eigenvectors of symbolic matrices for three CPU minutes. Seventeen garbage collections occur when the cons cell free-list runs out.

The FG demo computes the classical algebraic manipulation problem of the astronomer's f and g series up to index 20.

References

Alex75.
W. C. Alexander and D. B. Wortman, "Static and Dynamic characteristics of XPL Programs," *Computer* **8**(11) pp. 41-46 (November 1975).

Bake77.
Henry G. Baker, "A Note of the Optimal Allocation of Spaces in Maclisp," MIT AI Working paper 142 (March 1977).

Char80.
Bruce W. Char, "On Stieltjes' Continued Fraction for the Gamma Function," *Mathematics of Computation* **34**(150) pp. 547-551 (April 1980).

Clar77.
Douglas W. Clark and C. Cordell Green, "An Empirical Study of List Structure in Lisp," *CACM* **20**(2) pp. 78-86 (February 1977).

Fate80.
Richard J. Fateman, *Addendum to the Macsyma Reference Manual for the VAX.* 1980.

Fitc77.
J. P. Fitch and A. C. Norman, "Implementing LISP in a High-Level Language," *Software-Practice and Experience* **7** pp. 713-725 (1977).

Fode80.
John K. Foderaro and Keith L. Sklower, *The Franz Lisp Manual.* 1980.

Fost71.
C. C. Foster, R. H. Gonter, and E. M. Riseman, "Measures of Op-Code Utilization," *IEEE Transactions on Computers*, pp. 582-584 (May 1971).

Grou77.
Mathlab Group, *Macsyma Reference Manual*, Laboratory for Computer Science, MIT (1977).

Henr80.
Robert R. Henry, *Techniques to Measure Static and Dynamic Operator and Operand Statistics on the VAX.* 1980.

Joy81.
W. N. Joy, "Virtualizing a Swap-Based system to do Paging in an Architecture lacking Page-Referenced Bits," UCB CS Internal memorandum (1981).

Knut68.
D. E. Knuth, *The Art of Computer Programming*, Addison-Wesley (1968).

Stan80.
Thomas Standish, *Data Structure Techniques*, Addison-Wesley (1980).

Stee77.
Guy L. Steele, "Data Representations in PDP-10 Maclisp," *Proceedings of the 1977 MACSYMA Users' Conference*, pp. 203-214 (July 1977).

Stre78.
William D. Strecker, "VAX-11/780: A Virtual Address Extension to the DEC PDP-11 Family," *Proc NCC*, pp. 967-980 (June 1978).

SMP - A Symbolic Manipulation Program

Chris A. Cole and Stephen Wolfram
Physics Department, California Institute of Technology, Pasadena CA 91125.

SMP is a new general-purpose symbolic manipulation computer program which has been developed during the past year by the authors, with help from G.C.Fox, J.M.Greif, E.D.Mjolsness, L.J.Romans, T.Shaw and A.E.Terrano. The primary motivation for the construction of the program was the necessity of performing very complicated algebraic manipulations in certain areas of theoretical physics. The need to deal with advanced mathematical constructs required the program to be of great generality. In addition, the size of the calculations anticipated demanded that the program should operate quickly and be capable of handling very large amounts of data. The resulting program is expected to be valuable in a wide variety of applications.

Of the existing symbolic manipulation programs, SCHOONSCHIP was the only one designed to handle the very large expressions encountered, and MACSYMA the only one of any generality.

In this paper, we describe some of the basic concepts and principles of SMP. The extensive capabilities of SMP are described, with examples, in the "SMP Handbook" (available on request from the authors).

The basic purpose of SMP is to manipulate symbolic expressions. These expressions may represent algebraic formulae, on which mathematical operations are performed. By virtue of their symbolic nature, they may also represent procedures and actions.

The ability to manipulate symbolic expressions, as well as sets of numbers, allows for much greater generality and a much richer architecture than in numerical computer languages.

The structure of expressions in SMP is defined recursively as follows:

expr consists of *symbol*
or *expr*[*expr*, *expr*, ...] (projection)
or { [*expr*] : *expr*, [*expr*] : *expr*, ...} (list)

Symbols are the fundamental units. Projections represent extraction of a part in an expression. Lists allow expressions to be collected together.

These three fundamental forms suffice to represent all the objects, operations and procedures required in symbolic manipulations.

Symbols are labelled by a unique name (e.g. `x3` or `Mult`) which is used to represent them for input and output. Expressions may be assigned as values for symbols. A symbol to which a value has been assigned becomes essentially a short notation for its value, and is replaced by the value whenever it occurs. If no value has been assigned to a symbol, the results of (most) operations are such as would hold for any possible value of the symbol. The set of possible values represented by a symbol may be delimited by assigning a list of properties to the symbol.

The projection `f[`*expr*`]` represents the part of the expression `f` selected by the "filter" *expr*. If `f` is a list, the entry with index *expr* is selected. If `f` is a symbol with no value, operations performed on `f[`*expr*`]` hold for any value of `f`. Properties assigned to `f` may specify particular treatment. "System-defined" symbols stand for many fundamental operations. Projections from these symbols (e.g. `Plus`) yield expressions which represent the action of these operations on the filter expressions.

In projections such as `f[x1,x2,x3]` or `f[x1][x2]` with several filters, the filters are used successively or together ("curried" or "uncurried") in selection of a part of `f`.

Lists are ordered and indexed sets of expressions. The index and value of each entry in a list may be arbitrary expressions. A particular value in a list is extracted by projection with the corresponding index as a filter.

If the value of some symbol `f` is a list, the entries of the list describe parts of the "object" `f`: they give the values for projections of `f` with different filters. Entries may be introduced into such a list by assignment of values for projections of `f`.

For example, `f:{[1]:a}` (or `f[1]:a`) defines `f` to be an object whose projection with filter `1` is `a`. The values of other projections from `f` remain unspecified. `f[x+y]:(x+y)^2` then yields `{[x+y]:(x+y)^2, [1]:a}` and defines the value for a projection of `f` with filter `x+y` to be `(x+y)^2`.

Lists whose indices are successive integers (starting at `1` , and termed "contiguous") are used to represent vectors. They are analogous to "arrays" in numerical languages such as FORTRAN, ALGOL or APL. Lists whose indices are fixed symbols are analogous to C, COBOL or PL/1 "structures" or PASCAL "records".

SMP incorporates many list manipulation facilities. `Ar` generates a list with a specified structure and entries according to a given "template": it alone encompasses many of the list manipulation capabilities of APL. Projections may be defined to be distributed over entries of any lists appearing as their filters, allowing lists to be used to collect a set of expressions on which the same operations may be performed. `Flat` "unravels" sublists within lists. `Sort`, `Cat` (concatenate), `Cyc` (cycle), `Rev` (reverse), `Union` and `Inter` (intersection) are also provided.

An ordinary symbol (such as `x`) is taken to stand for the same expression whenever it appears. A "generic" symbol (such as `$x`) may represent any one of a possibly infinite class ("genus") of expressions. Different occurrences of a generic symbol may stand for different members of the class. "Generic expressions" or "patterns" may stand for any one of a class of expressions in which generic symbols are replaced by suitable expressions.

List entries whose indices and values are patterns define transformations for classes of expressions. Projection of the list with an expression which is a particular case of the index pattern yields a corresponding specialization of the value pattern. The necessary replacements of generic symbols in the index are performed in the value.

For example, `g:{[$x]:$x^2}` (or `g[$x]:$x^2`) defines `g` to be the operation of transforming an arbitrary expression into its square. Thus `g[2]` becomes `4` while `g[x+y]` becomes `(x+y)^2`. This is to be contrasted with the assignment `f[x+y]:(x+y)^2` given above, which defined a value only for projection with the specific filter `x+y`.

Lists with entries of the form `{[$x]:`*expr*`}`, where *expr* is some expression containing the generic symbol `$x`, correspond to "lambda functions" in LISP, with `$x` the "bound variable". Assignments such as `f[$x]:$x-2` parallel "function definitions" familiar from FORTRAN, C and so on.

When several occurrences of the same generic symbol appear explicitly in a particular pattern, they must correspond to the same expression. Generic symbols in different patterns may represent different expressions.

List entries whose indices are arbitrary patterns define transformations for expressions with particular structures. For example, the assignment `f[$x,1-$x]:h[$x]` defines `f[a,1-a]` to become `h[a]` and `f[5,-4]` to become `h[5]`.

Two patterns are "literally equivalent" if all of their parts are identical, possibly after properties of projections such as commutativity have been accounted for. A pattern *expr2* "matches" *expr1* if the simplified form of *expr2* after suitable replacements for generic symbols is literally equivalent to *expr1*. Replacements for generic symbols must be deducible by literal comparison from at least one of the occurrences of each generic symbol in *expr2*. Thus `f[$x+$y,$x,a+$y]` is determined to match `f[5,2,a+3]`, but `f[$x+$y,$x-$y]` is not determined to match `f[5,-1]`.

If assignments are made for several patterns with overlapping domains of applicability, the assignments for more specific patterns are used in preference to those for more general cases. Hence, with `g[0]:a; g[$x]:1/$x` the value of `g` becomes `{[0]:a, [$x]:1/$x}` so that projections of `g` with the specific filter `0` are `a` but projections with other filters are the reciprocals of those filters.

A Boolean condition may be associated with a pattern to restrict expressions which it matches. *pat*`_=`*cond* represents a pattern equivalent to *pat*, but constrained to match only expressions for which *cond* is determined to be "true" after necessary replacements for generic symbols. Hence, for example, `f[$x_=(5<$x<7)]:$x^2` defines values for projections of `f` whose filters lie between 5 and 7; thus `f[6]` becomes `36` but `f[1]` or `f[x]` remain unevaluated.

The arbitrary structure of patterns used as indices in lists allows definition of "functions" whose "arguments" are constrained to be of particular "types".

SMP incorporates standard logical and relational operations such as `~` (`Not`), `&` (`And`), `~=` (`Uneq` unequal), together with character determination projections such as `Natp` (natural number) and `Polyp` (polynomial). "False" and "true" are identified with `0` and non-zero numbers respectively.

The values of entries in a list may themselves be lists. The resulting form may be pictured as an "n-ary" or "multiway" tree. Each list is a node on the tree, with branches leading to the list entries and labeled by the list indices. A particular part of the tree is selected by a projection with a succession of filters specifying the branch to be taken at each node encountered in descent from the root of the tree.

Contiguous lists of lists (with successive integer indices) represent matrices and tensors. Lists of lists with fixed indices are analogous to hierarchical data bases. Lists of lists with patterns as indices represent "functions" with several parameters.

"Multi-generic" symbols (such as `$$x`) represents sequences ("null projections") of expressions. Thus for example, `f[$$x]` stands for projections of `f` with arbitrary sets of filters; in `f[a,b,c]` `$$x` represents `[a,b,c]`. The assignment `Log[$x $$x]:Log[$x]+Log[$$x]` (space indicates multiplication) defines a logarithms to be expanded, so that `Log[a b (x+y)]` becomes `Log[a]+Log[b]+Log[x+y]`.

All SMP expressions have the structure of n-ary trees. Projections are nodes whose branches (labeled by successive integer indices) lead to the filters of the projection. Symbols form the ultimate terminals ("leaves") of the tree. Parts of an expression may be selected by projections with suitable filters, and may be modified, added or removed by assignments or deassignments for these projections. With `t:f[a^2,b]` the value of the projection `t[1,1]` is `a` and the assignment `t[1,1]:x^2` causes `t` to become `f[x^4,b]`.

Expressions input to SMP are evaluated by replacing each of their parts (starting with the smallest) by any values assigned to them. This process is carried out to the maximum extent possible: unless further assignments are made, the resulting output expressions can be evaluated no further; if input again, they will be output unchanged.

Picturing an expression as an n-ary tree, the terminals (symbols) are evaluated first, followed by the successively larger parts encountered on ascending towards the root of the tree. In evaluating projections, any system-defined procedures are invoked first. Projections with properties such as associativity or antisymmetry are cast into a canonical form, in which mathematically equal expressions are rendered syntactically equivalent. Finally, values assigned to the objects ("projectors") of the projections are scanned. Any value assigned for the required part (projection) is used. Assigning `a:3` the expression `a^2` is first evaluated to `3^2` and then simplified by the system-defined procedure associated with `Pow` to `9`. The expression `0^0` would be left unchanged by this system-defined procedure. Its value may be specified by an assignment such as `0^0:1`.

The SMP procedure for evaluation is arranged so that maximal simplifications and cancellations occur at all stages. In this way, the complexity of intermediate expressions used in generating simple final results is kept to a minimum.

Values assigned for a projection may involve further projections from the same object ("projector"), thereby representing recursive or self-referential function definitions. Evaluation of such values is performed in a sequence of passes through the complete expressions involved; in each pass the recursion is carried only one step further. Thus with the definition `g[$x]:$x g[$x-1]` the expression `g[2]` is evaluated first to `1g[0]` then simplified to `g[0]` and then evaluated to `0g[-1]` and at this stage immediately simplified to `0`. In conventional recursive evaluation schemes, `g[-1]` would be evaluated before the product `0g[-1]` was established to be zero, and the evaluation of `g[2]` would not terminate. The direct recursive definition `f[0]:f[1]:1; f[$x]:f[$x-1]+f[$x-2]` of the Fibonacci series provides a further example. The simplest recursive evaluation of `f[`*n*`]` forms a binary tree requiring exponential time and memory space; in SMP, the time and space required grow only quadratically with *n*.

There are two possible kinds of assignment of a value *expr2* to an expression *expr1*. In "immediate assignment", *expr1*`:`*expr2* specifies the value of *expr1* to be the present value of *expr2*. The resulting value is maintained in an evaluated form, and updated by any subsequent assignments when used. In "delayed assignment", *expr1*`::`*expr2* specifies that whenever the value of *expr1* is requested, the value of *expr2* found at that time is to be given. In this case, the unevaluated form of *expr2* is maintained as the value of *expr1*, and is evaluated afresh when it is used.

With the assignments `b:c a:b` the value of `a` becomes `c`. A subsequent reassignment `b:d` leaves the value of `a` unchanged. On the other hand, with a delayed assignment `a::b` the value of `a` is always the value of `b` at the time of the request. Thus, with `b:c; a::b` a request for the value of `a` would give `c` : after the reassignment `b:d` a request for `a` would however give `d`.

Expressions assigned as "delayed values" for patterns are evaluated after replacement of the necessary generic symbols.

Delayed values may represent "procedures" to be re-executed whenever "called". (A feature which exemplifies the necessity for delayed assignment is that conditionals may depend on symbolic expressions of undetermined truth value.)

Projections may have the property that some or all of their filters are to be maintained in an unsimplified form. In this way, a "procedure" assigned as the value of a symbol may be "passed by name".

SMP incorporates constructs necessary for programming: local variables, `If`, `For`, `Do`, `Rpt` (repeat) and `Swtch` (switch), together with local and non-local returns and jumps.

Assignments define values for expressions to be used whenever the expressions appear. "Replacements" such as `x->y+1` are syntactic constructs which may be applied selectively in a particular expression by an `S` (substitution) projection. Replacements may involve patterns. Note that in substitutions, as in assignments, associativity and commutativity (or other filter reordering symmetries) of projections are accounted for, so that `S[a b c^2 d, $x^2 a -> 1-$x]` yields `(1-c) b d`.

In addition to standard arithmetic operations (such as `+` (`Plus`), `^` (`Pow`), `.` (`Dot` inner product) and `**` (`Omult` outer product)) and elementary functions (such as `Log`) SMP treats a large number of the special functions of mathematical physics (such as `Chg` (confluent hypergeometric function) and `Geg` (Gegenbauer function)). Numerical values of expressions involving such functions are obtained by `N[`*expr*`]`. Simplifications and transformations are not made automatically unless the results are very simple. The necessary formulae are contained in an extensive library of "external files", and given as replacements, to be applied selectively when required. Thus, for example, `S[`*expr*`,XTrig[2,6]]` applies the half-angle relations for trigonometric functions in *expr*. The name `XTrig[2,6]` of the set of relations to apply is found from the list of formulae in the "SMP Reference Manual".

SMP incorporates a variety of facilities for effecting structural simplifications and modifications on expressions. Parts in expressions may be reassigned to have modified values. The parts may be identified using the projection `Pos` which yields a list of positions of a particular pattern in an expression. The projection `Map` may be used to apply a "template" to a "domain" or set of parts in an expression. A template is an expression used to specify an action on a set of expressions. Application of the template `f` to for example `x` and `y` yields `f[x,y]`. The template `$x^($y+$x)` has two "slots" indicated by `$x` and `$y` into which expressions are inserted: application to `x` and `y` yields `x^(x+y)`. Other facilities include `Cb[`*expr*`,`*form*`]`, which combines coefficients of terms matching the pattern *form* in *expr*.

SMP incorporates projections for performing expansions using distributivity. It also contains facilities for factorization of polynomials, for forming partial fractions, and for other polynomial manipulations.

SMP performs derivatives, sums, products, integrals, series expansions and solves equations. Assignments may be made to define derivatives, integrals, inverses and so on, for new mathematical functions. For example, `D[f[$x],{$x,1,$y}]:g[$y]` defines the first derivative of the "function" `f` of one "argument" to be `g`.

Input and output in SMP conform as closely as possible to standard mathematical notation. Input syntax may be modified and new forms introduced. Arbitrary output forms may be defined. Numerical values of expressions may be plotted.

An Extension of Liouville's Theorem on Integration in Finite Terms

M.F. Singer
Department of Mathematics
North Carolina State University
Raleigh, NC 27650

B.D. Saunders[1]
Department of Mathematical Sciences
Rensselaer Polytechnic Institute
Troy, NY 12181

B.F.Caviness[1]
General Electric Research and Development Center
Schenectady, NY 12345
on sabbatical leave from
Department of Mathematical Sciences
Rensselaer Polytechnic Institute
Troy, NY 12181

1. Introduction

In 1969 Moses [MOSE69] first raised the possibility of extending the Risch decision procedure for indefinite integration to include a certain class of special functions. Some of his ideas have been incorporated as heuristic methods in Macsyma and Reduce. However little progress has been made on the theory necessary to extend the Risch algorithm. One step in this direction was the paper by Moses and Zippel [MOZI79] in which a weak Liouville theorem was given for special functions.

In this paper we give an extension of the Liouville theorem [RISC69, p. 169] and give a number of examples which show that integration with special functions involves some phenomena that do not occur in integration with the elementary functions alone.

Our main result generalizes Liouville's theorem by allowing, in addition to the elementary functions, special functions such as the error function, Fresnel integrals and the logarithmic integral to appear in the integral of an elementary function. The basic conclusion is that these functions, if they appear, appear linearly.

2. Statement of Results

We begin by defining a generalization for the elementary functions. Let F be a differential field of characteristic zero with field of constants C and derivation $'$. Let A and B be finite indexing sets and let

$$E = \{G_\alpha(exp\ R_\alpha(Y))\}_{\alpha \in A}$$
$$L = \{H_\beta(\log S_\beta(Y))\}_{\beta \in B}$$

be sets of expressions where

(1) $G_\alpha, R_\alpha, H_\beta, S_\beta$ are in $C(Y)$ for all $\alpha \in A$, $\beta \in B$; i.e., they are all rational functions with constant coefficients, and (2) for all $\beta \in B$, if $H_\beta(Y) = \frac{P_\beta(Y)}{Q_\beta(Y)}$; P_β, $Q_\beta \in C[Y]$; P_β, Q_β relatively prime, then $degP_\beta \leqslant degQ_\beta + 1$. We say that a differential extension E of F is an ***EL**-elementary extension of F* if there exists a tower of fields $F = F_0 \subset \cdots \subset F_n = E$ such that $F_i = F_{i-1}(\theta_i)$ where for each i, $1 \leqslant i \leqslant m$, exactly one of the following holds:

(i) θ_i is algebraic over F_{i-1}.
(ii) $\theta_i' = u'\theta_i$, $u \in F_{i-1}$. In this case we write $\theta_i = exp(u)$.
(iii) $\theta_i' = u'/u$, u a non-zero element of F_{i-1}. In this case we write $\theta_i = \log u$.
(iv) For some $\alpha \in A$, $\theta_i' = u'G_\alpha(\exp R_\alpha(u))$, $u \in F_{i-1}$ and $\exp R_\alpha(u) \in F_{i-1}$. In this case and the next, θ_i is a special function of u.
(v) For some $\beta \in B$, $\theta_i' = u'H_\beta(\log S_\beta(u))$, $u \in F_{i-1}$ and $\log S_\beta(u) \in F_{i-1}$.

We can now state the generalization of Liouville's theorem.

Theorem. Let F be a differential field with an algebraically closed subfield of constants C. Let γ be in F and assume there exists an ***EL***-elementary extension E of F with the same field of constants and an element y in E such that $y' = \gamma$. Then there exist constants a_i, $b_{i\alpha}$, $c_{i\beta}$ in C and $w_0, w_1, \ldots, w_n, u_{i\alpha}, u_{i\beta}$, $\exp R_\alpha(u_{i\alpha})$, $\log S_\beta(u_{i\beta})$ algebraic over F such that

$$\gamma = w_0' + \sum_{i=1}^{n} a_i \frac{w_i'}{w_i} + \sum_{\alpha \in A} \sum_{i \in I_\alpha} b_{i\alpha} u_{i\alpha}' G_\alpha(\exp R_\alpha(u_{i\alpha}))$$

$$+ \sum_{\beta \in B} \sum_{i \in J_\beta} c_{i\beta} u_{i\beta}' H_\beta(\log S_\beta(u_{i\beta}))$$

where I_α and J_β are finite sets of integers for all α, β.

3. Discussion of Hypotheses and Some Counter-Examples

The proof of the theorem will be given in a later paper. Now some comments about the definitions.

The definition of ***EL***-elementary functions is broad enough to include such functions as the error function, the Fresnel integrals and the logarithmic integral. The error function is defined by

$$erf(u) = \frac{2}{\sqrt{\pi}} \int u' e^{-u^2} dx$$

where $G_\alpha(exp\ R_\alpha(Y)) = exp(-Y^2)$ with $G_\alpha(W) = W$ and $R_\alpha(Y) = -Y^2$.

The Fresnel integrals are defined by

$$S(u) = \int u' \sin\left(\frac{\pi}{2} u^2\right) dx$$

and

$$C(u) = \int u' \cos\left(\frac{\pi}{2} u^2\right) dx.$$

For $S(u)$ we have that

$$G_\alpha(\exp R_\alpha(Y)) = \frac{\left[e^{i\frac{\pi}{2}Y^2}\right]^2 - 1}{2i\, e^{i\frac{\pi}{2}Y^2}}$$

where $G_\alpha(W) = \frac{W^2-1}{2iW}$ and $R_\alpha(Y) = i\frac{\pi}{2}Y^2$.

The logarithmic integral is defined by

$$li(u) = \int \frac{u'}{\log u} dx$$

with $H_\beta(W) = \frac{1}{W}$ and $S_\beta(Y) = Y$.

EL-elementary functions do not include the dilogarithm (or Spence function) defined by

1. Partially supported by National Science Foundation grant MCS-7909158

$$Li_2(u) = -\int \frac{u'\log u}{u-1}dx$$

nor the exponential integral

$$Ei(u) = \int \frac{u'e^u}{u}dx.$$

The condition (2) in the definition of **EL**-elementary extension seems artificial but the theorem is not true without it. For example, let $F = C(x,\log x)$ where C is the field of complex numbers, $E = \{\}$, and $L = \{(\log x(x+1))^2\}$. In this case, the index set B is a singleton and $H = Y^2$. This is excluded by condition (2) since degree(numerator H) = 2 > degree(denominator H)+1. Let $\gamma = \frac{\log x}{x+1}$.

Claim. (a) $\int \frac{\log x}{x+1}$ lies in an **EL**-elementary extension of F, but

(b) $\frac{\log x}{x+1} \neq w_0' + \sum c_i \frac{w_i'}{w_i} + \sum d_i u_i'(\log u_i(u_i+1))^2$

for any $w_0, w_1, \ldots, w_n, u_i, \log(u_i(u_i+1))$ algebraic over $C(x,\log x)$.

To verify (a), compute $\int (\log x(x+1))^2 dx$ by parts. First we have that

$$\int (\log x)^2 = x(\log x)^2 - 2x\log x + 2x;$$

$$\int (\log(x+1))^2 = (x+1)(\log(x+1))^2 - 2(x+1)\log(x+1) + 2(x+1);$$

and

$$\int (\log x)\log(x+1) = x(\log x)\log(x+1) - (x+1)\log(x+1) - x(\log x) + 2x + \int \frac{\log x}{x+1}.$$

Hence

$$\int (\log x(x+1))^2 = \int (\log x + \log(x+1))^2 = \int (\log x)^2 + 2\int (\log x)\log(x+1) + \int (\log(x+1))^2 = \text{elementary function} + 2\int \frac{\log x}{x+1}.$$

To verify (b) assume that

$$\frac{\log x}{x+1} = w_0' + \sum c_i \frac{w_i'}{w_i} + \sum_1^m d_i u_i'(\log u_i(u_i+1))^2$$

with $w_i, u_i, \log(u_i(u_i+1))$ algebraic over $C(x,\log x)$. We may assume, without loss of generality, that $u_i' \neq 0$. From the structure theorem [ROCA79] we have for each i, $1 \leqslant i \leqslant m$, that $u_i(u_i+1) = c_i x^{r_i}$ for some $r_i \in Q$, the field of rational numbers, and $c_i \in C$. Neither c_i nor r_i is zero for otherwise $u_i'=0$. Hence

$\log(u_i(u_i+1)) = r_i \log x + k_i$, $k_i \in C$.

Furthermore each u_i is algebraic over $K = C(x,\log x, x^{r_1}, \ldots, x^{r_m})$ and satisfies the irreducible equation $u(u+1) - c_i x^{r_i} = 0$. Consider the Galois group of $K(u_1, \ldots, u_m):K$. Then $tr(u_i)$ = integer x (-1) since $u_i^2 + u_i - c_i x^{r_i} = 0$. Therefore $tr(u_i') = \{tr(u_i)\}' = 0$. Apply the trace to both sides of

$$\frac{\log x}{x+1} = w_0' + \sum c_i \frac{w_i'}{w_i} + \sum_1^m d_i u_i'(r_i \log x + k_i)^2$$

to obtain

$$\mu \frac{\log x}{x+1} = (tr\ w_0)' + \sum c_i \frac{(Nw_i)'}{Nw_i}$$

where μ is a positive integer and Nw_i is the norm of w_i. But this contradicts the fact that $\int \frac{\log x}{x+1}$ is not elementary and hence (b) is verified.

One would also like a theory that includes functions like the dilogarithm and exponential integral, but this remains an open problem.

4. Algorithmic Considerations

To facilitate the discovery of a new decision procedure based on the theorem it would be desirable, as in the Risch algorithm, to have reasonable conditions on the field F to guarantee that $w_0, w_i, u_{i\alpha}, \exp R_\alpha(u_{i\alpha}), u_{i\beta}, \log S_\beta(u_{i\beta})$ are actually in F, not simply algebraic over F.

When F is a Liouvillian extension of its field of constants, the lemma on p. 338 of [ROSI77] implies that each $R_\alpha(u_{i\alpha})$ and $\log S_\beta(u_{i\beta})$ is in F and that some power of each $\exp R_\alpha(u_{i\alpha})$ and $S_\beta(u_{i\beta})$ is in F.

The following example shows that F being Liouvillian over C is not strong enough to guarantee that the $u_{i\alpha}$ and $\exp R_\alpha(u_{i\alpha})$ are in F. Let $E = \{\exp(-Y^2)\}$ and $L = \{\}$. Consider $F = C(x, e^x, \exp(-e^x + x/2))$. Let $u = \exp(x/2)$. Then $\left\{\frac{\sqrt{\pi}}{2} erf(u)\right\}' = u'\exp(-u^2) = \frac{1}{2}\exp(-e^x + x/2)$, but neither u nor $\exp(-u^2)$ is in F although both are algebraic over F. In this example we even have that F is a purely transcendental extension of C but this is not strong enough to conclude that u and $\exp(-u^2)$ are in F.

If $\int \gamma$ can be expressed in terms of elementary functions and the error function, one can prove that

$$\gamma = w_0' + \sum c_i \frac{w_i'}{w_i} + \sum d_i u_i' e^{-u_i^2}$$

where $w_0, w_i, u_i' e^{-u_i^2}, u_i^2$ and $(e^{-u_i^2})^2$ are in F. This may be sufficient to obtain a decision procedure for this case.

5. References

[MOSE69] J. Moses, The integration of a class of special functions with the Risch algorithm, *SIGSAM Bulletin*, No. 13 (December 1969), pp. 14-27.

[MOZI79] J. Moses and R. Zippel, An extension of Liouville's theorem, in *Symbolic and Algebraic Computation*, E.W. Ng (ed.), Springer-Verlag, 1979, pp. 426-430.

[RISC69] R.H. Risch, The problem of integration in finite terms, *Trans. Amer. Math. Soc.*, (1969), pp. 167-189.

[ROCA79] M. Rothstein and B.F. Caviness, A structure theorem for exponential and primitive functions, *SIAM J. Comput.*, *8* (August 1979), pp. 357-367.

[ROSI77] M. Rosenlicht and M.F. Singer, On elementary, generalized elementary, and liouvillian extension fields, *Contributions to Algebra* (1977), pp. 329-342.

Formal Solutions of Differential Equations in the Neighborhood of Singular Points

(REGULAR AND IRREGULAR)

J. DELLA DORA, E. TOURNIER

Laboratoire IMAG
Tour des Mathématiques
B.P. 53 X
38041 GRENOBLE *cédex*
FRANCE

1 - INTRODUCTION

This paper presents an algorithm for building a fundamental system of formal solutions in the neighbourhood of every singularity of a linear homogenous differential operator :

$$\Gamma = \sum_{i=1}^{n} q_i \frac{d^i}{dx^i} ,$$

where $q_i \in k(x)$ field of rational fractions with k of characteristic 0 and algebricaly closed.

This study requires the following steps :

1st - The exact localization of each root of q_n .

2nd - The study of the nature of the singularities in question, which are located at the roots.

3rd - The algorithm which gives the formal solutions in the neighbourhood of these singularities.

4th - The resummation of the formal solutions obtained.

We will discuss here the points 2 and 3.
Point 4 is the subject of publications [6,8].
Point 1 has not yet been studied. This explains why at present , we only study operators of type :

$$\text{(I)}\quad L = \sum_{i=0}^{n} p_i \frac{d^i}{dx^i}$$

localized at the origin ; that is to say

$p_i \in k[[x]]$ field of formal series in one indeterminate over k .

According to the works of J.P. RAMIS [6] and B. MALGRANGE [9] we can prove that the study of the 3rd point can be reduced (in a non trivial way [4]) to the study of *regular singularities* only. Thus, the heart of this program is the setting up of an algorithm (similar to the classic algorithm of Frobenius) which builds a fundamental system of formal solutions in the neighbourhood of a *regular singular point.*

2 - THE FROBENIUS ALGORITHM (GENERIC CASE)

First of all, we have to introduce some notation and theorems. We note v the valuation of a formal series which is its order.

(If $p(x) = \sum_{j=n}^{+\infty} p_j x^j$, with $p_n \neq 0$, $n \in \mathbb{Z}$, then $v(p) = n$)

Fuchs has proved [5] the following theorem :

THEOREM

0 is a regular singularity of the operator L, if and only if :

$$v(p_i) - i \geq v(p_n) - n , \ \forall i = 0,\ldots,n-1$$

With the Fuchs theorem L can be written as :

$$L = x^n R_0(x)\, \partial^n + x^{n-1} R_1(x)\, \partial^{n-1} + \ldots + R_n(x)$$

with $R_i(x) = \sum_{j=0}^{+\infty} r_j^i x^j$, $\partial = \frac{d}{dx}$ and $R_0(0) \neq 0$

We can therefore easily prove that :

(II) $L(x^\lambda) = x^\lambda f(x,\lambda)$, (for every λ belonging to k)

in which $f(x,\lambda) = \sum_{j=0}^{+\infty} f_j(\lambda) x^j$

with $f_j(\lambda) = \sum_{i=0}^{n} [\lambda]_i r_j^{n-i}$

and where $[\lambda]_i = \lambda(\lambda-1)\ldots(\lambda-i+1)$; $[\lambda]_0 = 1$

The Frobenius algorithm [4] shows that in the neighbourhood of 0. There exists n formal solutions of either the form

$$(III)_a \qquad y = x^\lambda \left(\sum_{j=0}^{+\infty} g_j x^j\right) \qquad \textit{(generic case)}$$

or

$$(III)_b \qquad y = x^\lambda (\phi_0 + \phi_1 \cdot \log(x) + \ldots \ldots \phi_s \cdot (\log(x))^s \qquad \textit{(general case)}$$

where $\phi_i \in k[[x]]$.

We can, therefore, prove (by substitution of y in $L(y) = 0$) the following theorem :

THEOREM

$y = x^\lambda \left(\sum_{i=0}^{+\infty} g_i x^i\right)$ is a formal solution of $L(y) = 0$ if and only if

$$\sum_{\nu=0}^{m} g_\nu f_{m-\nu}(\lambda+\nu) = 0 \text{ , } \forall m = 0,1,\ldots,+\infty.$$

We write this theorem in a form of a matrix expression.

Let A be an infinite matrix, and g an infinite vector such that :

$$A(m+1,j+1) = \begin{cases} f_{m-j}(\lambda+j) & \text{for } j \leq m \\ 0 & \text{otherwise} \end{cases}$$

$$g^t = (g_0, g_1, \ldots)$$

We have the infinite linear system $Ag = 0$.

The solution is expressed as a function of g_0 and the f_i , and has the form :

$$(IV) \quad g_m(\lambda) = - \frac{g_0}{\prod_{j=1}^{m} f_0(\lambda+j)} \times \det \begin{vmatrix} f_0(\lambda+1) & 0 & 0 & 0 & f_1(\lambda) \\ f_1(\lambda+1) & f_0(\lambda+2) & 0 & & \\ \vdots & & \ddots & 0 & \vdots \\ & & & f_0(\lambda+m-1) & \\ f_{m-1}(\lambda+1) & \ldots & & f_1(\lambda+m-1) & f_m(\lambda) \end{vmatrix}$$

for $m \geq 1$

If we substitue in $(III)_{a,b}$ the values of g_m obtained in (IV) and applying L we obtain :

$$L\left(x^\lambda \sum_{j=0}^{\infty} g_j x^j\right) = x^\lambda f_0(\lambda)$$

$f_0(\lambda) = 0$ is called the "*indicial equation*" associated with the operator L .

If λ is the root of the indicial equation and if $f_0(\lambda+j) \neq 0$, $\forall j = 0,1,\ldots,+\infty$, we have the "*generic case*" , and so we can determine the formal solution completely.

Now if λ is a multiple root of $f_0(\lambda)$ or if there exists an integer i such that $f_0(\lambda+i) = 0$, we have the "*general case*" .

In this case solutions of the form $(III)_b$ may appear, and to determine these solutions we are now going to present the general algorithm of Frobenius.

3 - The Frobenius Algorithm

(GENERAL CASE)

3.1. SOME NOTATIONS

To the indicial equation, we associate the array of its roots ordered as follows :

$$R_0 = \begin{vmatrix} \rho_0 & \rho_1 & \dots & \rho_{\alpha-1} \\ \rho_\alpha & \rho_{\alpha+1} & \dots & \rho_{\beta-1} \\ \dots & & & \\ \dots & & & \end{vmatrix}$$

For the 1st row we have :

Realpart $(\rho_0) \geq$ Realpart $(\rho_1) \geq \dots \geq$ Realpart $(\rho_{\alpha-1})$

$\rho_j - \rho_i \in N$ for $j \leq i$ and $0 \leq i,j \leq \alpha-1$

and likewise for the other rows. The elements of 2 different rows must not differ from an element of Z .

It is usual to "blow up" this array to distinguish the multiple roots from the others. So, for the 1st row we have :

$\rho_0 = \rho_1 = \dots = \rho_{i-1}$ roots of multiplicity i

$\rho_i = \rho_{i+1} = \dots = \rho_{j-1}$ " " j-i

. . .

$\rho_\nu = \dots = \rho_{\alpha-1}$ " " $\alpha-\nu$

and likewise for the other rows. We note $\Omega = \rho_0 - \rho_{\alpha-1}$. We show now how to deal with the 1st row of this array. The others may be treated in the same way.

3.2. DETERMINATION OF THE SOLUTIONS

There is no problem in determining the solution associated with the root with the largest real part as described in § 2. There are 2 cases in which the previous method has to be modified :

- either we are in the subset $\rho_0 = \rho_1 \dots = \rho_{i-1}$ and in that case we always have the same expression for the solution
- or we are in a different subset $\rho_i = \rho_{i+1} \dots = \rho_{j-1}$ but in that case the denominators of g_i are null after a certain rank.

We also notice that numerically, it is not obvious to determine if 2 roots of an algebraic equation differ by an integer. However, we will use an algorithm proposed by Watanabe [3,10].

To determine the other solutions, the idea is to substitute g_0 by

$$g_0 \prod_{i=1}^{\Omega} f_0(\lambda+i) \text{ with } \Omega = \rho_0 - \rho_{\alpha-1} .$$

In that case, our formulas become :

$$L(x^\lambda \Sigma g_i x^i) = g_0 \prod_{i=0}^{\Omega} \rho_0(\lambda+i)x^\lambda$$

and

$$\frac{\partial^s}{\partial\lambda^s}(L(x^\lambda \sum_{i=0}^{+\infty} g_i(\lambda)x^i)) = \frac{\partial^s}{\partial\lambda^s}(g_0 \prod_{i=0}^{\Omega} f_0(\lambda+i)x^\lambda)$$

$$L(\frac{\partial^s}{\partial\lambda^s}(x^\lambda \sum_{i=0}^{+\infty} g_i(\lambda)x^i))$$

If we note $F(\lambda) = \prod_{i=0}^{\Omega} f_0(\lambda+i)$, we have :

$$\frac{\partial^s}{\partial\lambda^s}(g_0F(\lambda)x\ ) = g_0(F^{(s)}(\lambda)x^\lambda + \lambda \operatorname{Log}(x)\ F^{(s-1)}(\lambda)\ x^\lambda + \dots + \lambda(\lambda-1)\dots(\lambda-s+1)\ F(\lambda)x^\lambda)$$

We now consider *the first subset* $\rho_0 = \rho_1 \dots = \rho_{i-1}$.

ρ_0 is a root of order i of $F(\lambda)$, then

$$\frac{\partial^s}{\partial\lambda^s}(F(\lambda)x^\lambda)\ |_{\lambda=\rho_0} = 0 \quad , \quad s=0,1,\dots,i-1$$

$$\text{then } L(\frac{\partial^s}{\partial\lambda^s}(x^\lambda \sum_{i=0}^{+\infty} g_i(\lambda)x^i\ |_{\lambda=\rho_0}) = 0$$

for $s=0,1,\dots,i-1$

this shows us that the *formal series* :

$$\frac{\partial^s}{\partial\lambda^s}(x^\lambda \sum_{i=0}^{+\infty} g_i(\lambda)x^i)\ |_{\lambda=\rho_0} \quad \text{for } s=0,1,\dots,i-1$$

are formal solutions of $L(y) = 0$.

If we note $g(x,\lambda) = \sum_{i=0}^{+\infty} g_i(\lambda)x^i$ moreover, we have :

$$\frac{\partial^s}{\partial\lambda^s}(x^\lambda g(x,\lambda)) = x^\lambda \sum_{j=0}^{s} C_n^j(\operatorname{Log} x)^j\ g^{(s-j)}(x,\lambda)$$

with C_n^j the binomial coefficient

$$\text{as } g^{(k)}(x,\lambda) = \sum_{i=0}^{+\infty} \frac{\partial^k g_i(\lambda)}{\partial\lambda^k} x^i$$

and if we note $w_i(x,\lambda) = x \times g^i(x,\lambda)$, we can now write the i solutions of the equation $L(y) = 0$ which are as follows :

$$\begin{cases} W_0 = w_0(x,\rho_0) \\ W_1 = w_0(x,\rho_0)\log x + w_1(x,\rho_0) \\ \vdots \\ W_{i-1} = w_0(x,\rho_0)(\log x)^{i-1} + C_{i-1}\, w_1(x,\rho_0)(\log x)^{i-2} \\ \quad + \ldots + w_{i-1}(x,\rho_0) \end{cases}$$

These solutions are linearly independant because of the logarithmic terms of different powers.
Now we have to consider the *following subset* of the roots :

$$\rho_i = \rho_{i+1} \ldots = \rho_{j-1}$$

With a similar proof, we can show that the solutions are :

$$\begin{cases} W_i = w_0(x,\rho_i)(\log(x))^i + i\, w_1(x,\rho_i)(\log x)^{i-1} + \\ \quad + \ldots + w_i(x,\rho_i) \\ \vdots \\ W_{j-1} = w_0(x,\rho_i)(\log(x))^{j-1} + \ldots + w_{j-1}(x,\rho_i) \end{cases}$$

As $w_i(x,\rho_i) \neq 0$ and W_{i+r} contains the term $w_i(x,\rho_i)(\log x)^r$ the solutions W_i , W_{i+1} , ... , W_{j-1} are linearly independent.

We can also show that $W_i,\ldots,W_{j-1}$ are linearly independent of $W_1,\ldots,W_{i-1}$ (see [5]).

So we determine a fundamental system of formal solutions for the equation $L(y) = 0$ in the neighbourhood of regular singularities, and the complete algorithm is written as :

ALGORITHM FROBENIUS

Input :

N : order of the differential equation $L(y) = 0$

A(N) : array of the coefficients of L .

Output :

SOLUTION (I,J) : array of the formal solutions of L

[1] Let Y(0) be the general form of the solution of L with (k+1) terms

[2] FO(λ) := the indicial equation of L

[3] solve FO(λ) = 0

[4] RO := the sorted array of the roots of FO(λ) = 0

[4.1] factorization over Z

[4.2] multiplicity order of roots of FO(λ) = 0

[4.3] WATANABE_algorithm to know if real part of 2 roots differ from one another by an integer

[5] for the row I = 1 until I such that RO(I,*) = nil do

[5.1] SOLUTION (I,0) := FROBENIUS_generic_case

[5.2] for the column J = 1 until J such that RO(I,J) = nil do SOLUTION(I,J) := FROBENIUS_general_case

CONCLUSION

A differential operator of the form :

$$L = \sum_{i=0}^{n} p_i\, \partial^i \ , \ p_i \in k[[X]]$$

has a set of formal solutions, linearly independant which can be devided into two groups.

-1- A set of *regular* solutions (of Fuchs type) which can be computed directly using the previous algorithm.

-2- A set of *irregular* solutions which have the

form :

$$y = e^{P(\frac{1}{t})} \varphi(t)$$

where $P \in k[X]$, φ is a regular solution in t, and t is defined as the following :
there exists a positive integer q well defined such that $X = t^q$.
The method to be used is the following :

2.1 to extract the irregular part of the solution y , i.e. $e^{P(\frac{1}{t})}$, that is done by using a non trivial algorithm which gives q and P in a finite number of steps.

2.2 The change of variable $y = e^{P(\frac{1}{t})} \varphi(t)$ transforms the operator into an operator which has φ as a regular solution. Then this solution can be computed by the Frobenius algorithm.

For a more complete description of these algorithms and some example see [3, 4].

Acknowledgments

The authors wish to thank Professors B. Malgrange and J.P. Ramis for their informative discussions and are most grateful to Dr. Y. Siret who has been of great help with his policy of development and maintenance of algebraic systems at the C.I.C.G.

REFERENCES

[1] J. DELLA DORA, E. TOURNIER
"Localisation formelle d'un opérateur différentiel". (preprint).

[2] J. DELLA DORA, E. TOURNIER
"Polygône de Newton-Ramis-Malgrange". (preprint).

[3] J. DELLA DORA, E. TOURNIER
"Algorithme d'obtention des solutions formelles d'une équation différentielle homogène au voisinage d'une singularité régulière".
(Rapport de Recherche du Laboratoire IMAG (Informatique et Mathématiques Appliquées de Grenoble , février 1981).

[4] J. DELLA DORA, E. TOURNIER
"Etude des solutions d'une équation différentielle homogène au voisinage d'une singularité irrégulière".
(Rapport de Recherche du Laboratoire IMAG, avril 1981).

[5] E.L. INCE
"Ordinary differential equations". (Dover).

[6] J.P. RAMIS
"Devissage Gevrey".
(Asterisque n° 59-60, pages 173-204, 1978).

[7] J.P. RAMIS
"Théorèmes d'indice dans les espaces de séries formelles Geuvrey" (preprint)
IRMA, Université de Strasbourg (1980).

[8] J.P. RAMIS, J. THOMAN :
"Fonctions k-sommables et leurs applications (séminaire IMAG, (janvier 1980)).

[9] B. MALGRANGE
"Sur la réduction formelle des équations différentielles à singularités régulières".
(preprint) Institut Fourier de Grenoble (1980).

[10] S. WATANABE
"Formula manipulations solving linear Ordinary differential equations (I) and (II)".
I : (Publ. RIMS Kyoto Univ. Vol. 6, 1970 , 71-111)
II : (Publ. RIMS Kyoto Univ. vol. 11, 1976 , 297-337).

Elementary First Integrals of Differential Equations

by

M. J. Prelle[1,2]
Department of Mathematical Sciences
Rensselaer Polytechnic Institute
Troy, New York 12187

M. F. Singer
Department of Mathematics
North Carolina State University
Raleigh, North Carolina 27650

1. Introduction

It is not always possible and sometimes not even advantageous to write the solutions of a system of differential equations explicitly in terms of elementary functions. Sometimes, though, it is possible to find elementary functions which are constant on solution curves, that is, elementary first integrals. These first integrals allow one to occasionally deduce properties that an explicit solution would not necessarily reveal. Consider the following example:

Example 1: The preditor-prey equations

$$\frac{dx}{dt} = ax - bxy$$

$$\frac{dy}{dt} = -cy + dxy \qquad a, b, c, d \text{ positive real numbers}$$

Although these cannot be solved explicitly in finite terms, one can show that

$$F(x,y) = dx + by - c \log x - a \log y$$

is constant on solution curves $(x(t),y(t))$. Using the function $F(x,y)$, one can show that all solution curves are closed, that is all solutions are periodic.

[1] Current address: Box 211, Vassar College, Poughkeepsie, New York 12601.

[2] Partially supported by National Science Foundation Grant MCS-7909158.

Note that in this example the first integral is of the form

$$w_0(x,y) = \sum c_i \log w_i(x,y)$$

where the c_i are constants and the w_i are rational functions of x and y. Roughly speaking, the main result of this note is that if a system of differential equations has an elementary first integral, it will then have one of this form. Corollaries of the main result will show that the theory presented in this paper unifies and generalizes a number of results originally due to Mordukhai-Boltovski, Ritt and others. An attempt to do this was made in [SING: 77], but the results presented here are more general and the techniques more to the point.

2. Main Result and Corollaries

To fix notation, we will let (K,Δ) denote a differential field of characteristic zero with a given set of derivations $\Delta = \{\delta\}_{\delta\in\Delta}$. The constants of (K,Δ), that is all those elements annihilated by all δ in Δ, will be denoted by $C(K,\Delta)$. We assume that the reader is familiar with the definitions of elementary and liouvillian extensions and related notions. For precise definitions see [ROS: 76].

The following is our main result. Note that for $\delta_1,\ldots,\delta_n$ in Δ, any K-linear combination $u_1\delta_1 + \ldots + u_n\delta_n$, $u_i \in K$, is a derivation on any Δ-differential extension of K.

Theorem. Let (L,Δ) be an elementary extension of the differential field (K,Δ) with $C(L,\Delta) = C(K,\Delta)$. Let D be in the K-span of Δ and assume that $C(L,\Delta)$ is a proper subset of $C(L,\{D\})$. Then there exist $w_0,w_1,\ldots,w_m$, algebraic over K and $c_1,\ldots,c_m$ in $C(K,\Delta)$ such that

$$Dw_0 + \sum_{i=1}^{m} c_i \frac{Dw_i}{w_i} = 0 \qquad (1)$$

and

$$\delta w_0 + \sum_{i=1}^{m} c_i \frac{\delta w_i}{w_i} \neq 0 \qquad (2)$$

for some $\delta \in \Delta$.

Furthermore if (K,Δ) is a liouvillian extension of $C(K,\Delta)$, we can find $w_0,w_1,\ldots,w_m$ in K and constants $c_1,\ldots,c_m$ in $C(K,\Delta)$ satisfying (1) and (2).

Let us see how Example 1 fits into this scheme:

Example 1 revisited. Let $K = \mathbb{C}(x,y)$, $\mathbb{C}$ being the complex numbers; x and y indeterminants. Let $\Delta = \{\delta_x,\delta_y\}$ where δ_x (resp. δ_y) is the partial derivative with respect to x (resp. y). Let $D = (ax + bxy)\delta_x + (-cy + dxy)\delta_y$. Let (L,Δ) be an elementary extension of (K,Δ). L then consists of elementary functions of two variables. For g in L, $Dg = 0$ is equivalent to g being constant on solutions of our system of differential equations. For g in L, $\delta_x g \neq 0$ or $\delta_y g \neq 0$ is equivalent to g being not identically constant. Therefore, the hypothesis that $C(L,\Delta)$ is properly contained in $C(L,\{D\})$ is equivalent to the existence of a non-trivial elementary function of two variables that is constant on solutions of our equation. Since $(\mathbb{C}(x,y),\Delta)$ is a liouvillian extension of $\mathbb{C}$, the conclusion states that there must exist $w_0,w_1,\ldots,w_m$ in $\mathbb{C}(x,y)$ such that $w_0 + \sum_{i=1}^{m} c_i \log w_i$ is constant on solutions of our system (since $D(w_0 + \sum_{i=1}^{m} c_i \log w_i) = Dw_0 + \sum_{i=1}^{m} c_i \frac{Dw_i}{w_i} = 0$) and such that $w_0 + \sum_{i=1}^{m} c_i \log w_i$ is not identically constant. Notice that letting $m = 2$, $w_0 = dx + by$, $w_1 = x$ and $w_2 = y$ illustrates the conclusion of the Theorem.

The proof of the Theorem relies on results from [ROS: 76] and [ROSSIN: 77]. We can deduce the following generalization of a theorem of Mordukhai-Boltovski, [M. -B.: 06]. (Also see [RITT: 48].):

Corollary 1. Let K be a differential field of functions of two variables x,y with derivations δ_x and δ_y and L an elementary extension of $(K,\{\delta_x,\delta_y\})$ with $C(K,\{\delta_x,\delta_y\}) = C(L,\{\delta_x,\delta_y\})$. Let f be in K and assume there exists a non-constant g in L which is a first integral of $y' = f(x,y)$. Then there exist $w_0,\ldots,w_m$ algebraic over K and constants $c_1,\ldots,c_m$ such that

$$w_0(x,y) + \sum_{i=1}^{m} c_i \log w_i(x,y)$$

is a non-constant first integral of $y' = f(x,y)$.

Proof: Let $D = \delta_x + f\delta_y$. Note $Dh = 0$ if and only if h is a first integral. Our hypotheses imply $Dg = 0$ while $\delta_x g \neq 0$ and $\delta_y g \neq 0$. The conclusion now follows from the Theorem.

Mordukhai-Boltovski proved this theorem when K is an algebraic extension of $\mathbb{C}(x,y)$. In this case (or more generally when K is a liouvillian extension of $\mathbb{C}(x,y)$) we can strengthen the conclusion to state that the w_i are in K. One can easily state a version of this corollary which also includes systems of partial differential equations.

We can also deduce the following generalization of a result of Ritt. Let k be an ordinary differential field with derivation δ and let $u_1,\ldots,u_n$ be elements of k. Define a new differential field (K,Δ) as follows: Let $K = k(U_1,\ldots,U_n)$ where $U_1,\ldots,U_n$ are indeterminants. Let $\Delta = \{\delta_0,\ldots,\delta_n\}$ where:

1) δ_0 restricted to k is δ and $\delta_0 U_i = 0$ for $i = 1,\ldots,n$.
2) For $i = 1,\ldots,n, \delta_i a = 0$ for all a in L, $\delta_i U_j = 0$ if $i \neq j$ and $\delta_i U_i = 1$.

Using our Theorem and other considerations we can show:

Corollary 2. Let (K,Δ) be as above and let (L,Δ) be an elementary extension of (K,Δ) so that $C(L,\Delta) = C(K,\Delta)$. Let $D = \delta_0 + \sum_{i=1}^{n} u_i\delta_i$ and assume $C(L,\Delta)$ is properly contained in $C(L,\{D\})$. Then there exist

$w_0,\ldots,w_m$ in k, and constants $c_1,\ldots,c_n$, $d_1,\ldots,d_m$ in $C(k,\{\delta\})$, not all the c_i's being zero, so that

$$\sum_{i=1}^{n} c_i u_i = \delta w_0 + \sum_{i=1}^{m} d_i \frac{\delta w_i}{w_i}$$

Loosely speaking, this result says that if there is an elementary function $g(x,U_1,\ldots,U_n)$ so that $g(x,\int u_1,\ldots,\int u_n)$ is a constant, then some non-trivial linear combination with constant coefficients of the $\int u_i$ is elementary over k. For n = 1, this result is due to RITT, [RITT: 23], [RITT: 48]. The result also appears in [SING: 77] and includes the result of [MOZI: 79].

For the next corollary, we let K be a differential field of functions of one variable y with derivation δ_y. Adjoin a new variable x to K and extend δ_y to K(x) so that $\delta_y(x) = 0$. Define a new derivation δ_x on K(x) by $\delta_x(a) = 0$ for all a in K and $\delta_x(x) = 1$.

Corollary 3. Let $(L,\{\delta_y,\delta_x\})$ be an elementary extension of $(K(x),\{\delta_y,\delta_x\})$ with $C(K,\{\delta_y\}) = C(L,\{\delta_y,\delta_x\})$. Let $f \neq 0$ be in K and $D = \delta_x + f\delta_y$. Assume there exists a g in L so that $Dg = 0$ but $\delta_x g \neq 0$. Then there exist $w_0,\ldots,w_m$ algebraic over K and constants $c_0,\ldots,c_m$ such that

$$D(c_0 x + w_0 + \sum_{i=1}^{m} c_i \log w_i(y)) = 0$$

but

$$\delta_x(c_0 x + w_0 + \sum_{i=1}^{m} c_i \log w_i(y)) \neq 0$$

Furthermore, we must have

$$\frac{1}{f(y)} = -\frac{1}{c_0}[w_0 + \sum_{i=1}^{m} c_i \frac{\delta_y w_i}{w_i}].$$

Roughly speaking, this corollary states that $y' = f(y)$ has an elementary first integral if and only if $\int \frac{1}{f(y)}\,dy$ is elementary. In [SING: 75], Singer dealt with a special case of this result.

3. Algorithmic Considerations

This work was motivated by our desire to find a decision procedure for finding elementary first integrals. It is our hope that the above theorem will be a starting point, by showing that if such integrals exist then there must exist ones of a prescribed form. Persuing this idea leads to interesting problems:

Example 1 revisited again. This time consider the system of differential equations:

$$\frac{dx}{dt} = P(x,y)$$

$$\frac{dy}{dt} = Q(x,y) \tag{3}$$

where P and Q are polynomials in x and y with constant coefficients. F(x,y) is a first integral if and only if DF = 0 where $D = P\frac{\partial}{\partial x} + Q\frac{\partial}{\partial y}$. One can show that there exists a first integral of the form $w_0 + \sum_{i=1}^{m} c_i \log w_i$ with w_i in $\mathbb{C}(x,y)$ if and only if there exists a rational integrating factor R(x,y) for the form Qdx - Pdy. Such a function satisfies

$$DR = -\left(\frac{\partial P}{\partial x} + \frac{\partial Q}{\partial y}\right)R \tag{4}$$

Therefore to decide if (3) has an elementary first integral, we must decide if (4) has a rational solution R(x,y) in $\mathbb{C}(x,y)$.

Assume that such an R exists and write $R = \prod_{i=1}^{m} f_i^{n_i}$ with f_i irreducible in $\mathbb{C}[x,y]$ and n_i in Z. From (4) we can conclude that for each i, f_i must divide Df_i. Results of Darboux ([JOU: 79], Thrm. 3.3, p. 102 and Lemma 3.53, p. 112) imply that the degree of each f_i is bounded. No effective bound is given. This suggests the following problem:

<u>Problem</u>: Given P,Q in $\mathbb{C}[x,y]$, let $D = P\frac{\partial}{\partial x} + Q\frac{\partial}{\partial y}$. Give an effective procedure to find an integer N so that if f is irreducible in $\mathbb{C}[x,y]$ and f divides Df, then the degree of f is less than N.

The next problem would be to bound m in the expression of R. The results of Darboux imply that $m \leq \frac{(d+1)d}{2} + 2$ where d = max(deg P,deg Q) or (3) has a <u>rational</u> first integral. Once one can bound the degrees of the f_i, the set of coefficients of f so that f divides Df forms a projective variety which we can construct. We can then decide if this has fewer than $\frac{(d+1)d}{2} + 2$ points and if it does find them. If it has more, Darboux's results show us how to construct a <u>rational</u> first integral. Otherwise we can find all f so that Df divides f. Equation (4) can then be written as

$$\frac{DR}{R} = \sum n_i \frac{Df_i}{f_i} = -\left(\frac{\partial P}{\partial x} + \frac{\partial Q}{\partial y}\right).$$

This now is equivalent to a system of linear equations in the n_i and we must check to see if these can be solved with n_i in Z.

Therefore, the problem of finding elementary first integrals is reduced to the above open problem. When max(deg P,deg Q) = 1 a solution appears in ([JOU: 79], pp. 8-19). For max(deg P,deg Q) > 1, no solution seems to be known. Partial results appear in ([PAIN: 72] V. I, pp. 173-218, V. II, pp. 433-458) and ([POINC: 34] V. III, pp. 32-97).

References

[JOU: 79] J. P. Jouanolou, Equations de Pfaff algébriques, Lecture notes in mathematics, V. 708, Springer-Verlag, 1979.

[M. -B: 06] D. Mordukhai-Boltovski, Researches on the integration in finite terms of differential equations of the first order, Communications de la Societé Mathematique de Kharkov, X(1906-1909), 34-64, 231-69. (In Russian)

[MOZI: 79] J. Moses, R. Zippell, An extension of Liouville's Theorem, in Symbolic and Algebraic Computation, E. W. Ng (Ed.), Springer-Verlag, 1979, 426-430.

[PAIN: 72] Oeuvres de Paul Painlevé, V. I and II, C.N.R.S., 1972.

[POINC: 34] Oeuvres de Henri Poincaré, V. III, Gauthier-Villars, 1934.

[RITT: 23] J. F. Ritt, On the integrals of elementary functions, Trans. Amer. Math. Soc., 25, (1923), 211-222.

[RITT: 48] J. F. Ritt, Integration in finite terms. Liouville's Theory of Elementary Methods, Columbia University Press, N.Y., 1948.

[ROS: 76] M. Rosenlicht, On Liouville's theory of elementary functions, Pacific J. Math., 65(1976), 485-492.

[ROSSIN: 77] M. Rosenlicht, M. Singer, On elementary, generalized elementary, and liouvillian extension fields, in Contributions to Algebra, Academic Press, 1977, 329-342.

[SING: 75] M. Singer, Elementary solutions of differential equations, Pacific J. Math., 59 No. 2, 535-547.

[SING: 77] M. Singer, Functions satisfying elementary relations, Trans. Amer. Math. Soc., 227, 1977, 185-206.

A Technique for Solving Ordinary Differential Equations Using Riemann's P-functions

Shunro Watanabe
Department of Mathematics, Tsuda College
Kodaira-shi Tokyo Japan

Abstract

This paper presents an algorithmic approach to symbolic solution of 2nd order linear ODEs. The algorithm consists of two parts. The first part involves complete algorithms for hypergeometric equations and hypergeometric equations of confluent type. These algorithms are based on Riemann's P-functions and Hukuhara's P-functions respectively. Another part involves an algorithm for transforming a given equation to a hypergeometric equation or a hypergeometric equation of confluent type. The transformation is possible if a given equation satisfies certain conditions, otherwise it works only as one of heuristic methods. However our method can solve many equations which seem to be very difficult to solve by conventional methods.

1. Introduction

We consider in this paper the following 2nd order linear ODE with rational function coefficients,

(1.1) $y'' + F(x)y' + G(x)y = H(x).$

As is well-known, the general solution of (1.1) is the sum of the special solution of (1.1) and the general solution of the corresponding homogeneous equation,

(1.2) $y'' + F(x)y' + G(x)y = 0.$

We will confine ourselves to only (1.2) in this paper. The class of ODEs we are considering covers many equations which appear in mathematical physics. Therefore, although we consider very restricted class of functions mathematically, our work will have a good significance practically.

We have, at present, no general method for solving (1.2). However, for several subclasses of (1.2), mathematicians gave the solutions. One such subclass is the class of hypergeometric equations :

(1.3) $$y''+ \frac{ax+b}{(x-\alpha_1)(x-\alpha_2)}y'+ \frac{cx^2+dx+e}{(x-\alpha_1)^2(x-\alpha_2)^2}y = 0,$$

and another such subclass is the class of hypergeometric equations (h.g.e.s) of confluent type:

(1.4) $$y''+ \frac{ax+b}{x-\alpha}y'+ \frac{cx^2+dx+e}{(x-\alpha)^2}y = 0.$$

The equation in these two subclasses has a remarkable property that the general solution is completely characterized by a finite number of parameters of several local solutions. This remarkable property is found only in the above two subclasses at present. The general solution represented by a few local parameters is called Riemann's P-function or the P-function of confluent type.

The solutions of some h.g.e.s are representable in algebraic functions, and the condition for this was discovered in 1872 by H.A.Schwarz. In 1941-52, Hukuhara investigated the transformation rules of P-functions and discovered the conditions for that the solution of h.g.e. was representable in elementary functions. Our algorithm for obtaining the representation for the P-function explicitly is derived by the researches of Hukuhara.

Hukuhara already found a method of solving

(1.2) by transforming the equation to either (1.3) or (1.4). This transformation was done by removing the apparent singular points. His method has, however, a defect that the removal of an apparent singular point often causes an appearance of other singular points.

In 1970, Kimura discovered a method of transforming a Fuchs type DE to a h.g.e. which has several apparent singular points except for singular points at 0, 1, and infinity. Kimura's method seems to be extended to the h.g.e. of confluent type.

This paper describes the methods of Fuchs, Gauss, Riemann, Hukuhara, Kimura, et.al. for solving the equation (1.2) and the implementation of their methods into REDUCE 2.

2. P-functions and Apparent Singular Points

The solution of ODE (1.2) is characterized much by the number and the rank of the singular points of the equation. Because (1.2) is linear in y, the singular points of the solution y coincide with those of the coefficients of the equation, namely with the zero points of the denominators of F and G. The singularity at infinity is defined by the singularity at t=0 with the transformation t=1/x. The simplest singular points are called regular singular points. The rank of the singularity of (1.2) at infinity is the maximum of f and g/2 plus one, where f is the degree of F and g is the degree of G. If the rank is less than or equal to zero, infinity is a regular singular point. If $x=\alpha$ is a pole of F as well as G, and if the order of the pole is at most one for F and at most two for G, then $x=\alpha$ is a regular singular point of (1.2). This is well-known Fuchs' theorem. According to these definitions, all the singular points of (1.3), i.e. α_1, α_2, and infinity are regular singular points. Furthermore, in (1.4), $x=\alpha$ is a regular singular point and infinity is an irregular singular point of rank one.

At each regular singular point $x=\alpha$, (1.2) has two independent power series solutions of the form

(2.1a) $y_1=(x-\alpha)^{\lambda_1}\{ g_0 + g_1(x-\alpha)^1+\dots \},\ g_0 \neq 0,$

(2.1b) $y_2=(x-\alpha)^{\lambda_2}\{h_0+h_1(x-\alpha)^1+\dots\}+c_0y_1\log(x-\alpha),$

where $h_0\neq 0$, and λ_1 and λ_2 are two solutions of the characteristic equation of (1.2) at $x=\alpha$,

(2.2) $$\begin{cases} \lambda(\lambda-1)+F_1(\alpha)\lambda+G_1(\alpha)=0, \\ F_1(x)=(x-\alpha)F(x),\ G_1(x)=(x-\alpha)^2G(x). \end{cases}$$

The λ_1 and λ_2 are called exponents of (1.2) at α. Equation (1.3) has six exponents λ_1, λ_2, μ_1, μ_2, and ν_1,ν_2 at α_1, α_2, and infinity respectively. There exists a connection among them, which is known as Fuchs' relation,

(2.3) $\lambda_1+\lambda_2+\mu_1+\mu_2+\nu_1+\nu_2 = 1.$

Conversely, for given values of λ_1, λ_2, μ_1, μ_2, and ν_1, ν_2 satisfying the relation (2.3), there is only one equation of the form (1.3) having the same exponents. The general solution of this equation is written as

(2.4) $$y = P\left\{\begin{matrix} \alpha_1 & \alpha_2 & \infty & \\ \lambda_1 & \mu_1 & \nu_1 & x \\ \lambda_2 & \mu_2 & \nu_2 & \end{matrix}\right\},$$

and it is called Riemann's P-function. We must note here that Riemann's P-function is not a function but a class of functions which satisfy (1.3). The general solution of (1.3) is characterized by only values of parameters in the braces of (2.4).

Similarly, we can construct P-functions for (1.4). Using the following linear fractional transformation with respect to the independent variables,

(2.5) $t = (Ax+B)/(Cx+D),\ AD-BC \neq 0,$

we can transform (1.4) to its standard form

(2.6) $$y'' + \frac{ax+b}{x}y' + \frac{cx^2+dx+e}{x^2}y = 0.$$

The characteristic equation at x=0 is

(2.7) $\sigma(\sigma-1)+b\sigma+e = 0,$

whose two roots σ_1 and σ_2 give the exponents of (2.6) at x=0. When $a^2\neq 4c$ in (2.6), (2.6) has two independent solutions at infinity. The solutions have the following form known as asymptotic expansions:

(2.8) $y_i=e^{\lambda_i x}x^{-\mu_i}\{c_{i0}+c_{i1}x^{-1}+c_{i2}x^{-2}+\dots\},c_{i0}\neq 0,$

where λ_i and μ_i are defined by

(2.9) $\lambda_i^2+a\lambda_i+b=0,\ \mu_i=(b\lambda_i+d)/(2\lambda_i+a),\ i=1,2.$

There is a relation connecting the values σ_1, σ_2, μ_1, and μ_2 which is similar to Fuchs' relation, and known as Hukuhara's relation:

(2.10) $\sigma_1+\sigma_2+\mu_1+\mu_2 = 1$.

Conversely, for given values of σ_1, σ_2, μ_1, and μ_2 satisfying the relation (2.10), there is only one equation of the form (2.6) that has σ_i and σ_2 as two exponents at x=0 and has two independent solutions of the form (2.8) at infinity. The general solution in this case is written as

$$(2.11) \qquad y = P\left\{\begin{matrix} \overbrace{\quad\infty\quad} & 0 & \\ \lambda_1 \quad \mu_1 & \sigma_1 & x \\ \lambda_2 \quad \mu_2 & \sigma_2 & \end{matrix}\right\},$$

and it is called Hukuhara's P-function. We must note here that Hukuhara's P-function is not a function but a class of functions which satisfy (2.6). The general solution of (2.6) is characterized by only values of parameters in the braces of (2.11).

When $a^2=4c$ in (2.6), the equation has two independent solutions at infinity. The solutions have the following form known as asymptotic expansions:

$$(2.12) \quad y_i= e^{\lambda x+\beta_i t}\, t\, \{c_{i0}+c_{i1}t^{-1}+..\}\ , c_{i0}\neq 0,\ i=1,2,$$

where t, λ, and β_i are defined by

$$(2.13) \quad t=\sqrt{x},\ \lambda^2+a\lambda+b=0,\ \ \beta_i^2 + 4(d+b\lambda)=0,\ i=1,2.$$

Conversely, for given values of σ_1, σ_2 , λ , and $\beta(=|\beta_i|)$ satisfying (2.13), there is only one equation of the form (2.6) which has σ_1 and σ_2 as two exponents at x=0, and has two asymptotic expansions (2.12) at infinity. The general solution in this case is written as

$$(2.14) \qquad y = P\left\{\begin{matrix} \overbrace{\quad\infty^*\quad} & 0 & \\ \lambda \quad \beta & \sigma_1 & x \\ \lambda \quad -\beta & \sigma_2 & \end{matrix}\right\},$$

and it is called Hukuhara's P-function.

When $x=\alpha$ is a regular singular point, we can rewrite (1.2) as follows

$$(2.15) \quad (x-\alpha)^2y'' + (x-\alpha)F_1(x)\ y' + G_1(x)y = 0,$$

where $F_1(x)$ and $G_1(x)$ have no singular point at $x=\alpha$. At a regular singular point α, we can calculate the power series solutions by substituting y in (2.15) by $(x-\alpha)^\lambda\ \{g_0+g_1(x-\alpha)^1+\dots+g_m(x-\alpha)^m+\dots\}$. It follows that the coefficients g_m must satisfy the following conditions

$$(2.16) \quad g_m f_0(\lambda+m)+g_{m-1}f_1(\lambda+m-1)+\dots+g_0 f_m(\lambda)=0,$$

where m=0,1,2,... and f_ℓ is defined by

$$(2.17) \quad f_\ell(\lambda)= \frac{1}{\ell!}\frac{\partial^\ell}{\partial x^\ell}\{\lambda(\lambda-1)+F_1(x)\lambda+G_1(x)\}\Big|_{x=\alpha},$$

$$\ell=0,1,2,\dots ,$$

and λ is the solution of the characteristic equation $f_0(\lambda)=0$ at $x=\alpha$. Let $\lambda_1-\lambda_2$, with λ_1 and λ_2 roots of $f_0(\lambda)=0$, be a positive integer k. If $f_0(\lambda+m)\neq 0$, g_m is determined uniquely by (2.16). When $f_0(\lambda+k)=0$, which happens for the case $\lambda=\lambda_2$ and $\lambda+k=\lambda_1$, for example, g_k cannot be calculated unless the following condition is satisfied,

$$(2.18) \quad g_{k-1}f_1(\lambda-k+1)+\dots+g_0 f_k(\lambda)=0.$$

The α is called an apparent singular point of (2.15) if (2.18) is satisfied. In other words, if $\lambda_1-\lambda_2$ is a positive integer and $c_0=0$ in the solution (2.1), then α is an apparent singular point.

3. Riemann's P-functions

We can determine whether or not a given Riemann's P-function (2.4) has an explicit expression by investigating the so-called Riemann's indices which are three differences of characteristic roots at the singular points α_1, α_2, and infinity

(3.1) (λ, μ, ν).

Here, the sign and the order of the differences are unessential. Let the points α_1, α_2, and infinity be transformed to β_1, β_2, and β_3 respectively, by the transformation (2.5). Then we have

$$(3.2) \quad P\left\{\begin{matrix} \alpha_1 & \alpha_2 & \infty & \\ \lambda_1 & \mu_1 & \nu_1 & x \\ \lambda_2 & \mu_2 & \nu_2 & \end{matrix}\right\} = P\left\{\begin{matrix} \beta_1 & \beta_2 & \beta_3 & \\ \lambda_1 & \mu_1 & \nu_1 & t \\ \lambda_2 & \mu_2 & \nu_2 & \end{matrix}\right\},\ t=\frac{Ax+B}{Cx+D}.$$

The definition of P-function gives us

$$(3.3) \quad (x-\alpha_1)^p(x-\alpha_2)^q y=P\left\{\begin{matrix} \alpha_1 & \alpha_2 & \infty & \\ \lambda_1+p & \mu_1+q & \nu_1-p-q & x \\ \lambda_2+p & \mu_2+q & \nu_2-p-q & \end{matrix}\right\},$$

where y is the P-function in the left hand side of (3.2). Setting the values p and q to $-\lambda_2$ and $-\mu_2$ respectively in (3.3), we get

$$(3.4) \quad (x-\alpha_1)^{-\lambda_2}(x-\alpha_2)^{-\mu_2}\ y=P\left\{\begin{matrix} \alpha_1 & \alpha_2 & \infty & \\ \lambda & \mu & \nu+\xi & x \\ 0 & 0 & \xi & \end{matrix}\right\},$$

where y is the P-function in the left hand side of (3.2) and $\lambda=\lambda_1-\lambda_2$, $\mu=\mu_1-\mu_2$, $\nu=\nu_1-\nu_2$, and $\xi=\nu_2-\lambda_2-\mu_2$. When ξ, $\xi+\nu\neq 0,-1,-2,\dots,-(n-1)$, with n a positive integer, we have the following differentiation fomula for the Riemann's P-function in the right hand side of (3.4):

$$(3.5)\quad \frac{d^n}{dx^n} P\begin{Bmatrix} \alpha_1 & \alpha_2 & \infty & \\ \lambda & \mu & \nu+\xi & x \\ 0 & 0 & \xi & \end{Bmatrix} = P\begin{Bmatrix} \alpha_1 & \alpha_2 & \infty & \\ \lambda-n & \mu-n & \nu+\xi+n & x \\ 0 & 0 & \xi+n & \end{Bmatrix}.$$

<u>Theorem</u> (Hukuhara). Let $p=\lambda-\lambda'$, $q=\mu-\mu'$, and $r=\nu-\nu'$ be integers such that $p \geq q \geq r$ and $p+q+r=$ an even integer. Then, for two Riemann's P-functions

$$y=P\begin{Bmatrix} 0 & 1 & \infty & \\ \lambda & \mu & \nu+\xi & x \\ 0 & 0 & \xi & \end{Bmatrix},\quad z=P\begin{Bmatrix} 0 & 1 & \infty & \\ \lambda' & \mu' & \nu'+\xi' & x \\ 0 & 0 & \xi & \end{Bmatrix},$$

we have the following relation

$$(3.6)\quad z=(x-1)^{\mu-m-r-n} D_x^n \left[(x-1)^{-\mu+r+m} D_x^m \{x^{-\xi-r} D_t^r (x^{\xi}y)\}\right],$$

where $t=1/x$, $m=(p+q-r)/2$, $n=(p-q+r)/2$, $D_x^n = d^n/dx^n$, and $D_t^r = d^r/dt^r$. Here, the following conditions are assumed to make the differentiation possible:

$$(3.7)\quad \begin{cases} \xi,\ \xi+\lambda \neq 0,-1,-2,\dots,-(r-1), \\ \xi+r, \xi+\nu \neq 0,-1,-2,\dots,-(m-1), \\ \xi+\mu,\ \xi+\mu+\nu \neq 0,-1,-2,\dots,-(n-1). \end{cases}$$

The proof is straightforward under the conditions (3.7).

If one of the conditions (3.7) is not satisfied, which is the case for $\xi=-k$, $0 \leq k \leq r-1$, for example, then $x^{\xi}y$ can be expressed by elementary functions. Now we investigate the conditions for that the DE can be solved by the integration, which is defined as follows.

<u>Definition</u>. A h.g.e. (1.3) is said to be solvable if and only if the equation is transformed to

$$(3.8)\quad z'' + H(x)\, z' = 0,$$

after a finite number of the following transformations

$$(3.9)\quad z=(x-\alpha)^{\rho}y,\ (\ \alpha=\alpha_1,\ \alpha_2,\ \text{or } 0\ ),\ \text{and } z=y'.$$

<u>Theorem</u>. The h.g.e. (1.3) or corresponding Riemann's P-function (2.4) is solvable if and only if one of four values $\lambda\pm\mu\pm\nu$ is an odd integer, where $\lambda=\lambda_1-\lambda_2$, $\mu=\mu_1-\mu_2$, and $\nu=\nu_1-\nu_2$. For example, in the case $\lambda+\mu+\nu=-2m+1$, $m \geq 1$, we have

$$(3.10)\quad y=P\begin{Bmatrix} \alpha_1 & \alpha_2 & \infty & \\ \lambda_1 & \mu_1 & \nu_1 & x \\ \lambda_2 & \mu_2 & \nu_2 & \end{Bmatrix} = (x-\alpha_1)^{\lambda_2}(x-\alpha_2)^{\mu_2}\frac{d^{m-1}}{dx^{m-1}}$$

$$\left[(x-\alpha_1)^{\lambda+m-1}(x-\alpha_2)^{\mu+m-1}\{A+B\int (x-\alpha_1)^{-\lambda-m}(x-\alpha_2)^{-\mu-m}dx\}\right],$$

using (3.3) and (3.5).

The transformation $x^r=t$ transforms

$$(3.11)\quad x^2y'' + x\,F_0(x^r)y' + G_0(x^r)y = 0$$

to

$$(3.12)\quad r^2t^2\ddot{y} + r\,t\{F_0(t)+r-1\}\dot{y} + G_0(t)y = 0,$$

where r is a positive integer, and the dot denotes the differation with respect t. From (3.11) and (3.12) we get the two famous formulas

$$(3.13)\quad P\begin{Bmatrix} 1 & -1 & \infty & \\ 0 & 0 & \nu & x \\ \lambda & \lambda & \nu' & \end{Bmatrix} = P\begin{Bmatrix} 0 & 1 & \infty & \\ 0 & 0 & \nu/2 & t \\ 1/2 & \lambda & \nu'/2 & \end{Bmatrix},\ x^2=t,$$

$$(3.14)\quad P\begin{Bmatrix} 1 & \omega & \omega^2 & \\ \lambda & \lambda & \lambda & x \\ \lambda' & \lambda' & \lambda' & \end{Bmatrix} = P\begin{Bmatrix} 0 & 1 & \infty & \\ 0 & \lambda & 0 & t \\ 1/3 & \lambda' & 1/3 & \end{Bmatrix},\ x^3=t,$$

where $\omega=(-1+\sqrt{3})/2$.

In 1872, H.A.Schwarz published a paper giving criteria on whether or not the general solution of a h.g.e. is algebraic, and the criteria were made into a table in terms of Riemann's indices, known as Schwarz' table. Later in 1949, Hukuhara and Ohashi published a paper written in Japanese giving criteria on whether or not the general solution of a h.g.e. can be expressed in elementary functions, allowing transformations of independent variable of the form $t=x^r$, $r=1,2$, and $t=(Ax+B)/(Cx+D)$, $AD-BC\neq0$, as well as the transformation of dependent variable of the form $z=(x-\alpha)^{\rho}y$ and $z=y'$. The criteria were also made into a table in terms of Riemann's indices. The following table summarizes the results of Hukuhara and Ohashi (type 0, type 1, ..., and type 5) and those of Schwarz (type 1 with r a rational number, type 2, ..., and type 15).

type	λ	μ	ν	
0	λ	μ	ν	$\lambda\pm\mu\pm\nu$ = an odd integer
1	1/2+k	1/2+m	r	r=any complex number
2	1/2+k	1/3+m	1/3+n	k+m+n=an even integer
3	2/3+k	1/3+m	1/3+n	
4	1/2+k	1/3+m	1/4+n	k+m+n=an even integer
5	2/3+k	1/4+m	1/4+n	
6	1/2+k	1/3+m	1/5+n	k+m+n=an even integer
7	2/5+k	1/3+m	1/3+n	"
8	2/3+k	1/5+m	1/5+n	"
9	1/2+k	2/5+m	1/5+n	"
10	3/5+k	1/3+m	1/5+n	"
11	2/5+k	2/5+m	2/5+n	"
12	2/3+k	1/3+m	1/5+n	"
13	4/5+k	1/5+m	1/5+n	"
14	1/2+k	2/5+m	1/3+n	"
15	3/5+k	2/5+m	1/3+n	"

In 1952, Hukuhara and Ohashi published another paper written in Japanese describing an integration method for equations of type 0,..., and type 5. The integration method for equations of type 0 has already shown in (3.10). For equations of type 1,..., and type 5, we explain only equations of type 1 to save the paper. For the case $k=m=0$, the Riemann's P-function is expressed explicitly as

$$(3.15)\quad P\begin{Bmatrix} 0 & 1 & \infty & \\ 0 & 0 & r/2 & x \\ 1/2 & 1/2 & -r/2 & \end{Bmatrix} = A(\sqrt{x}+\sqrt{x-1})^r + B(\sqrt{x}-\sqrt{x-1})^r.$$

The general case of equations type 1 can be reduced to the above case by Hukuhara's theorem (3.6). The integration methods for the equations of other types are quite similar to that of type 1.

4. Hukuhara's P-functions

In 1941, Hukuhara introduced P-functions for h.g.e.s of confluent type as was mentioned in §2. He obtained many theorems and formulas similar to those of Riemann's P-functions. Let us apply the transformations of independent variables

$$(4.1)\quad x+\alpha = t, \text{ and } x = \delta t,$$

to (2.6). Then we get

$$(4.2)\quad P\begin{Bmatrix} \overbrace{\quad\infty\quad} & & 0 & \\ \lambda_1 & \mu_1 & \sigma_1 & x \\ \lambda_2 & \mu_2 & \sigma_2 & \end{Bmatrix} = P\begin{Bmatrix} \overbrace{\quad\infty\quad} & & \alpha & \\ \lambda_1 & \mu_1 & \sigma_1 & t \\ \lambda_2 & \mu_2 & \sigma_2 & \end{Bmatrix},\quad x+\alpha=t,$$

$$(4.3)\quad P\begin{Bmatrix} \overbrace{\quad\infty\quad} & & 0 & \\ \lambda_1 & \mu_1 & \sigma_1 & x \\ \lambda_2 & \mu_2 & \sigma_2 & \end{Bmatrix} = P\begin{Bmatrix} \overbrace{\quad\infty\quad} & & 0 & \\ \delta\lambda_1 & \mu_1 & \sigma_1 & t \\ \delta\lambda_2 & \mu_2 & \sigma_2 & \end{Bmatrix},\quad x=\delta t.$$

Simiarly we get the following relation by the transformation $z=e^{\lambda x}x^{\sigma}y$:

$$(4.4)\quad e^{\lambda x}x^{\sigma}\, P\begin{Bmatrix} \overbrace{\quad\infty\quad} & & 0 & \\ \lambda_1 & \mu_1 & \sigma_1 & x \\ \lambda_2 & \mu_2 & \sigma_2 & \end{Bmatrix} = P\begin{Bmatrix} \overbrace{\quad\infty\quad} & & 0 & \\ \lambda_1+\lambda & \mu_1-\sigma & \sigma_1+\sigma & x \\ \lambda_2+\lambda & \mu_2-\sigma & \sigma_2+\sigma & \end{Bmatrix}.$$

Setting $\lambda_2+\lambda$ and $\sigma_2+\sigma$ to zero in (4.4), we get

$$(4.5)\quad e^{-\lambda_2 x}x^{-\sigma_2} P\begin{Bmatrix} \overbrace{\quad\infty\quad} & & 0 & \\ \lambda_1 & \mu_1 & \sigma_1 & x \\ \lambda_2 & \mu_2 & \sigma_2 & \end{Bmatrix} = P\begin{Bmatrix} \overbrace{\quad\infty\quad} & & 0 & \\ \lambda & \mu_1+\sigma_2 & \sigma & x \\ 0 & \mu_2+\sigma_2 & 0 & \end{Bmatrix},$$

where $\lambda=\lambda_1-\lambda_2$ and $\sigma=\sigma_1-\sigma_2$. When $\lambda_1(\mu_2+k)\neq 0,1,2,\ldots,n-1$, with n a positive integer. We get the following differentiation formula for the P-function in right hand side of (4.5):

$$(4.6)\quad \frac{d^n}{dx^n} P\begin{Bmatrix} \overbrace{\quad\infty\quad} & & 0 & \\ \lambda_1 & \mu_1 & \sigma_1 & x \\ 0 & \mu_2 & 0 & \end{Bmatrix} = P\begin{Bmatrix} \overbrace{\quad\infty\quad} & & 0 & \\ \lambda_1 & \mu_1 & \sigma_1-n & x \\ 0 & \mu_2+n & 0 & \end{Bmatrix}.$$

Relations (4.5) and (4.6) lead us to the following theorem which gives the condition for that one P-function is transformed to another P-function.

<u>Theorem</u> (Hukuhara). Let y and z be the following P-functions:

$$(4.7)\quad y = P\begin{Bmatrix} \overbrace{\quad\infty\quad} & & 0 & \\ \lambda_1 & \mu_1 & \sigma_1 & x \\ \lambda_2 & \mu_2 & \sigma_2 & \end{Bmatrix},\quad z = P\begin{Bmatrix} \overbrace{\quad\infty\quad} & & 0 & \\ \lambda_1' & \mu_1' & \sigma_1' & x \\ \lambda_2' & \mu_2' & \sigma_2' & \end{Bmatrix},$$

where we assume that $\lambda_1-\lambda_2=\lambda_1'-\lambda_2'$ and both p and q defined by

$$(4.8)\quad p = \mu_1'-\mu_1+\sigma_1'-\sigma_1,\quad q = \mu_2'-\mu_1+\sigma_1'-\sigma_1,$$

are integers. Then, if integers m, m', n, and n' satisfy the relations $n'-n=p$ and $m'-m=q$, y and z are related to each other as

$$(4.9)\quad z = e^{\lambda_2' x}x^{\sigma_1'}\, D^{m'}\{e^{\lambda x}\, D^{n'}\{x^{-\sigma+m+n}\, D^{m}\{e^{-\lambda x}\, D^{n}\{e^{-\lambda_2 x}\, x^{-\sigma_2}y\}\}\}\},$$

where $\lambda=\lambda_1-\lambda_2$, $\sigma=\sigma_1-\sigma_2$, and $D=d/dx$.

This theorem is easily proved by the straight substitution. Note that the condition $\lambda_1-\lambda_2=\lambda_1'-\lambda_2'$ is unessential, because the transformation $x=\delta t$ breaks the condition (cf.(4.3)).

Note further that, in some cases, the differentiation is impossible in (4.9), for example, when $\mu_2+\sigma_2=1-k$ or equivalently $\mu_1+\sigma_1=k$. In such a case, y is representable in elementary functions. Let us show this for the case $\mu_2+\sigma_2=1-k$:

$$(4.10)\quad y = P\begin{Bmatrix} \overbrace{\quad\infty\quad} & & 0 & \\ \lambda_1 & \mu_1 & \sigma_1 & x \\ \lambda_2 & \mu_2 & \sigma_2 & \end{Bmatrix}$$

$$= e^{\lambda_1 x}x^{\sigma_1}D^k\left\{A+B\int e^{(\lambda_2-\lambda_1)x}x^{\sigma_2-\sigma_1+k-1}dx\right\}.$$

5. Methods for Removing Apparent Singular Points

A simple method for removing of apparent singular points was known to mathematicians before 1941. If $x=\alpha$ is an apparent singular point of (1.2), we have two power series solutions at $x=\alpha$:

$$(5.1)\quad \begin{cases} y_1 = (x-\alpha)^{\lambda+k}\,\{g_0+g_1(x-\alpha)^1+\ldots\},\ g_0\neq 0, \\ y_2 = (x-\alpha)^{\lambda}\,\{h_0+h_1(x-\alpha)^1+\ldots\},\ h_0\neq 0, \end{cases}$$

where k is a positive integer. For the particular case that k=1 and $\lambda=0$ in (5.1), then $x=\alpha$ is a regular point as is well-known. By making the transformation $y=(x-\alpha)^{\lambda}z$ in (1.2), we obtain an equation which has the following two solutions at $x=\alpha$,

$$(5.2)\quad \begin{cases} z_1=(x-\alpha)^{k}\{g_0+g_1(x-\alpha)^{1}+\ldots\},\ g_0\neq 0, \\ z_2=(x-\alpha)^{0}\{h_0+h_{\ell}(x-\alpha)^{\ell}+\ldots\},\ h_0\neq 0,\ h_{\ell}\neq 0. \end{cases}$$

Therefore, if k=1 in (5.2), then $x=\alpha$ is a regular point of the equation. When $k\geq 2$, the transformation $w=z'$ gives the following two solutions,

$$(5.3)\quad \begin{cases} w_1=(x-\alpha)^{k-1}\{g_0+g_1(x-\alpha)^{1}+\ldots\},\ g_0\neq 0, \\ w_2=(x-\alpha)^{\ell-1}\{h_0+h_1(x-\alpha)^{1}+\ldots\},\ h_0\neq 0. \end{cases}$$

We, thus reach again at the start solutions (5.1), and the repetition of this process removes the singular point. The above method is summarized in the transformations of the forms

$$(5.4)\quad z=(x-\alpha)^{\lambda}y,\ \text{ or } z=y'.$$

The latter transformation may cause, however, new singular points. This is a defect of the above method.

When all the singuar points are regular, the equation is said to be of Fuchs' type. As is well-known, if two equations of Fuchs' type

$$(5.5)\quad y''+A_1(x)\,y'+A_2(x)\,y=0$$

and

$$(5.6)\quad z''+B_1(x)\,z'+B_2(x)\,z=0$$

have the same set of singular points and the same monodromy group, there are rational functions $R_0(x)$ and $R_1(x)$ such that (5.5) is transformed from (5.6) by

$$(5.7)\quad y=R_0(x)\,z+R_1(x)\,z'.$$

In 1970, Kimura found that, for equations in proper subclasses of (5.5) and (5.6), their monodromy groups coincide. He showed that $R_0(x)$ and $R_1(x)$ are polynomials and can be calculated by the method of undetermined coefficients in that case.

Theorem (Kimura). Consider the equation

$$(5.8)\quad y''+\left\{\frac{\gamma}{x}+\frac{\delta}{x-1}+\sum_{i=1}^{k}\frac{\varepsilon_i}{x-a_i}\right\}y'+\frac{\alpha\beta x^{k}+\rho_1x^{k-1}+\ldots}{x(x-1)\prod_{i=1}^{k}(x-a_i)}y=0,$$

which has regular singular points $0,1,a_1,\ldots,a_k$, and infinity. Note that we can assume, without loss of generality, all of ε_i are negative integers. Assume that the singular points $a_1,\ldots,$ and a_k are apparent, and (5.8) is irreducible. Then if $\alpha-\beta\neq 1,2,\ldots,n-1$, with $n=-(\varepsilon_1+\ldots+\varepsilon_k)$, (5.8) is obtained from

$$(5.9)\quad z''+\left\{\frac{\gamma}{x}+\frac{\delta}{x-1}\right\}z'+\frac{\alpha(\beta+n)}{x(x-1)}z=0,$$

by a transformation of the form

$$(5.10)\quad y=(x^{n}+p_1x^{n-1}+\ldots+p_n)z+x(x-1)(x^{n-1}+q_2x^{n-2}+\ldots+q_n)z'.$$

6. Algorithm for Solving Linear ODEs

From the results described above, we get an integration method for (1.2). First, we judge whether or not a regular singular point is apparent by calculating finite coefficients of the power series solution at that point. The classification of all the singular points allows us to decide the type of the equation.

If the equation has the form (5.8), we apply the transformation (5.10) to obtain an equation of the standard form (1.3). Then, we apply the integration methods described in section 3 for the equation. If the given equation is of the type other than that of (5.8), we try to remove their apparent singular points by successive application of the transformation (5.4), and we try to integrate the reduced equation.

7. Examples

We have implemented the above algorithm into REDUCE 2 and tested our program on some fifty linear ODEs of the form (1.2). We will show only four examples with intermediate calculations.

Example 1.

$$(7.1)\quad y''+\frac{2x(x^2-5)}{x^4-1}y'+\frac{16(x^4+1)}{(x^4-1)^2}y=0.$$

The singular points of this equation are 1, -1, i, -i, and infinity. All of the singular points of this equation are regular. Putting the power series solution at x=1 as

$$y=(x-1)^{\lambda}\{g_0+g_1(x-1)^{1}+\ldots\},\ g_0\neq 0,$$

we obtain the following conditions on $\lambda, g_0, g_1, \ldots,$

$$g_0f_0(\lambda)=0,\ g_1f_0(\lambda+1)+g_0f_1(\lambda)=0,\ldots,$$

where $f_0(\lambda)=(\lambda-1)(\lambda-2)$, $f_1(\lambda)=2(\lambda-1)$. For the value $\lambda=1$ the conditions are satisfied, therefore x=1 is an apparent singular point. The transformation $y=(x-1)z$ transforms the equation (7.1) to

$$(7.2)\quad z'' + \frac{2(2x^2+3x-1)}{x^3+x^2+x+1}z' + \frac{2(x^4+3x^3+9x^2+11x+8)}{(x^3+x^2+x+1)^2}z=0.$$

The singular points of (7.2) are -1, i, -i, and infinity, the apparent singular point x=1 was removed by the transformation $y=(x-1)z$. Putting the power series solution of (7.2) at x=-1 as

$$z = (x+1)^{\lambda}\{g_0 + g_1(x+1)^1+\dots\},\ g_0\neq 0,$$

we obtain the following conditions on λ, $g_0, g_1, \dots,$

$$g_0 f_0(\lambda)=0,\ g_1 f_0(\lambda+1)+g_0 f_1(\lambda)=0,\dots,$$

where $f_0(\lambda)=(\lambda-1)(\lambda-2)$, $f_1(\lambda)=-3(\lambda-1)$. For the value $\lambda=1$ the conditions are satisfied for arbitrary g_0 and g_1, therefore x= -1 is an apparent singular point. The transformation $z=(x+1)w$ transforms the equation (7.2) to

$$(7.3)\quad w'' + \frac{6x}{x^2+1}w' + \frac{2(3x^2+7)}{(x^2+1)^2}w = 0.$$

The singular points of (7.3) are i, -i, and infinity, and all these are regular. Therefore the general solution of (7.3) can be written as

$$(7.4)\quad w = P\begin{Bmatrix} i & -i & \infty & \\ -1+\sqrt{3} & -1-\sqrt{3} & 3 & x \\ -1-\sqrt{3} & -1+\sqrt{3} & 2 & \end{Bmatrix}.$$

The indices of (7.4) are $(2\sqrt{3}, -2\sqrt{3}, 1)$, and the type of the P-function in (7.4) is 0. Therefore it has the following explicit expression

$$(7.5)\quad w=(x+i)^{-1+\sqrt{3}}(x-i)^{-1-\sqrt{3}}\{A+B\int(x-i)^{-2\sqrt{3}}(x+i)^{2\sqrt{3}}dx\}.$$

Consequently the general solution of (7.1) is given explicitly as

$$(7.6)\quad y = (x^2-1)w.$$

Example 2.

$$(7.7)\quad y'' - \frac{x^2+n^2}{x(x^2-n^2)}y' + \frac{x^2-n^2}{x^2}y = 0.$$

The singular points of (7.7) are 0, n, -n, and infinity. The first three are regular singular points, and infinity is an irregular singular point of rank one. The characteristic equations of (7.7) at 0 and n are $\lambda^2-n^2=0$ and $\lambda(\lambda-2)=0$, respectively. Putting the power series solution of (7.7) at x=n as

$$y = (x-n)^n\{g_0+g_1(x-n)^1+g_2(x-n)^2+\dots\},\ g_0\neq 0,$$

we obtain the following conditions on λ, $g_0, \dots,$

$$(7.8)\quad g_m f_0(\lambda+m)+g_{m-1}f_1(\lambda+m-1)+\dots+g_0 f_m(\lambda)=0,\quad m=0,1,2,\dots,$$

where $f_0(\lambda)=\lambda(\lambda-2)$, $f_1(\lambda)=-\lambda/(2n)$, $f_2(\lambda)=-3\lambda/(4n^2)$, $f_3(\lambda)=(-7\lambda-16n^2)/(8n^3)$. For the value $\lambda=0$ conditions (7.8) are satisfied, therefore x=n is an apparent singular point. We get the two solutions:

$$\begin{cases} y_1=(x-n)^2\{h_0+h_1(x-n)^1+\dots\},\ h_0\neq 0,\\ y_2=(x-n)^0\{g_0+\frac{2}{3n}g_0(x-n)^3+\dots\},\ g_0\neq 0.\end{cases}$$

By differentiating these functions, we obtain

$$\begin{cases} y_1'=(x-n)^1\{2h_0+3h_1(x-n)^1+\dots\},\ h_0\neq 0,\\ y_2'=(x-n)^2\{\frac{2}{n}g_0+\dots\},\ g_0\neq 0.\end{cases}$$

The transformation $y'=(x-n)z$ transforms the equation (7.7) to

$$(7.9)\quad z'' + \frac{x+3n}{x(x+n)}z' + \frac{(x-n)\{(x+n)^3-n\}}{x^2(x+n)^2}z = 0.$$

The singular points of (7.9) are 0, -n, and infinity. The first two are regular singular points, and infinity is an irregular singular point of rank one. The characteristic equations at 0 and -n are $(\lambda+1)^2-n^2=0$ and $(\lambda-1)(\lambda-2)=0$, respectively. The power series solution of (7.9) at x=-n,

$$z = (x+n)^{\lambda}\{g_0+g_1(x+n)^1+\dots\},\ g_0\neq 0,$$

must satisfy the following conditions,

$$g_0 f_0(\lambda)=0,\ g_1 f_0(\lambda+1)+g_0 f_1(\lambda)=0,\dots,$$

where $f_0(\lambda)=(\lambda-1)(\lambda-2)$, $f_1(\lambda)=-3\lambda(\lambda-1)/n$. For the value $\lambda=1$ the conditions are satisfied, therefore x=-n is an apparent singular point. The transformation $z=(x+n)w$ transforms the equation (7.9) to

$$(7.10)\quad x^2w'' + 3x\,w' + (x^2+1-n^2)\,w = 0.$$

The general solution of (7.10) can be written, by the relations (2.6), (2.7), (2.9), and (2.11), as follows

$$(7.11)\quad w=P\begin{Bmatrix} \overbrace{\infty} & 0 & \\ i\ 3/2 & n-1 & x \\ -i\ 3/2 & -n-1 & \end{Bmatrix} = \frac{1}{x}P\begin{Bmatrix} \overbrace{\infty} & 0 & \\ i\ 1/2 & n & x \\ -i\ 1/2 & -n & \end{Bmatrix} = \frac{1}{x}B_n(x),$$

where $B_n(x)$ is the general solution of Bessel's equation of order n ;

$$(7.12)\quad x^2y'' + x\,y' + (x^2-n^2)\,y = 0.$$

From the above sequence of transformations, and using simple calculations, we get the explicit expression of the general solution of (7.7) as

$$(7.13)\quad y = \frac{d}{dx}B_n(x).$$

Example 3.

$$(7.14)\quad y'' + \frac{2x+1}{x(x-1)}y' + \frac{64x^4+1}{4x^2(x-1)^2}y = 0.$$

The singular points of this equation are 0, 1, and infinity. The first two are regular singular points, and infinity is an irregular singular point of rank one. Putting the power series solution of (7.14) at x=0 as

$$y = x^{\lambda} \{g_0 + g_1 x^1 + \dots\}, \ g_0 \neq 0 ,$$

we obtain the following conditions on λ, g_0, $g_1, \dots$,

$$g_0 f_0(\lambda)=0, \ g_1 f_0(\lambda+1) + g_0 f_1(\lambda)=0, \dots,$$

where $f_0(\lambda)=\lambda^2 - \frac{1}{4}$, $f_1(\lambda) = \lambda - \frac{1}{6}$. For the value $\lambda = -\frac{1}{2}$ the conditions are not satisfied, hence x=0 is not the apparent singular point. The characteristic equation at x=1 is $4\lambda^2 - 16\lambda - 65 = 0$. The difference of two characteristic roots is 9. The power series solution of (7.14) at x=1 must satisfy (7.8) on $\lambda, g_0, \dots$, where $f_0(\lambda)=\lambda^2 - 4\lambda - \frac{65}{4}$, $f_1(\lambda)=\lambda - \frac{63}{2}$, $f_2(\lambda) = -\lambda - \frac{67}{4}$, $f_3(\lambda)=\lambda + 1$, $f_4(\lambda) = -\lambda - \frac{5}{4}$, $f_5(\lambda) = \lambda + \frac{3}{2}$, $f_6(\lambda) = -\lambda - \frac{7}{4}$, $f_7(\lambda)=\lambda + 2$, $f_8(\lambda) = -\lambda - \frac{9}{4}$, $f_9(\lambda) = \lambda + \frac{5}{2}$. For the values $g_0=1$ and $\lambda = -\frac{5}{2}$, we can calculate $g_0, g_1, \dots$, and g_8 as follows:

$$\begin{cases} g_0=1, \ g_1=-17/4, \ g_2=9, \ g_3=-3595/288, \\ g_4=70297/5760, \ g_5=-18991/2560, \ g_6=-74467/41472, \\ g_7=211505672/11612160, \ g_8=-79283484/1146880 . \end{cases}$$

These values do not satisfy the condition

$$g_8 f_1(\lambda+8) + g_7 f_2(\lambda+7) + \dots + g_0 f_9(\lambda) = 0,$$

therefore x=1 is not the apparent singular point. From the reasoning mentioned above, we can conclude that the general solution of (7.14) has no explicit expression.

Example 4.

$$(7.15) \quad y'' + \{\frac{1}{x} + \frac{3}{x-1} - \frac{1}{x-\frac{1}{4}}\}y' + \frac{1}{(x-1)(x-\frac{1}{4})} y = 0$$

The singular point of this equation are 0, 1, $\frac{1}{4}$, and infinity. The singular points 0, 1, and infinity are regular and $\frac{1}{4}$ is apparent. Equation (7.15) is obtained from

$$(7.16) \quad z'' + \{\frac{1}{x} + \frac{1}{x-1}\} z' + \frac{2}{x(x-1)} z = 0,$$

by the following transformation

$$(7.17) \quad y = (x + \frac{1}{2}) z + x(x-1) z' .$$

Equation (7.16) has the general solution representable in elementary functions:

$$(7.18) \quad z = P \begin{Bmatrix} 0 & 0 & \infty & \\ 0 & 0 & 1 & x \\ 0 & -2 & 2 & \end{Bmatrix} = (x-1)^{-1} \{A + B \int (x-1) dx\}.$$

8. Conclusion

We have described in this paper an algorithmic integration method for the equations belonged to a few subclasses of the class of the Eq. (1.2). Also we have shown that for the other class of Eq. (1.2), only the heuristic algorithm is possible. The number of the subclasses which may be integrated by an algorithmic method have increased in the past, and also will increase in the future by the efforts of many people. However, even now, it must be valuable to implement the algorithm on some proper symbolic computation systems.

9. References

(1) BIEBERBACH, L., Theorie der gewöhnlichen Differentialgleichungen, Springer 1965.

(2) CAYLEY, A., On the Schwarzian Derivative, and the Polyhedral Functions, Trans. Cambr. Phil. Soc. (1881) 5-68.

(3) HEARN, A.C., REDUCE 2 User's Manual, Univ. of Utah 1973.

(4) HUKUHARA, M., Integration Methods of Linear Ordinary Differential Equations, Iwanami 1941 (in Japanese).

(5) HUKUHARA, M. and OHASHI, S., On the determination of types of Riemann's P-function which can be expressed by elementary functions, Sugaku (1949) 227-230 (in Japanese).

(6) HUKUHARA, M. and OHASHI, S., On Riemann's P-function which is expressed by elementary functions, Sugaku (1952) 27-29 (in Japanese).

(7) KIMURA, T., On Riemann's equations which are solvable by quadratures, Funkcial. Ekvac. 12 (1969) 269-281.

(8) KIMURA, T., On Fuchsian DEQs reducible to h.g.e.s by linear transformations, Funkcial. Ekvac. 13 (1970) 213-232.

(9) LAFFERTY, E.L., Hypergeometric Function Reduction, MACSYMA User's Conf. 1979 465-481.

(10) POOLE, E.G.C., Intoduction to the Theory of Linear Differential Equations, Oxford 1936 and Dover Publ. 1967.

(11) SCHWARZ, H.A., Über diejenigen Fälle, in welchen die Gaussische hypergeometrische Reihe eine algebraische Function ihres vierten Elementes darstellt, J. für die reine und angew. Math. 75 (1872) 292-335.

(12) WATANABE, S., Formula Manipulations Solving Linear ODEs II, Publ. RIMS Kyoto Univ. 11 no.2 (1976) 297-337.

Using Lie Transformation Groups to Find Closed Form Solutions to First Order Ordinary Differential Equations

Bruce Char*
Applied Mathematics Division
Argonne National Laboratory
Argonne, Illinois 60439

1. Introduction

Most work on computer programs to find closed form solutions to ordinary differential equations (o.d.e.s) has concentrated on implementing a catalog of those methods often cited in textbooks and reference works (see e.g.Kam61a, Inc44a]): algorithms of certain, easily recognized cases (e.g. separable, exact, homogeneous equations) and a useful guessing framework for changes of variable. This approach has been followed by Moses, Schmidt, and Lafferty, among others [Mos67a, Sch76a], [Sch79a, Laf80a].

We present here a different approach to cataloguing, using the relation between differential equations and Lie transformation groups. When presented with a given a first order o.d.e., we shall be concerned with finding continuous transformations (of the plane) which map the solution curves of the o.d.e. into each other. When a *group* of such transformations is found, it is possible to construct the solution to the o.d.e. via quadratures. We shall find that for many cases of interest, there are succinct algorithms for finding the transformations without knowing the solution curves beforehand. The guiding relationships for the catalogue search, and the justification for the quadrature formula used, has been known for about a century. Pioneering work was done by Sophus Lie and others in the 19th century (see e.g [Lie75a]). Its emplacement within a symbolic/algebraic computational setting is, of course, modern-day.

*This work was supported in part by the Applied Mathematical Sciences Research Program (KC-04-02) of the Office of Energy Research of the U.S. Department of Energy under Contract W-31-109-Eng-38, and in part by U.S. Dept. of Energy contract DE-AM03-76SF00034. Many of the results reported within were computed and processed into readable form by Ernie Co-Vax of the University of California at Berkeley, supported in part by the National Science Foundation under grant MCS#78-07291.

2. Lie Transformation Groups and Differential Equations

In this section, we review a few key definitions and theorems of Lie theory which we need to state the theorem crucial to our work. See [Dic24a, Coh11a], [Mar60a, Fra28a, Bey79a, Blu74a, Inc44a] for a fuller explanation of matters.

A *one-parameter Lie transformation group* on a differentiable manifold $\mathbf{M}$ is a C^2 function $T:\mathbf{R}\times\mathbf{M}\to\mathbf{M}$ where for all $t\in\mathbf{R}$, $T(t,\cdot)$: $\mathbf{M}\to\mathbf{M}$ is a homeomorphism which obeys the composition rule $T(t_2,T(t_1,\cdot)) = T(t_2+t_1,\cdot)$. A *local one-parameter Lie transformation group* on $\mathbf{M}$ is a C^2 function T obeying the composition rule, where for each compact $K\subset\mathbf{M}$, there exists an open interval I_K containing 0 such that T is defined on $I_k\times K$, and for each $t\in I_K$, $T(t;\cdot)$ is a homeomorphism of K onto some $K_t\in\mathbf{M}$.

An *infinitesimal generator* for a one parameter local transformation group on $\mathbf{R}^2$ is a differential operator U: $\xi\frac{\partial}{\partial x}+\eta\frac{\partial}{\partial y}$, where $\xi,\eta:\mathbf{R}^2\to\mathbf{R}$. We often refer to a group on $\mathbf{R}^2$ as the pair of functions (ξ,η). (See, e.g. [Coh11a, Mar60a] for an explanation of the unique correspondence between the (ξ,η) and T descriptions of a local transformation group.)

The *manifold* $\mathbf{L}(\mathbf{O})$ *of line elements* of an open set $\mathbf{O}\subset\mathbf{R}^2$ is a particular differentiable 3-manifold, diffeomorphic with $\mathbf{O}\times\mathbf{S}^1$, the product of $\mathbf{O}$ with the circle, informally. Since each point (x_0,y_0), of a (differentiable) curve in $\mathbf{O}$, has a tangent p_0 at that point, we can "lift" the curve, considered as a set of points $\{(x_0,y_0)\}$ in $\mathbf{O}$, to a set of points in $\mathbf{L}(\mathbf{O})$ $\{(x_0,y_0,p_0)\}$. The solution curves in $\mathbf{O}$ to a first order ordinary differential equation $M(x,y)$ + $N(x,y)p$ = 0, where $p=\frac{dy}{dx}$, thus defines a surface in $\mathbf{L}(\mathbf{O})$.

By looking at the action on (tangents to) curves in $\mathbf{O}$ of a local one-parameter transformation group on $\mathbf{O}$, the group (ξ,η) itself can be lifted to a local one-parameter transformation group on $\mathbf{L}(\mathbf{O})$ (ξ,η,η'), where

Theorem 2 The algorithm can recognize (and solve) any separable, linear, general homogeneous, or Bernoulli equation, or equations which can be reduced to separable form by the changes of variables $u = x$, $v = V(x,y)$, where $\frac{\partial V}{\partial x} / \frac{\partial V}{\partial y}$ is of the form $c_{21}x^{l_2}y^{k_2} + c_{22}$. ∎

There are other equations which can be solved beyond the claims of the theorems. For example, the group (x^2+1, y^{-1}) admits the differential equation

$$\tan^{-1}(x)y = y'f\left(-\frac{(y^2-2\tan^{-1}(x))}{2}\right)+(x^2+1)yf\left(-\frac{(y^2-2\tan^{-1}(x))}{2}\right).$$

Additionally, IF2 has limited abilities to find solutions for o.d.e.s which can be reduced to linear, separable, etc. through other changes of variables.

3.3.2. Groups of the form $(0,\psi(x)\rho(y))$, or $(\psi(x)\rho(y),0)$.

The algorithm for search through groups of these forms, generates conditions on a parameterized set of first order linear differential equations ψ and ρ must satisfy, from $U'f \equiv 0$ condition. This generates a system of linear equations whose solution (or lack thereof) determines the existence and form of the group.

Theorem The general form of the equation which admits a group of the form $(0,\varphi(x)\rho(y))$ is

$$\psi'(x)\int\frac{dy}{\rho(y)} + g(x) + \frac{y'}{\varphi(x)\rho(y)} - \frac{\psi'(x)}{\varphi(x)^2}\int\frac{dy}{\rho(y)} + \frac{y'}{\varphi(x)\rho(y)}$$

$$= f\left(\frac{\int\frac{dy}{\rho(y)}}{\varphi(x)}\right),$$

where f and g are an arbitrary functions. ∎

The general form of the equation which admits groups of the form $(\varphi(x)\rho(y),0)$ is

$$\frac{\frac{1}{\varphi(x)\rho(y)} + \left(\frac{\int\frac{dx}{\varphi(x)}\rho'(y)}{\rho(y)^2} + g(y)\right)y'}{\frac{1}{\varphi(x)\rho(y)} + \frac{\int\frac{dx}{\varphi(x)}\rho'(y)}{\rho(y)^2}y'} = f\left(\frac{\int\frac{dx}{\varphi(x)}}{\rho(y)}\right)$$

where f and g are arbitrary functions.

3.3.3. Groups of the form $(\psi(x),\varphi(y))$, or $(\varphi(y),\psi(x))$.

The techniques involved here are similar to those of the previous section

Theorem For groups of the form $(\varphi(y),\varphi(x))$, if we define the solution to $\frac{\varphi(x)}{\varphi(y)} = \frac{dy}{dx}$ in terms of y as $u(x,const.) = y$, and in terms of x as $w(y,const.) = x$, then $(\varphi(y),\varphi(x))$ leave invariant equations of the form

$$\frac{y'}{\varphi(x)+\varphi(y)y'} = \psi(w(y,const.))\, f\left(\int\frac{dx}{\varphi(x)} - \int\frac{dy}{\varphi(y)}\right)$$

as well as

$$\frac{1}{\varphi(x)+\varphi(y)y'} = \varphi(u(x,const.))\, f\left(\int\frac{dx}{\varphi(x)} - \int\frac{dy}{\varphi(y)}\right)$$ ∎

We refer to the methods of sections 3.3.2 and 3.3.3 collectively as Algorithm SEP1234. SEP1234 can, among other things, find integrating factors for those equations reduced to linear, or Bernoulli, by the change of variables $(x,y)\rightarrow(u,v)$ through either $u=f(x)$, $v=g(y)$, or $u=f(y)$, $v=g(x)$, (f, g arbitrary). This is straightforward to verify using the change of variables formula.

3.3.4. Summary of results, comparison with other techniques

Table 3 (see Appendix) summarizes the abilities of the two Algorithms, and compares them to the abilities of symbol manipulation programs already implemented: Moses' original SOLDIER [Mos67a], Lafferty's MACSYMA equation solver ODE (Summer, 1980 version), and Peter Schmidt's work on EULE, and change-of-variable extensions to EULE [Sch79a]. (See [Cha80a] for a discussion and comparison of these programs.)

4. Implementations of IF2 and SEP1234

Algorithms IF2 and SEP1234 were implemented as MACSYMA programs named *food2* and *food4* respectively. Table 4 gives comparable times for *food2* , *food4* , and Edward Lafferty's o.d.e. solver ODE (which incorporates a number of special techniques, case-by-case trials, and simple changes of variables), on a number of test equations, all on a VAX 11/780 system running Berkeley Virtual Memory UNIX [Fod81a].

As can be seen from the table, when ODE solves an equation, typically reaches a solution faster than *food2* or *food4*, regardless of whether it uses its ODE2, Riccati, integrating factor, or non-linear transform methods. *food2* is seen to be much slower than either *food4* or ODE. Its use seems to be precluded on equations with more than three or four terms when expanded, even on systems with as large a virtual address space as the standard Vaxima system (approximately 2 megabytes for data and user program). The fifth entry of Table 4 is an example of one of the broad class of equations solvable in principle by SEP1234 and IF2, which ODE did not solve.

We ran *food4* on the 367 first order, first degree Kamke [Kam61a] equations, and found that it could find closed-form solutions to approximately 58% of the equations (ODE can find solutions approximately 70% of the time, Schmidt claims a success rate of 85%), given a 90 minute cpu execution time limit. It is worth mentioning that over 70% of the equations solved via ODE's exact and elementary integrating factor methods also admit one of the groups found by *food4*, since no *a priori* predictions could be made about the efficacy of *food4* in that area. Even though the performance of ODE was in general superior to *food2* and *food4*, *food4* was able to find solutions to eight Kamke equations (nos. 77, 78, 320, 337, 120, 321, 350, 357) that ODE was unable to solve.

p 45 on next page

$$\eta' \equiv (\frac{\partial \eta}{\partial x} + (\frac{\partial \eta}{\partial y} - \frac{\partial \xi}{\partial x})p - \frac{\partial \xi}{\partial y}p^2).$$

The infinitesimal generator for the group induced on **L(O)** by U is

$$U' \equiv \xi\frac{\partial}{\partial x} + \eta\frac{\partial}{\partial y} + \eta'\frac{\partial}{\partial p}.$$

Note that U' is a differential operator on functions $f(x,y,p)$.

A function h:**M**→**R** is *invariant* under a local one-parameter transformation group T on **O**, if h is constant along any given curve $\{T(t,x_0,y_0),\ t \text{ varying}\}$ in **O**. If the infinitesimal generator for T is (ξ,η), then $Uh = \xi\frac{\partial h}{\partial x} + \eta\frac{\partial h}{\partial y} = 0$ on **O** iff h is invariant under the transformation. A set S is said to be invariant under T when every point in S is mapped to another element of S under the transformation. If the set is described implicitly through an equation $h(x,y) = 0$, then S is invariant under a transformation group iff $Uh = 0$ for all points (x,y) such that $h(x,y) = 0$.

Similar notions of invariance occur with extended transformation groups (which are still one-parameter transformation groups, but on **L(O)**). In particular, a surface in **L(O)** implicitly described by a differential equation $w(x,y,p) = M(x,y) + N(x,y)p = 0$ is invariant under a given extended transformation group, iff $U'w = 0$, for all (x,y) such that $w(x,y,p) = 0$. When this is the case, it is said that "the transformation group admits the differential equation" or "the equation is kept invariant by the group".

We can now state Lie's result which we rely upon for our algorithms:

Theorem Suppose the local one-parameter transformation group (ξ,η) on **O** admits the first order ordinary differential equation $M(x,y) + N(x,y)p = 0$. Suppose that for all $(x,y)\in$**O**, $\xi M + \eta N \neq 0$. Then $\frac{1}{\xi M + \eta N}$ is an integrating factor for the equation,

$$\frac{M(x,y) + N(x,y)p}{\xi M + \eta N} = 0$$

is an *exact* differential equation. ▪ (See e.g [Dic24a,p.304])

3. An approach to finding solutions: "group search"

3.1. The relation between elementary methods and transformation groups

The notion of searching for transformation groups which are admitted by an ordinary differential equation would not be appealing unless groups which are associated with many elementary equations were also succinct. However, the proponents of Lie theory found simple groups for the elementary types of equations, as well as many others, as summarized by Table 1 (see Appendix).

3.2. Reduction via change of variables

Many of the techniques used successfully by Schmidt [Sch76a, Sch79a] involved reduction of an equation through change of variables to one of the "elementary types" (i.e. homogeneous, linear, separable). Through the use of the implicit function theorem, it is possible to describe (the infinitesimal generator of) a group which admits such an equation. We computed the change of variables formula in two fundamental situations, summarized in Table 2 (see Appendix).

These evidence convinced us that many equations of interest admit groups of relatively simple structure. This, along with the integrating factor result of the previous section, suggested to us that a directed search through groups with simple structure would be a useful way to organize an extensive catalog search. The remainder of this section outlines some of the methods we devised for realization of this idea. We do *not* claim to have solved the problem of finding a general method for finding a group for an arbitrary first order o.d.e. without knowing its general solution beforehand*.

3.3. Algorithms to search particular classes of groups

For the purposes of this section, we assume that "all equations" refer to the class of equations which the programs like those in MACSYMA [Mos74a] can canonically simplify in principle , e.g. Brown's class of rational exponentials [Bro69a]. We found algorithms for three interesting classes:

3.3.1. Groups of the form $\left[\varphi(x)(c_{11}x^{l_1}y^{k_1}+c_{12}),\ \varphi(x)(c_{21}x^{l_2}y^{k_2} + c_{22})\right]$.

Briefly, the algorithm searches groups of this form by differentiating the expression $U'f = 0$ for a group of such form with respect to x, and looking through the (finite) number of possibilities for the c_{ij}, l_i, and k_i φ is then found via quadratures. In the sequel, we refer to this method as Algorithm IF2. Using classical results about one-parameter transformation groups (see e.g [Coh11a,p.50]), it is straightforward to verify

Theorem 1 Every equation of the form

$$f(\frac{\alpha x^{m+q+1}}{m+q+1} - \frac{\beta y^n}{n+1}) = \frac{\varphi(x)x^m}{\alpha(m+q+1)x^{n+q} - c2(n+1)y^n y'}$$

where f is an arbitrary function, and α and β are arbitrary constants, is admitted by some group of the above form. ▪

A similar procedure finds groups of the form $\left[\varphi(y)(c_{11}x^{l_1}y^{k_1}+c_{12}),\ \varphi(y)(c_{21}x^{l_2}y^{k_2} + c_{22})\right]$. Through application of the change of variables formulae for groups, we find

*In general, every first order ordinary differential equation is admitted by some one-parameter transformation group. This is not the case for higher order equations (see [Coh11a] for details).

5. Conclusions

What is presented here is obviously just one way to apply Lie theory in a computational setting. (Beyer [Bey79a] has pursued other directions.) While a transformation group search is just as surely a "toolkit" of cases as the methods of EULE or ODE, the group search paradigm unifies much of what ODE or EULE achieve through their many techniques. Certainly by inclusion of more types of groups, group search programs can be made arbitrarily "expert" in the sense of "expert" that EULE or ODE strive for. It is an open question whether transformation group techniques can be made efficient enough for it to be worthwhile to do so; our *food2/focd4* implementations were intended only to be first words, not last ones, on that matter. We foresee improvements in the efficiency and sophistication of the present ideas, and exploration of extensions to additional groups, to higher order equations, and to systems of equations.

Appendix – Tables 1-4

equation type	equation form	admits group(s)
exact	$M(x,y) + N(x,y)y' = 0$, where $\frac{\partial N}{\partial x} = \frac{\partial M}{\partial y}$	no general form -- depends on equation
separable	$y' = \varphi(x)\psi(y)$	$(\frac{1}{\varphi(x)}, 0)$ $(0, \psi(y))$
linear	$y' + p(x)y + q(x) = 0$	$(0, e^{-\int p(x)dx})$
Bernoulli	$y' + p(x)y + q(x)y^s = 0$	$(0, e^{\int (s-1)p(x)dx}y^s)$
general homogeneous	$y' = \frac{y}{x}f(\frac{y}{x^n})$	(x, ny)
	$y' = f(rx + sy)$	$(s, -r)$
	$xy' - ny = g(x)f(\frac{y}{x^n})$	$(\frac{x^{n+1}}{g(x)}, \frac{x^n ny}{g(x)})$
	$\frac{1}{g(x)}(x \pm yy') = F(x^2 \pm y^2)$	$(\frac{1}{g(x)}, \frac{\pm x}{g(x)y})$
	$x^{1-r}y^{s-1}y' = f(\frac{x^r}{r} \pm \frac{y^s}{s})$	$(x^{1-r}, \pm y^{1-s})$
	$\frac{x^r y^s(xy'-ny)}{sxy' + ry} = f(\frac{y}{x^n})$	$(x^r y^s x, x^r y^s ny)$
	$\frac{y - xy'}{x - nyy'} = f(x^2 - ny^2)$	(ny, x)
	$\varphi_1(x)(\mu y' + u'y + v') = F(\mu y + v)$ where $\mu \equiv \exp(\int \varphi_2(x)dx)$ and $v \equiv \int \mu\varphi_3(x)dx$	$(\varphi_1(x), -\varphi_1(x)(\varphi_2(x)y + \varphi_3(x)))$

Table 1*
Groups associated with elementary types of ODEs

*Note: these tables are a summary of various tables in the Lie transformation group literature, as found in [Fra28a, Coh11a, Blu74a, Dic24a]. The literature lists corresponding groups and equations where the roles of x and y are interchanged, e.g. $(\frac{1}{g(y)}, \frac{\pm y}{g(y)x})$.

If transformation $u = U(x,y),\ v = V(x,y)$ reduces o.d.e. to		then o.d.e. admits (ξ,η) group(s)
separable	$\frac{dv}{du} = \varphi(u)\psi(v)$	$\xi = \frac{1}{\varphi(U(x,y))}\frac{\partial X}{\partial u},\ \eta = \frac{1}{\varphi(U(x,y))}\frac{\partial Y}{\partial u}$ $\xi = \psi(V(x,y))\frac{\partial X}{\partial v},\ \eta = \psi(V(x,y))\frac{\partial Y}{\partial v}$
linear	$\frac{dv}{du} + P(u)v + Q(u) = 0$	$\xi = \Phi(U(x,y))\frac{\partial X}{\partial v},\ \eta = \Phi(U(x,y))\frac{\partial Y}{\partial v}$ ($\Phi(u)$ is of the form $e^{-\int P(u)du}$)

Table 2
Groups for changes of variable reducing to separable/linear ODE

type of equation: name	change of var.	Program/Algorithm: SOLDIER	ODE	Schmidt EULE, heuristics	IF2	SEP1234
exact		X	X	X	A	A
linear		X	X	X	X	X
separable		X	X	X	X	X
homogeneous		X	X	X	X	X
general homog.		X	X	P	X	X
Bernoulli		X	X	X	X	X
"almost linear"		X		X		X
elem. int. factors		X	X	X	A	A
Riccati			P	H,P		
reversed variables	$u = y$ $v = x$	P	P	l,gh,b,r	l,gh,b	l,gh,b
explicit func. $f(x,y)$	$u = x$ $v = f(x,y)$			H		
explct. trig func. of $f(y)$	$u = x$ $v = trig(y)$			H		l,gh
explct. log or exp func. of $f(y)$,	$u = x$ $v = \log(f(y))$, $v = \exp(f(y))$			H		l,gh
explct. func. $f(x)$	$u = f(x)$ $v = y$			H	l,gh	l,gh
explct. SUM of x & y subterm in eq.	$u = x$ $v = SUM$			H		
explct. SUM of y subterm in eq.	$u = x$ $v = SUM^{const.}$			H		l,gh
explct. SUM of x subterm in eq.	$u = SUM^{const.}$ $v = y$			H		l,gh
linear coefficients	$u = ax+b$ $v = cy+d$	A	A	H	l,gh	l,gh
linear coefficients	$u = ay+b$ $v = cx+d$	A	A	H	l,gh	l,gh
linear ch. of var.	$u = ax+by$ $v = cx+dy$		P	H	•	
double ch. of var.	$u = x^a y^b$ $v = x^c y^d$; $u = x^2+ay^2$ $v = xy^{-1}$; $x = v\cos(u)$ $y = v\sin(u)$; etc.		P	H	†	
ch. of var.	$u = x$ $v = yx^c$		P	H	s,gh	
ch. of var.	$u = x$ $v = y + kx^p$		P	H	s,gh	

Table 3

Notes on Table 3

Key: X - solved by program; H - heuristic which solves some equations of this type if reducible to one of the elementary types, or a Riccati equation which can be solved by the other techniques of the program; A - many equations of this class solved, but no methods included specifically to handle this type; P - algorithmic techniques which solve some equations of this type; l - can solve equations reduced to linear; b - can solve equations reduced to Bernoulli; gh - can solve equations reduced to (general) homogeneous, s - can solve equations reduced to separable, r - can solve equations reduced to Riccati.

*The only equation Schmidt found which prescribed the use of this method is Murphy equation 534 [Mur60a,p.262]. This equation, $[a + x(x+y)]y' = b+(x+y)y$, is reduced to a separable equation $f(u) = \frac{v'}{v}$ via the change of variables. Thus, the equation admits a group of the form $\xi = ax+by$, $\eta = cx + dy$, which is of a form found by IF2.

†Most of the equations Schmidt gives [Sch79a,p.168] have groups which are found by IF2. For example, Kamke equations 1-366 and 1-367 are from [Dic24a] who generates the equations from Lie transformation groups). Murphy equation 711, also given as an example of this type of equation, has a group discoverable by IF2, as shown in an example in section 4.1.

equation	*food2*	*food1*	ODE
$y' + p(x)y = q(x)y^s$ (Bernoulli)	2053 sec exec. 564 sec g.c.	530 sec. exec. 151 sec g.c.	15 sec exec 4 sec g.c.
$xy' = y\log(x^a y^b)$ (general homog.)	924 sec 224 sec g.c.	150 sec 42 sec g.c.	15 sec 4 sec g.c.
$y' = \log(x)(y^2 - e^y)$ (separable)	storage exhausted after 1476 sec.	73 sec exec 20 sec g.c.	11 sec exec 4 sec g.c.
$y' = (y+x)^2$ (admits group (1,1))	1930 sec. 484 sec. g.c.	273 sec. 77 sec. g.c.	77 sec. 32 sec. g.c.
$y' + y\log(y)+yf(x) = 0$ ($u=x$, $v=\log(y)$ reduces to linear)	1789 sec. 450 sec. g.c.	205 sec. 58 sec. g.c.	56 sec. (no solution found)
$\frac{y^{s-1}}{x^{r-1}}y' = \frac{x^r}{r} + \frac{y^s}{s}$ (admits group $\xi=x^{1-r}$, $\eta=y^{1-s}$, has elementary integrating factor in this form)	failed to find answer within 5400 sec. (90 min.)	526 sec. 150 sec. g.c.	135 sec. 35 sec. g.c.

Table 4

exec = total execution time, including garbage collection time (g.c.)

References

Bey79a. William A. Beyer, "Lie-Group Theory for Symbolic Integration of First-Order Ordinary Differential Equations," in *Proceedings of the 1979 MACSYMA User's Conference*, (1979).

Blu74a. G.W. Bluman and J.D. Cole, *Similarity Methods for Differential Equations*, Springer-Verlag, New York (1974).

Bro69a. W.S. Brown, "Rational Exponential Expressions and a Conjecture Concerning π and e," *American Mathematical Monthly* **76** pp. 28-34 (1969).

Cha80a. Bruce W. Char, *Algorithms Using Lie Transformation Groups to Solve First Order Ordinary Differential Equations Algebraically*, University of California, Berkeley (1980). (Ph.D thesis, Computer Science Division).

Coh11a. Abraham Cohen, *An Introduction to the Lie Theory of One-parameter Groups*, D.C. Heath, Boston (1911).

Dic24a. L.E. Dickson, "Differential Equations from the Group Standpoint," *Annals of Mathematics Second Series* **25** pp. 287-378 (1923-24).

Fod81a. John K. Foderaro and Richard J. Fateman, "Characterization of VAX Macsyma," *These Proceedings.*, (1981).

Fra28a. P. Franklin, "The Canonical Form of a One-parameter group," *Annals of Mathematics, 2nd Series* **29**(2) pp. 113-122 (Apr. 1928).

Inc44a. E.L. Ince, *Ordinary Differential Equations*, Dover, New York (1944).

Kam61a. E. Kamke, *Differentialgleichungen-Losungsmethoden und Losungen, 7th ed. 1. Gewohnliche Differentialgleichungen*, Akad Verl. Ges. Geest & Portig K.G., Leipzig, Germany (1961). (also published in 3rd edition (1944) by Chelsea Publishing Co., New York, 1959).

Laf80a. Edward L. Lafferty, *ODE*, Personal communication. 1980.

Lie75a. Sophus Lie, *Sophus Lie's 1880 Transformation Group Paper*, Math Sci Press, 18 Gibbs St., Brookline MA 02146 (1975). Translated by Michael Ackerman. Comments by Robert Hermann.

Mar60a. Lawrence Markus, *Group Theory and Differential Equations*, Lecture Notes at the University of Minnesota 1959-60.

Mos67a. Joel Moses, "Symbolic Integration," Project Mac report MAC-TR-47, Massachusetts Institute of Technology (1967). (Ph.d thesis, Department of Electrical Engineering and Computer Science).

Mos74a. Joel Moses, "MACSYMA - The Fifth Year," in *Proceedings of the Eurosam 74 Conference*, , Stockholm (August 1974).

Mur60a. G.M. Murphy, *Ordinary Differential Equations*, D. Van Nostrand, Princeton (1960).

Sch76a. Peter Schmidt, "Automatic Symbolic Solution of Differential Equations of the First Order and First Degree," pp. 114-125 in *Proceedings of the 1976 ACM Symposium of Symbolic and Algebraic Computation*, (1976).

Sch79a. Peter Schmidt, "Substitution Methods for the Automatic Symbolic Solution of Differential Equations of First Order and First Degree," pp. 164-176 in *Proceedings of Eurosam79*, ed. Edward Ng, (1979).

The Computational Complexity of Continued Fractions

V. Strassen *)

Seminar für Angewandte Mathematik
Universität Zürich, Freiestr.36, CH-8032 Zürich

Abstract. The Knuth-Schönhage algorithm for expanding a quolynomial into a continued fraction is shown to be essentially optimal with respect to the number of multiplications/divisions used, uniformly in the inputs.

1. Introduction Let k be a field, k[x] the polynomial ring over k in the indeterminate x. Let A_1, A_2 be polynomials, $A_2 \neq 0$. Applying the division algorithm successively (Euclid's algorithm) we get

$$
\begin{aligned}
A_1 &= Q_1 A_2 + A_3 \\
A_2 &= Q_2 A_3 + A_4 \\
&\cdots\cdots \\
A_{t-1} &= Q_{t-1} A_t,
\end{aligned} \tag{1.1}
$$

where $A_i \neq 0$, deg $A_{i+1} < \deg A_i$ for i>1. The sequence $(Q_1, \ldots, Q_{t-1})$ depends only on the quolynomial A_1/A_2 and is called the continued fraction of A_1/A_2. (For its significance in several branches of mathematics see [13], [26].) The name comes from the identity

$$A_1/A_2 = Q_1 + 1/(Q_2 + 1/(\ldots + 1/(Q_{t-2} + 1/Q_{t-1})\ldots)),$$

valid in k(x), which follows from (1.1) by dividing the i-th equation by A_{i+1} and eliminating all A_i/A_{i+1} with i>1. $(Q_1, \ldots, Q_{t-1})$ determines A_1/A_2 uniquely.

Knuth [8] associates with (A_1, A_2) the extended sequence $(Q_1, \ldots, Q_{t-1}, A_t)$, which he calls the euclidean representation of (A_1, A_2). It represents the pair (A_1, A_2) uniquely. In fact, one has a bijection between pairs of polynomials (A_1, A_2) such that $\deg A_1 \geq \deg A_2 \geq 0$ and finite sequences of polynomials $(Q_1, \ldots, Q_{t-1}, A_t)$ such that $t \geq 2$, $\deg Q_1 \geq 0$, $\deg Q_i > 0$ for $1 < i < t$ and $\deg A_t \geq 0$. If we put $n = \deg A_1$, $m = \deg A_2$, then obviously $\sum_1^{t-1} \deg Q_i + \deg A_t = n$, $\sum_2^{t-1} \deg Q_i + \deg A_t = m$.

*)The results of this paper have been announced in [25].

©1981 ACM O-89791-047-8/81-0800-0051 $00.75

The euclidean representation is rather informative. It contains the continued fraction of A_1/A_2 and the gcd A_t of A_1, A_2. Brown and Collins [4] have exhibited the resultant of A_1, A_2 essentially as a power product of the leading coefficients of the Q_i and A_t. In particular, if $A_2 = \frac{d}{dx} A_1$, one gets the discriminant of A_1. If in addition $k = \mathbf{R}$, one can read off from the euclidean representation the number of zeros of A_1 in any real interval in linear time, since Sturm's algorithm may be carried out using the values of the Q_i and A_t at the endpoints of the interval.
In the sequel we will exclusively work with the euclidean representation, however our results will apply mutatis mutandis also to continued fractions.

How fast can we compute the sequence $(Q_1, \ldots, Q_{t-1}, A_t)$ from (A_1, A_2)? For simplicity and elegance we will allow in this paper additions, subtractions and multiplications by fixed scalars (which are thought to be stored in the program) for free and will thus only count "nonscalar" multiplications and divisions (Ostrowski's measure). For n=m, Euclid's algorithm requires in the worst and "normal" case, when $\deg Q_i = 1$ for $1 < i < t$ and $\deg A_t = 0$, about n^2 mult/div. The algorithm cannot be essentially improved if one insists on computing the A_i in addition to the Q_i (use a linear independence argument).

Lehmer [10] suggested to employ the fact that for small $\deg Q_i$ only a small initial segment of A_i, A_{i+1} is needed to compute Q_i. Taking up this idea, Knuth [8] and Schönhage [16] constructed an ingenious $O(n \log n)$ algorithm for computing the euclidean representation. Actually all three authors were concerned with the number theoretic analogue of our situation ($\mathbb{Z}$ instead of $k[x]$). The translation to the somewhat simpler polynomial setting is due to Moenck [11].

In the present paper we will show that the Knuth-Schönhage algorithm is optimal up to a multiplicative constant. In fact we will prove this not only for the worst case, but in a strong sense uniformly over the set of input polynomials A_1, A_2, at least when k is algebraically closed.

We use the model of a computation tree, allowing tests of the form

"if a = 0 goto i else goto j"

free of charge. A computation tree computes a "collection" (φ,π), where φ is a function on the set J of inputs and π a finite partition of J (see section 5). Fixing n and m, we have in the case of the euclidean representation

$$J = \{(A_1,A_2): \deg A_1 = n, \deg A_2 = m\}$$
$$= \{(a_o,\ldots,a_n,b_o,\ldots,b_m) \in k^{n+m+2}: a_o b_o \neq 0\}$$

(with the identification $A_1 = \sum_o^n a_i t^{n-i}$, $A_2 = \sum_o^m b_j t^{m-j}$),

$$\varphi(A_1,A_2) = (Q_1,\ldots,Q_{t-1},A_t)$$

(also represented by its sequence of coefficients of total length n+t),

$$\pi = \{D(n_1,\ldots,n_t): t \geq 2,\ n_1,\ n_t \geq 0,\ n_i > 0 \text{ for } 1<i<t,\ \sum_1^t n_i = n,\ \sum_2^t n_i = m\},$$

where

$$D(n_1,\ldots,n_t) = \{(A_1,A_2) \in J: (\deg Q_1,\ldots,\deg Q_{t-1},\deg A_t) = (n_1,\ldots,n_t)\}$$

is the set of inputs, whose euclidean representation has the "format" $(n_1,\ldots,n_t)$. Our main result is the following (see sections 4 and 6):

1. The Knuth-Schönhage algorithm computes the euclidean representation with cost

$$\leq 30n(H(n_1,\ldots,n_t)+6.5)$$

on $D(n_1,\ldots,n_t)$ (H is the entropy function, see (2.1))

2. Let k be algebraically closed. Any algorithm that computes the euclidean representation has cost

$$\geq n(H(n_1,\ldots,n_t)-2)$$

on a Zariski dense open subset of $D(n_1,\ldots,n_t)$. (Any algorithm may of course be speeded up on particular inputs by a table look up procedure.)

In particular, for n=m, the order n log n of the worst case of the Knuth-Schönhage algorithm cannot be improved. The above result 2. remains true also for nonclosed fields, if the algorithm is assumed to yield the euclidean representation over the algebraic closure as well. If this condition is not satisfied, we still get order-sharp lower bounds on those $D(n_1,\ldots,n_t)$ with $t \geq (\frac{1}{2}+\varepsilon)m$. ($\varepsilon>0$; this of course covers the worst case.) Similar results hold in the important situation of polynomials over a field $\mathbb{Z}_p$, where p is not known in advance (see Brown [3]).

For proving lower bounds we employ the geometric degree method (Strassen [23], see also Borodin-Munro [2], Schönhage [17], Schnorr [15]). For this reason the paper assumes some knowledge of the language of classical algebraic geometry (see Mumford [12], Shafarevich [18], Samuel [14]). Let k be algebraically closed. The degree of a closed irreducible set $X \subset k^n$ is the typical number of points of intersection of X with an affine subspace of k^n of complementory dimension. (This coincides with the degree of the closure of X in $\mathbf{P}^n$). The degree of a closed, but reducible subset of k^n is the sum of the degrees of its components. The degree of a locally closed set X is the degree of its closure $\overline{X}$ (thus also the sum of the degrees of its components). We have Bezout's inequality

$$(1.2) \qquad \deg(X \cap Y) \leq \deg X \deg Y$$

for closed $X,Y \subset k^n$. We will use this inequality mainly in the case, when Y is an affine subspace of k^n, where it becomes $\deg(X \cap Y) \leq \deg X$. If $f_1,\ldots,f_r \in k(x_1,\ldots,x_n)$ are rational functions, we denote by $\deg(f_1,\ldots,f_r)$ the degree of the locally closed graph $W \subset k^{n+r}$ of the rational map $k^n \to k^r$ defined by $(f_1,\ldots,f_r)$. Then if $L(f_1,\ldots,f_r)$ is the complexity of $f_1,\ldots,f_r$ with respect to the cost measure introduced above (see [23], [2]), we have the degree bound

$$(1.3) \qquad L(f_1,\ldots,f_r) \geq \log \deg(f_1,\ldots,f_r).$$

(In this paper log always means $\log_2$.)

2. Symbolic multiplication of several polynomials

The results of this section will be used later, but they are also of independent interest. Let $n_1,..,n_t$ be nonnegative integers, $n=\sum_i n_i$. We denote the entropy of the probability vector $(n_1/n,..,n_t/n)$ by $H(n_1,...,n_t)$, i.e.

$$(2.1) \qquad H(n_1,...,n_t) = -\sum_{n_i>o} (n_i/n)\log(n_i/n).$$

(In case n=0 we set $H(n_1,...,n_t)=0$.) Obviously the entropy does not change if we remove from $(n_1,..,n_t)$ all n_i which are =0. We list a few properties of the entropy, some of which will be used in later sections (for detailed proofs see Fano [5]). We have $0\leq H(n_1,...,n_t)\leq \log n$ for n>0, with both bounds attained. $nH(n_1,...,n_t)$ is monotonic in each argument n_i (as one sees by differentiating): If $n_i\leq n_i'$ for all i and if $n'=\sum_1^t n_i'$ then

$$(2.2) \qquad nH(n_1,...n_t) \leq n'H(n_1',...,n_t').$$

Since inserting zeroes into the sequence $(n_1,..,n_t)$ does not change the entropy, this implies

$$(2.3) \qquad (\sum_1^{t-1} n_i)H(n_1,...,n_{t-1}) \leq nH(n_1,...,n_t).$$

The following crucial property is easily checked:

$$(2.4) \qquad (\sum_1^s n_i)H(n_1,...,n_s)+(\sum_{s+1}^t n_i)H(n_{s+1},...,n_t)$$
$$= n(H(n_1,...,n_t)-H(\sum_1^s n_i, \sum_{s+1}^t n_i)).$$

It is convenient to extend the definition (2.1) by allowing nonnegative real numbers p_i in place of n_i. Since the entropy is invariant under scaling, we can reformulate (2.4) as follows: If $\sum_1^s n_i=pn$ then

$$(2.5) \qquad pH(n_1,...,n_s)+(1-p)H(n_{s+1},...,n_t)$$
$$= H(n_1,...,n_t)-H(p,1-p).$$

In the sequel we will often write H for $H(n_1,...,n_t)$.

Lemma 2.1 $\log(n!/(n_1!...n_t!)) \geq n(H(n_1,...,n_t)-2)$.

Proof We may assume that all n_i are positive. By Stirlings formula with error estimate we have
$n!/(n_1!...n_t!)\geq 2^{nH}(2\pi n)^{1/2} \prod_{j=1}^t (2\pi n_j)^{-1/2} e^{-1/(12n_j)}$.
(See e.g. Fano [5], (8.87); notice that Fano works with natural logarithms.) Because of $n_1...n_t\leq(n/t)^t$ this implies

$\log(n!/n_1!...n_t!)) \geq$

$\geq nH+(1/2)\log(2\pi n)-(t/2)\log(2\pi n/t)-(t/12)\log e$

$\geq nH-(t/2)\log(2\pi n/t)-(t/12)\log e.$

Given n, the sum of the absolute values of the last two terms is maximal for

$t = 2\pi n e^{-5/6}$

and for this value of t we have

$(t/2)\log(2\pi n/t)+(t/12)\log e \leq 2n.$

Let k be an infinite field and let $x,p_{11},...,p_{1n_1},...,p_{t1},...,p_{tn_t}$ be indeterminates over k. We put

$$(2.6) \qquad P_i = x^{n_i}+p_{i1}x^{n_i-1}+...+p_{in_i}$$

and

$$(2.7) \qquad A = P_1...P_t.$$

The polynomial A has the form

$$A = x^n+a_1x^{n-1}+...+a_n,$$

where the a_i are polynomials in the p_{js}, i.e. $a_i\in k[p_{11},...,p_{tn_t}]=k[\underline{p}]$. The following theorem holds true irrespective of whether we interpret the complexity $L(a_1,...,a_n)$ in $k[\underline{p}]$ (not allowing division) or in the field of rational functions $k(\underline{p})$. In either case linear operations are not to be counted.

Theorem 2.2

$$n(H(n_1,...,n_t)-2) \leq L(a_1,...,a_n) \leq n(H(n_1,..,n_t)+1).$$

In particular, as $H \longrightarrow \infty$

$$L(a_1,...,a_n) \sim nH(n_1,...,n_t).$$

Proof We may assume that all n_i are positive. Left inequality: W.l.o.g. k algebraically closed. Choose $\alpha_1,...,\alpha_n\in k$ such that the polynomial $x^n+\alpha_1x^{n-1}+...+\alpha_n$ has n simple roots in k, say $\theta_1,...,\theta_n$. We will determine the number of solutions of the system of equations

$$(2.8) \qquad a_1 = \alpha_1,...,a_n = \alpha_n.$$

A point $(\kappa_{11},...,\kappa_{1n_1},...,\kappa_{t1},...,\kappa_{tn_t})\in k^n$ is a solution of (2.8) if and only if

$$(x-\theta_1)...(x-\theta_n) = (x^{n_1}+\kappa_{11}x^{n_1-1}+...+\kappa_{1n_1})...$$
$$...(x^{n_t}+\kappa_{t1}x^{n_t-1}+...+\kappa_{tn_t}).$$

Therefore we have a bijection of the set of solutions of (2.8) and the set of partitions of

$\{\theta_1,\ldots,\theta_n\}$ into t classes with $n_1,\ldots,n_t$ elements respectively. Thus they are exactly $n!/(n_1!\ldots n_t!)$ solutions. By (1.2) and Lemma 2.1 this implies

$$(2.9) \qquad \text{logdeg}(a_1,\ldots,a_n) \geq$$
$$\geq \log(n!/(n_1!\ldots n_t!)) \geq n(H-2)$$

and therefore by (1.3)

$$L(a_1,\ldots,a_n) \geq n(H-2).$$

Right inequality: A word is a finite (possibly empty) sequence from the set $\{0,1\}$. An s-code is a sequence $w_1,\ldots,w_s$ of s words such that for any $i\neq j$ the word w_i is not an initial segment of the word w_j. We will first show by induction on t (t being the number of polynomials to be multiplied symbolically, see (2.7)) that for any t-code $w_1,\ldots,w_t$

$$(2.10) \qquad L(a_1,\ldots,a_n) \leq \sum_1^t n_i \text{ length}(w_i).$$

This is clear for t=1. Now let t>1. It suffices to show (2.10) for a t-code $w_1,\ldots,w_t$ for which $\sum n_i$ length(w_i) is as small as possible. Because of t>1 the code does not contain the empty word. We partition $\{1,\ldots,t\}$ into the set E of those i for which w_i begins with 0 and its complement F. E is nonempty. Otherwise all w_i would begin with 1. Deleting the initial 1 in each w_i would still leave us with a t-code, in contradiction to the assumption of minimality above. Similarly F is nonempty. We assume w.l.o.g. $E=\{1,\ldots,s\}$. Deleting the initial zeroes in $w_1,\ldots,w_s$ and the initial ones in $w_{s+1},\ldots,w_t$ we obtain an s-code $\tilde{w}_1,\ldots,\tilde{w}_s$ and a (t-s)-code $\tilde{w}_{s+1},\ldots,\tilde{w}_t$. We put $m=n_1+\ldots+n_s$ and define $b_1,\ldots,b_m,c_1,\ldots,c_{n-m} \in k[\underline{p}]$ by

$$x^m+b_1x^{m-1}+\ldots+b_m = P_1\ldots P_s$$
$$x^{n-m}+c_1x^{n-m-1}+\ldots+c_{n-m} = P_{s+1}\ldots P_t.$$

Obviously

$$x^n+a_1x^{n-1}+\ldots+a_n =$$
$$= (x^m+b_1x^{m-1}+\ldots+b_m)(x^{n-m}+c_1x^{n-m-1}+\ldots+c_{n-m}).$$

Our induction hypothesis implies

$$L(b_1,\ldots,b_m) \leq \sum_1^s n_i \text{ length}(\tilde{w}_i)$$
$$L(c_1,\ldots,c_{n-m}) \leq \sum_{s+1}^t n_i \text{ length}(\tilde{w}_i).$$

Since the symbolic multiplication of two monic polynomials of degrees m and n-m can be achieved with n nonlinear operations (cf. [23], p. 244) we conclude

$$L(a_1,\ldots,a_n) \leq L(b_1,\ldots,b_m)+L(c_1,\ldots,c_{n-m})+n$$
$$\leq \sum_1^s n_i \text{ length}(\tilde{w}_i) + \sum_{s+1}^t n_i \text{ length}(\tilde{w}_i)+n$$
$$= \sum_1^s n_i(\text{length}(\tilde{w}_i)+1)+ \sum_{s+1}^t n_i(\text{length}(\tilde{w}_i)+1)$$
$$= \sum_1^s n_i \text{ length}(w_i) + \sum_{s+1}^t n_i \text{ length}(w_i)$$
$$= \sum_1^t n_i \text{ length}(w_i).$$

Thus we have proved (2.10) for an arbitrary t-code $w_1,\ldots,w_t$. Now we can always choose a t-code $w_1,\ldots,w_t$ such that

$$\sum_1^t n_i \text{length}(w_i) \leq n(H+1)$$

(see Fano [5], section 3.5). Therefore

$$L(a_1,\ldots,a_n) \leq n(H+1).$$

This completes the proof of the theorem.

3. Conversion of a continued fraction into a rational fraction

We need the following

Lemma 3.1 Let k be algebraically closed and let

$$f_1(y,x_1,\ldots,x_n)$$
$$\cdots\cdots\cdots$$
$$f_r(y,x_1,\ldots,x_n)$$

be polynoials. For $\mu\in k$ let $W_\mu \subset k^{n+r}$ be the graph of the map

$$(\alpha_1,\ldots,\alpha_n) \longmapsto (f_1(\mu,\underline{\alpha}),\ldots,f_r(\mu,\underline{\alpha})).$$

Then the function

$$\mu \longmapsto \deg(W_\mu)$$

is Zariski lower semicontinuous (i.e. it equals its maximum value except on finitely many points).

Proof Let W be the graph of the map

$$(\mu,\alpha_1,\ldots,\alpha_n) \longmapsto (f_1(\mu,\underline{\alpha}),\ldots,f_r(\mu,\underline{\alpha})).$$

W is an (n+1)-dimensional closed subvariety of $k\times k^{n+r}$. For any $\mu\in k$ the graph W_μ is an n-dimensional closed subvariety of k^{n+r} and we have

$$(3.1) \qquad \{\mu\}\times W_\mu = W\cap(\{\mu\}\times k^{n+r}) = W.(\{\mu\}\times k^{n+r})$$

since W and $\{\mu\}\times k^{n+r}$ intersect transversally (see van der Waerden [27], Samuel [14], Hartshorne [6], Mumford [12]). Let $\overline{W}$ be the closure of W in $k\times\mathbb{P}^{n+r}$,

$\overline{W}_\mu$ the closure of W_μ in $\mathbb{P}^{n+r}$. Since

$$(3.2) \qquad \dim(\overline{W}\cap(\{\mu\}\times\mathbf{P}^{n+r})) \le \dim((\{\mu\}\times W_\mu)\cup(\overline{W}-W)) = \max\{\dim W_\mu, \dim(\overline{W}-W)\} = n,$$

$\overline{W}$ and $\{\mu\}\times\mathbf{P}^{n+r}$ intersect properly. By conservation of number (see van der Waerden [27], Samuel [14]) the quantity

$$\deg(\overline{W}.(\{\mu\}\times\mathbf{P}^{n+r}))$$

is independent of μ, say $=c$.

On the other hand we have

$$(3.3) \qquad \{\mu\}\times\overline{W}_\mu = \overline{W}.(\{\mu\}\times\mathbf{P}^{n+r})$$

for all but finitely many points $\mu_1,\ldots,\mu_v$. For if we restrict both sides of (3.3) to affine space $k\times k^{n+r}$, we get equality by (3.1). So (3.3) can be invalidated only by the appearance of components of $\overline{W}\cap(\{\mu\}\times\mathbf{P}^{n+r})$ disjoint from $k\times k^{n+r}$. Now any such component lies in $\overline{W}-W$ and has dimension n by (3.2), so it is a component of $\overline{W}-W$. But $\{\mu\}\times\mathbf{P}^{n+r}$ can contain a component of $\overline{W}-W$ for only finitely many μ. (3.3) implies

$$\deg(W_\mu) = \deg(\overline{W}_\mu) = \deg(\overline{W}.(\{\mu\}\times\mathbf{P}^{n+r})) = c$$

for $\mu\notin\{\mu_1,\ldots,\mu_v\}$. Also we have

$$\deg(W_{\mu_i}) = \deg(\overline{W}_{\mu_i}) \le \deg(\overline{W}.(\{\mu_i\}\times\mathbf{P}^{n+r})) = c$$

since $\overline{W}_{\mu_i}$ is always a component of $\overline{W}.(\{\mu_i\}\times\mathbf{P}^{n+r})$. These two statements prove the lemma.

Now let k be an arbitrary infinite field, $t\ge 2$ and $n_1,\ldots,n_t$ be nonnegative integers such that $n_i>0$ for $1\le i<t$. Let x and

$$q_{1o},\ldots,q_{1n_1},$$
$$\cdots\cdots$$
$$q_{to},\ldots,q_{tn_t}$$

be indeterminates over k. Put

$$Q_i = q_{io}x^{n_i}+\ldots+q_{in_i}$$

for $1\le i<t$ and

$$A_t = q_{to}x^{n_t}+\ldots+q_{tn_t}.$$

Then the system of polynomial equations

$$A_1 = Q_1A_2+A_3$$
$$A_2 = Q_2A_3+A_4$$
$$\cdots\cdots$$
$$A_{t-1} = Q_{t-1}A_t$$

uniquely determines polynomials $A_1,\ldots,A_{t-1}$. We have

$$A_1 = a_ox^n+\ldots+a_n$$
$$A_2 = b_ox^m+\ldots+b_m,$$

where $n = \sum_1^t n_i$, $m = \sum_2^t n_i$ and $a_o,\ldots,a_n,b_o,\ldots,b_m\in k[\underline{q}]$. In the following theorem we can interpret $L(a_o,\ldots,a_n,b_o,\ldots,b_m)$ either in $k[\underline{q}]$ (not allowing division) or in $k(\underline{q})$. As usual, linear operations are free.

<u>Theorem 3.2</u> Let $n>0$. Then $n(H(n_1,\ldots,n_t)-2) \le L(a_o,\ldots,a_n,b_o,\ldots,b_m) \le 8n(H(n_1,\ldots,n_t)+7)$.

<u>Proof</u> Left hand inequality: W.l.o.g. k algebraically closed. By induction one easily sees

$$\begin{pmatrix}A_i\\A_{i+1}\end{pmatrix}=\begin{pmatrix}Q_i & 1\\1 & 0\end{pmatrix}\cdots\begin{pmatrix}Q_{t-1} & 1\\1 & 0\end{pmatrix}\begin{pmatrix}A_t\\0\end{pmatrix},$$

in particular

$$(3.4) \qquad \begin{pmatrix}A_1\\A_2\end{pmatrix}=\begin{pmatrix}Q_1 & 1\\1 & 0\end{pmatrix}\cdots\begin{pmatrix}Q_{t-1} & 1\\1 & 0\end{pmatrix}\begin{pmatrix}A_t\\0\end{pmatrix}.$$

Let $\mu\in k$ be different from 0. In (3.4) we make the substitution

$$q_{io}\mapsto 1/\mu$$
$$q_{ij}\mapsto (1/\mu)p_{ij}$$

(where $1\le i\le t$, $1\le j\le n_i$ and the p_{ij} are new indeterminates) and multiply both sides of (3.4) by μ^t. We get

$$(3.5) \qquad \begin{pmatrix}a_o(\mu)x^n+a_1(\mu)x^{n-1}+\ldots+a_n(\mu)\\b_o(\mu)x^m+b_1(\mu)x^{m-1}+\ldots+b_m(\mu)\end{pmatrix} = \begin{pmatrix}P_1 & \mu\\\mu & 0\end{pmatrix}\cdots\begin{pmatrix}P_{t-1} & \mu\\\mu & 0\end{pmatrix}\begin{pmatrix}P_t\\0\end{pmatrix},$$

where

$$P_i = x^{n_i}+p_{i1}x^{n_i-1}+\ldots+p_{in_i}$$

and where $a_j(\mu)$, $b_j(\mu)$ are obtained from a_j, b_j by the above substitution and subsequent multiplication by μ^t. Since the graph of the polynomial map defined by $a_i(\ ),\ldots,a_n(\ )$ is essentially a hyperplane section of the graph of $a_1,\ldots,a_n$, the degree of the former is less or equal than the degree of the latter. Thus for any $\mu\neq 0$

$$(3.6)\qquad \log \deg(a_o,\ldots,a_n,b_o,\ldots,b_m) \geq \log \deg(a_1,\ldots,a_n) \geq \log \deg(a_1(\mu),\ldots,a_n(\mu)).$$

(3.5) shows that the $a_j(\mu)$ are polynomials in $\underline{p}$ which depend polynomially on the parameter μ. In particular $a_j(\mu)$ make sense for $\mu=0$ and (3.5) remains correct in this case, i.e.

$$(3.7)\qquad x^n+a_1(0)x^{n-1}+\ldots+a_n(0) = P_1\ldots P_t.$$

By Lemma 3.1 we have

$$(3.8)\qquad \log \deg(a_1(\mu),\ldots,a_n(\mu)) \geq \log \deg(a_1(0),\ldots,a_n(0))$$

for all but finitely many μ. (3.7) together with (2.9) imply

$$(3.9)\qquad \log \deg(a_1(0),\ldots,a_n(0)) \geq n(H-2).$$

(3.6), (3.8) and (3.9) yield

$$(3.10)\qquad \log \deg(a_o,\ldots,a_n,b_o,\ldots,b_m) \geq n(H-2).$$

Now (1.3) gives the left hand inequality of the theorem. Right hand inequality: For any 2 by 2 matrix

$$G = \begin{pmatrix} g_{11} & g_{12} \\ g_{21} & g_{22} \end{pmatrix}$$

whose coefficients g_{ij} are polynomials in x with coefficients in $k(\underline{q})$ we put

maxdeg G = maximum of the degrees with respect to x of the g_{ij}

L(G) = L (the set of the coefficients with respect to x of the g_{ij}).

Let

$$G_i = \begin{pmatrix} Q_i & 1 \\ 1 & 0 \end{pmatrix}$$

for $i \leq t-1$ and

$$G_t = \begin{pmatrix} A_t & 0 \\ 0 & 0 \end{pmatrix}.$$

Then maxdeg $G_i=n_i$ and by (3.4) it suffices to show

$$(3.11)\qquad L(G_1\ldots G_t) \leq 8n(H(n_1,\ldots,n_t)+7).$$

The problem of computing the matrix product $G_1\ldots G_t$ is similar to the problem of computing the product of polynomials $P_1\ldots P_t$ as in Theorem 2.2 (of course in both cases we are dealing with symbolic computations, i.e. computations on coefficients), the main difference being that matrices do not commute. We replace (3.11) by

$$(3.12)\qquad L(G_1\ldots G_t) \leq cn(H(n_1,\ldots,n_t)+d)+7(t-1),$$

where we will choose c, $d \geq 1$ at the end of the proof, which is by induction on t.

The start (t=2) being clear, let t>2 and therefore n>0. There is a unique s ($1 \leq s \leq t$) such that

$$\sum_1^{s-1} n_i \leq n/2, \quad \sum_1^{s} n_i > n/2.$$

Define p, p' by

$$pn = \sum_1^{s-1} n_i, \quad p'n = \sum_1^{s} n_i.$$

Then we have $p \leq 1/2 < p'$. Choose $0<\varepsilon<1/2$.

Case $p' \leq 1-\varepsilon$: We first compute $G_1\ldots G_s$ and $G_{s+1}\ldots G_t$ and then by one matrix multiplication $G_1\ldots G_t$. Using the matrix multiplication algorithm of [20] together with the fact that

$$\text{maxdeg}(G_1\ldots G_t) = \text{maxdeg}\begin{pmatrix} A_1 & 0 \\ A_2 & 0 \end{pmatrix} \leq n,$$

then the induction hypothesis and property (2.5) of the entropy function, we get

$$L(G_1\ldots G_t) \leq L(G_1\ldots G_s)+L(G_{s+1}\ldots G_t)+7(n+1)$$
$$\leq cp'n(H(n_1,\ldots,n_s)+d)+7(s-1)$$
$$+ c(1-p')n(H(n_{s+1},\ldots,n_t)+d)+7(t-s-1)+7(n+1)$$
$$\leq cn(H(n_1,\ldots,n_t)+d)+7(t-1)+7n-cnH(p',1-p').$$

Now $H(p',1-p') \geq H(1-\varepsilon,\varepsilon)$. Thus if the condition

$$(3.13)\qquad 7 \leq cH(\varepsilon,1-\varepsilon)$$

is satisfied we have (3.12).

Case $p'>1-\varepsilon$, $p \geq \varepsilon$: We first compute $G_1\ldots G_{s-1}$ and $G_s\ldots G_t$ and then $G_1\ldots G_t$. Again (3.13) implies (3.12).

Case $p<\varepsilon$, $p'>1-\varepsilon$: We first compute $G_1\ldots G_{s-1}$ and $G_{s+1}\ldots G_t$ and then by two matrix multiplications $(G_1\ldots G_{s-1})G_s(G_{s+1}\ldots G_t)$. Using induction hypothesis and properties (2.3) and (2.5) of the entropy function we get

$$L(G_1\ldots G_t) \le L(G_1\ldots G_{s-1})+L(G_{s+1}\ldots G_t)+14(n+1)$$
$$\le cpn(H(n_1,\ldots,n_{s-1})+d)+7(s-2)+$$
$$+ c(1-p')n(H(n_{s+1},\ldots,n_t)+d)+7(t-s-1)+14(n+1)$$
$$\le cn(p'H(n_1,\ldots,n_s)+(1-p')H(n_{s+1},\ldots,n_t)+(p+1-p')d)+$$
$$+ 7(t-1)+14n$$
$$\le cn(H(n_1,\ldots,n_t)+2\varepsilon d)+7(t-1)+14n.$$

Thus in this case the condition

$$(3.14) \qquad 14 \le (1-2\varepsilon)cd$$

implies (3.12). Now we choose $\varepsilon=0.325$, $c=8$ and $d=5$. Then (3.13) and (3.14) are satisfied and the theorem follows from (3.12), because $n_i>0$ for $1<i<t$ implies $t-1\le n+1$.

4. Conversion of a rational fraction into a continued fraction: Analysis of the Knuth-Schönhage Algorithm

Let k be an infinite field, $n\ge m$ nonnegative integers. Given univariate polynomials A_1, A_2 over k with $\deg A_1=n$, $\deg A_2=m$, there are unique nonzero polynomials $Q_1,\ldots,Q_{t-1}$, $A_3,\ldots,A_t$ such that

$$(4.1) \qquad \begin{aligned} A_1 &= Q_1 A_2 + A_3 \\ A_2 &= Q_2 A_3 + A_4 \\ &\cdots\cdots \\ A_{t-1} &= Q_{t-1} A_t \end{aligned}$$

and $\deg A_{i+1}<\deg A_i$ for $i\ge 2$ (Euclid's algorithm). The sequence

$$(Q_1,\ldots,Q_{t-1},A_t)$$

is called the euclidean representation of (A_1,A_2) (Knuth [8]). We have $t\ge 2$. If we put

$$(4.2) \qquad \begin{aligned} n_i &:= \deg Q_i \quad (1\le i\le t-1) \\ n_t &:= \deg A_t, \end{aligned}$$

then

$$n_i \ge 0$$

for all i and

$$n_i > 0$$

for $1<i<t$. Furthermore

$$n = \sum_1^t n_i, \quad m = \sum_2^t n_i.$$

We define $A_{t+1}=0$ and

$$(4.3) \qquad M_i = \begin{pmatrix} 0 & 1 \\ 1 & -Q_i \end{pmatrix}.$$

Then

$$(4.4) \qquad \begin{pmatrix} A_s \\ A_{s+1} \end{pmatrix} = M_{s-1}\ldots M_1 \begin{pmatrix} A_1 \\ A_2 \end{pmatrix}$$

for $s\le t$. (The reader will notice that (4.4) for $s=t$ is just the inverse relationsship of (3.4), if there the indeterminates q_{ij} are replaced by elements of k.)

In this section we will show, using an algorithm that is essentially due to Knuth [8] and Schönhage [16] (see also Moenck [11]), how to compute the coefficients of A_1, A_2 in a rather efficient way. Since the length of the output (number of coefficients of the Q_i) depends on the input (A_1,A_2), it is clear that our computational model has to be extended by allowing branching instructions, say of the form

"if f = 0 then goto i else goto j".

In this way we get the wellknown model of a computation tree. For the purpose of proving lower bounds for the complexity of our problem we will discuss this model in some detail in the next section. In the present section we will analyse the cost (number of nonlinear multiplications/divisions as a function of the input) of the Knuth-Schönhage algorithm and implicitly present the algorithm using an informal approach. The style is such, however, that it can easily be formalized with the help of Propositions 5.2 and 5.3. For completeness we will give an ALGOL-like formulation of the algorithm at the end of this section.

We need a preliminary result. Given a polynomial $A=a_o x^q+\ldots+a_q \in k[x]$ of degree $q\ge 0$ and an integer ℓ, we set

$$(4.5) \qquad A|\ell = \begin{cases} 0 & \text{if } \ell<0 \\ a_o x^\ell+\ldots+a_\ell & \text{if } 0\le\ell\le q \\ a_o x^\ell+\ldots+a_q x^{\ell-q} & \text{if } \ell>q. \end{cases}$$

($A|\ell$ consists, so to speak, of the significant part of A of length $\ell+1$.) Obviously

$$(Ax^j)|\ell = A|\ell.$$

Given two pairs of polynomials (A,B) and (A',B') such that

$$\deg A \ge \deg B \ge 0$$
$$\deg A' \ge \deg B' \ge 0,$$

and an integer ℓ, we say that (A,B) and (A',B') coincide up to ℓ iff

$$A|\ell = A'|\ell$$

$$B|(\ell-(\deg A-\deg B)) = B'|(\ell-(\deg A'-\deg B')).$$ Coincidence up to ℓ is an equivalence relation. (A,B) and (Ax^j,Bx^j) coincide up to ℓ for every $j\geq 0$ (given that $\deg A\geq\deg B\geq 0$). If (A,B) and (A',B') coincide up to ℓ and $\ell\geq\deg A-\deg B$ then $\deg A-\deg B=\deg A'-\deg B'$. The qualitative idea of the following lemma is due to Lehmer [10].

Lemma 4.1 Besides (4.1) consider Euclid's algorithm for another pair $A_1',A_2'\in k[x]$ with $\deg A_1'\geq\deg A_2'$:

$$A_1' = Q_1' A_2' + A_3'$$
$$A_2' = Q_2' A_3' + A_4'$$
$$\cdots\cdots$$
$$A_{t'-1}' = Q_{t'-1}' A_{t'}'.$$

Let ℓ be a nonnegative integer and $1\leq s\leq t$ be such that $\sum_1^{s-1} n_i\leq\ell$ and either $s=t$ or $\sum_1^s n_i>\ell$. Define s' similarly (using A_1', A_2' instead of A_1, A_2). Then if (A_1, A_2) and (A_1', A_2') coincide up to 2ℓ we have $s=s'$ and $Q_i=Q_i'$ for $1\leq i\leq s-1$.

Proof We show by induction on $1\leq j\leq s$:

$$j \leq s',$$
$$Q_i = Q_i' \quad \text{for all } i < j,$$

and either $j=s$ or (A_j, A_{j+1}) and (A_j', A_{j+1}') coincide up to $2(\ell-\sum_1^{j-1} n_i)$. (This implies the lemma by symmetry.) The start of the induction is clear and the induction step is a consequence of the following statement:

Let (A,B) and (A',B') coincide up to 2ℓ, where $\ell\geq\deg A-\deg B$, and let

$$(4.6)\qquad \begin{array}{ll} A = Q B + C, & \deg C < \deg B \\ A' = Q'B' + C', & \deg C' < \deg B'. \end{array}$$

Then $Q=Q'$ and either $C=0$ or $\ell-\deg Q<\deg B-\deg C$ or (B,C) and (B',C') coincide up to $2(\ell-\deg Q)$.

To prove this statement we may assume

$$\deg A = \deg A' > 2\ell$$

(by multiplying (A,B) and (A',B') with appropriate powers of x) and therefore

$$\deg(A-A') \leq \deg A - 2\ell - 1,$$
$$\deg B = \deg B'$$
$$\deg(B-B') \leq \deg A - 2\ell - 1.$$

Subtracting the equations (4.6) we get

$$(4.7)\qquad A-A' = Q(B-B')+(Q-Q')B'+C-C'.$$

The polynomials $A-A'$, $Q(B-B')$ and $C-C'$ all have degrees $<\deg B$. Therefore

$$\deg(Q-Q')B' < \deg B,$$

which implies $Q=Q'$. But then (4.7) gives

$$(4.8)\qquad \deg(C-C') < \deg Q + \deg A-2\ell.$$

Now assume $C\neq 0$, $\ell-\deg Q\geq\deg B-\deg C$. Comparing this with (4.8) we get $\deg C=\deg C'$ (in particular $C'\neq 0$). But then (4.8) implies

$$C|2(\ell-\deg Q)-(\deg B-\deg C)$$
$$= C'|2(\ell-\deg Q)-(\deg B'-\deg C').$$

This proves the statement and the lemma.

In the sequel when we speak of computing polynomials from polynomials, we always think of computing their coefficients from the coefficients of the given polynomials. At the end of such a computation one will of course also know the degrees of the output polynomials. Similar remarks apply to matrices built up from polynomials.

Lemma 4.2 Let $n\geq m\geq 0$ and $\ell\geq 0$. The function that assigns to any pair of polynomials A_1, A_2 with $\deg A_1=n$, $\deg A_2=m$ the sequence

$$(Q_1,\ldots,Q_{s-1},M_{s-1}\cdots M_1)$$

with $\sum_1^{s-1} n_i\leq\ell$ and either $s=t$ or $\sum_1^s n_i>\ell$, is computable in time

$$c\ell(H(n_1,\ldots,n_{s-1},\ell-\sum_1^{s-1} n_i)+d)+e(s-1)+1,$$

where $c=30$, $d=5$ and $e=16$.

Proof Induction on ℓ. The cases $\ell=0$ or $\ell<n-m$ being clear, assume $1\leq\ell\leq n$ (w.l.o.g.) and $\ell\geq n-m$. By working with the initial segments $A_1|2\ell$ and $A_2|2\ell-(n-m)$ instead of A_1 and A_2 resp. and using Lemma 4.1 we may assume w.l.o.g. that

$$\deg A_1 \leq 2\ell$$
$$\deg A_2 \leq 2\ell.$$

By induction hypothesis (applied to $\lfloor\ell/2\rfloor$ instead of ℓ) and (2.2) we can compute

$$(A_1,A_2) \longmapsto (Q_1,\ldots,Q_{r-1},M_{r-1}\cdots M_1)$$

in time

$$c(\ell/2)(H(n_1,\ldots,n_{r-1},(\ell/2)-\sum_1^{r-1} n_i)+d)+e(r-1)+1,$$

where $\sum_1^{r-1} n_i\leq\ell/2$ and either $r=t$ or $\sum_1^r n_i>\ell/2$. By

(4.4) we can compute

$$(A_1,A_2,M_{r-1}\cdots M_1) \longmapsto (A_r,A_{r+1})$$

in time

$$4((\sum_1^{r-1} n_i)+2\ell+1) \le 10\ell+4.$$

Therefore

(4.9) $$(A_1,A_2) \longmapsto (Q_1,\ldots,Q_{r-1},M_{r-1}\cdots M_1,A_r,A_{r+1})$$

is computable in time

$$c(\ell/2)(H(n_1,\ldots,n_{r-1},(\ell/2)-\sum_1^{r-1} n_i)+d)+e(r-1)+10\ell+5.$$

Now if r=s (this can be tested at no cost since r=s iff $A_{r+1}=0$ or $\sum_1^{r-1} n_i+\deg A_r-\deg A_{r+1}>\ell$) no further computation is necessary. Otherwise we can compute

$$(A_r,A_{r+1}) \longmapsto (Q_r,A_{r+1},A_{r+2})$$

in time

$$6n_r+1+2\ell+1 \le 8\ell+2$$

(by a division with remainder, using Sieveking [19], Strassen [23], Kung [9]). If $A_{r+2}\neq 0$ we apply the induction hypothesis to (A_{r+1},A_{r+2}) instead of (A_1,A_2) and $\ell-\sum_1^r n_i$ instead of ℓ. Thus

$$(A_{r+1},A_{r+2}) \longmapsto (Q_{r+1},\ldots,Q_{s-1},M_{s-1}\cdots M_{r+1})$$

is computable in time

$$c(\ell-\sum_1^r n_i)(H(n_{r+1},\ldots,n_{s-1},(\ell-\sum_1^r n_i)-\sum_{r+1}^{s-1} n_i)+d)+e(s-r-1)+1.$$

If $A_{r+2}=0$ we have r=s-1. So in any case (if r<s) we can compute

(4.10) $$(Q_1,\ldots,Q_{r-1},M_{r-1}\cdots M_1,A_r,A_{r+1}) \longmapsto (Q_1,\ldots,Q_{s-1},M_{s-1}\cdots M_{r+1},M_{r-1}\cdots M_1)$$

in time

$$c(\ell-\sum_1^r n_i)(H(n_{r+1},\ldots,n_{s-1},\ell-\sum_1^{s-1} n_i)+d)+e(s-r-1)+8\ell+3.$$

By two matrix multiplications taking into account the special form (4.3) of M_r we can compute

$$(M_{s-1}\cdots M_{r+1},M_r,M_{r-1}\cdots M_1) \longmapsto (M_{s-1}\cdots M_1)$$

in time

$$2(\ell+1)+7(\ell+1) = 9(\ell+1).$$

Together with (4.10) we see that

$$(Q_1,\ldots,Q_{r-1},M_{r-1}\cdots M_1,A_r,A_{r+1}) \longmapsto (Q_1,\ldots,Q_{s-1},M_{s-1}\cdots M_1)$$

is computable in time

$$c(\ell-\sum_1^r n_i)(H(n_{r+1},\ldots,n_{s-1},\ell-\sum_1^{s-1} n_i)+d)+e(s-r-1)+17\ell+12,$$

when r<s. Finally we combine this with (4.9). Thus (in any case)

$$(A_1,A_2) \longmapsto (Q_1,\ldots,Q_{s-1},M_{s-1}\cdots M_1)$$

is computable in time

$$c(\ell/2)(H(n_1,\ldots,n_{s-1},(\ell/2)-\sum_1^{s-1} n_i)+d)+e(s-1)+10\ell+5=:t_1$$

if r=s, or in time

$$c(\ell/2)(H(n_1,\ldots,n_{r-1},(\ell/2)-\sum_1^{r-1} n_i)+d)$$
$$+\ c(\ell-\sum_1^r n_i)(H(n_{r+1},\ldots,n_{s-1},\ell-\sum_1^{s-1} n_i)+d)$$
$$+\ e(s-1)+27\ell+1 =: t_2$$

if r<s. (We have used e=16.) To complete the induction, we have to show

(4.11) $$t_i \le c\ell(H(n_1,\ldots,n_{s-1},\ell-\sum_1^{s-1} n_i)+d)+e(s-1)+1.$$

Now by (2.2)

$$t_1 \le c\ell(H(n_1,\ldots,n_{s-1},\ell-\sum_1^{s-1} n_i)+d)$$
$$-\ c(\ell/2)d+e(s-1)+10\ell+5$$

implying (4.11) easily for c=30, d=5. To estimate t_2, choose $0<\varepsilon<1/2$.

Case $\sum_1^r n_i<(1-\varepsilon)\ell$: By (2.2) and (2.4) we have

$$t_2 \le c\sum_1^r n_i(H(n_1,\ldots,n_r)+d)+$$
$$+\ c(\ell-\sum_1^r n_i)(H(n_{r+1},\ldots,n_{s-1},\ell-\sum_1^{s-1} n_i)+d)+$$
$$+\ e(s-1)+27\ell+1$$
$$\le c\ell(H(n_1,\ldots,n_{s-1},\ell-\sum_1^{s-1} n_i)+d)+e(s-1)+1$$
$$+\ \ell(27-cH(1-\varepsilon,\varepsilon)).$$

Thus in this case (4.11) is a consequence of the condition

(4.12) $$27 \le cH(1-\varepsilon,\varepsilon).$$

Case $\sum_1^r n_i\ge(1-\varepsilon)\ell$: Here we have

$$t_2 \le c\sum_1^r n_i(H(n_1,\ldots,n_r)+d)+c(\varepsilon-\tfrac{1}{2})\ell d+$$
$$+\ c(\ell-\sum_1^r n_i)(H(n_{r+1},\ldots,n_{s-1},\ell-\sum_1^{s-1} n_i)+d)+$$
$$+\ e(s-1)+27\ell+1$$
$$\le c\ell(H(n_1,\ldots,n_{s-1},\ell-\sum_1^{s-1} n_i)+d)+e(s-1)+1+$$
$$+\ 27\ell+c(\varepsilon-\tfrac{1}{2})\ell d.$$

In this case (4.11) is implied by

(4.13) $$27 \le (\tfrac{1}{2}-\varepsilon)cd.$$

Now (4.12) and (4.13) are satisfied for $\varepsilon=0.35$, $c=30$, $d=5$.

Theorem 4.3 Let $n\geq m\geq 0$, $n>0$. The function that assigns to any pair of polynomials A_1,A_2 with $\deg A_1=n$, $\deg A_2=m$ their euclidean representation $(Q_1,\ldots,Q_{t-1},A_t)$ is computable in time

$$30n(H(n_1,\ldots,n_t)+6.5).$$

Proof Lemma 4.2 with $\ell=n$ shows that

$$(A_1,A_2)\longmapsto(Q_1,\ldots,Q_{t-1},M_{t-1}\cdots M_1)$$

is computable in time

$$30n(H(n_1,\ldots,n_t)+5)+16(t-1)+1.$$

By (4.4)

$$(A_1,A_2,M_{t-1}\cdots M_1)\longmapsto A_t$$

is computable in time $2(n+1)$. Now use $t-1\leq n+1$.

For completeness we now give an ALGOL-like procedure for the function that appears in Lemma 4.2.

```
procedure SCH(A,B,u;z,Q,M):
 if B=0 or u<deg A-deg B then
    begin  z := 1;
           M := (1 0
                 0 1)
    end
else
    begin  F := A|2u;
           G := B|(2u-(degA-degB));
           v := ⌊u/2⌋;
           SCH(F,G,v;z,Q,M);
           (F) := M(F);
           (G)     (G)
           if ( ∑   degQ[i])+degF-degG ≤ u
               i<z
                    and G≠0 then
             begin Q[z] := div(F,G);
                   H := rem(F,G);
                   F := G;
                   G := H;
                   v := u- ∑ deg Q[i];
                          i≤z
                   SCH(F,G,v;z',Q',M');
                   for i:=1 to z'-1 do Q[i+z]:=Q'[i];
                   M := M'(0    1   )M;
                          (1 -Q[z])
                   z := z+z'
             end
    end
```

Remarks: 1. u,v,z,z',i are variables for numbers, A,B,F,G,H are variables for polynomials, M,M' are variables for 2×2 matrices of polynomials, Q,Q' are variables of sequences of polynomials.

2. Given A,B such that $B\neq 0$, we have
   $A=\text{div}(A,B)\cdot B+\text{rem}(A,B)$
such that $\deg \text{rem}(A,B) < \deg B$.

3. Let the contents of A,B,u be A_1,A_2,ℓ respectively (see Lemma 4.2). Then after running the procedure SCH the contents of A,B,u will be unchanged, the content of z will be s, the content of $Q[i]$ will be Q_i for $1\leq i<s$ and the content of M will be $M_{s-1}\cdots M_1$ (hopefully).

5. The computational model

We will discuss here the notion of a computation tree in some generality. Let Ω,P be disjoint types (sets together with arity-functions). The $\omega\in\Omega$ are called operational symbols, the $\rho\in P$ relational symbols. A structure of type (Ω,P) is a set A together with an interpretation for each $m\geq 0$ of any m-ary $\omega\in\Omega$ as an m-ary partial operation in A and of any m-ary $\rho\in P$ as an m-ary relation in A. Notationally we will not distinguish between a symbol and its interpretation.

Example: $\Omega=\{0,1,+,-,*,/\}$, $P=\{\leq\}$, $A=\mathbf{R}$. (See also [21], [22].)

Let $s_1,s_2,\ldots,$ be variables (symbols which denote storage locations). A computation tree of type (Ω,P) is a binary tree B (see [1]) together with

1. A function that assigns:
— to any vertex with exactly one son (simple vertex) an operational instruction of the form
$$s_i := \omega(s_{j_1},\ldots,s_{j_m}),$$
where $m\geq 0$, $i,j_1,\ldots,j_m>0$ and $\omega\in\Omega$ m-ary,
— to any vertex with two sons (branching vertex) a test instruction of the form
$$\rho(s_{j_1},\ldots,s_{j_m}),$$
where $m\geq 0$, $j_1,\ldots,j_m>0$ and $\rho\in P$ m-ary,
— to any leaf an output instruction of the form
$$(s_{j_1},\ldots,s_{j_q}),$$
where $q\geq 0$, $j_1,\ldots,j_q>0$;

2. A partition σ of the set of leaves such that the length q of the assigned output instructions is

constant on σ-classes.

The purpose of the partition is to collect the relevant part of the information gathered by the various tests of the tree, as is most easily visualized in the case of a decision tree, i.e. a computation tree with $\Omega=\emptyset$. Computation trees will have inputs as well as outputs of the form

$$(A,a_1,\ldots,a_n),$$

where A is a structure of type (Ω,P) and $a_1,\ldots,a_n \in A$. n is called the length of the input (output). Let J be a set of inputs of length n (A may vary).

A <u>collection</u> for J is a pair (φ,π), where φ is a function that assigns to any input $\underline{a}=(A,a_1,\ldots,a_n)\in J$ an output $\varphi(\underline{a})=(A,b_1,\ldots,b_q)$ (the structure of $\varphi(\underline{a})$ is the same as that of $\underline{a}$, q may vary as a function of $\underline{a}$), and where π is a finite partition of J such that the length of $\varphi(\underline{a})$ is constant on π-classes.

<u>Examples</u> (If it is clear from the context, which structure is in front of an input (or output), we will often neglect to write it.)

(5.1) Matrix inversion: Let k be a field, $J=k^{n\times n}$,

$$\varphi((a_{ij})) = \begin{cases} (a_{ij})^{-1} & \text{if } \det(a_{ij})\neq 0 \\ \emptyset & \text{otherwise} \end{cases}$$

$\pi = \{\{\text{regular matrices}\}, \{\text{singular matrices}\}\}$. Then (φ,π) is a collection for J. Other interesting input sets for the same problem are e.g.

$$J = \bigcup_{p \text{ prime}} \{\mathbf{Z}_p\} \times \mathbf{Z}_p^{n\times n},$$

$$J = \{(a_{ij}) \in k^{n\times n}: (a_{ij}) \text{orthogonal}\}.$$

(5.2) Knapsack: Consider $\mathbf{R}$ as an ordered field and let $J=\mathbf{R}^n$,

$$\varphi(\underline{a}) = \emptyset \quad \text{for all } \underline{a}\in\mathbf{R}^n,$$

$$\pi = \{E,\mathbf{R}^n\setminus E\}, \text{ where}$$

$$E = \{\underline{a}\in\mathbf{R}^n: \exists I \subset \{1,\ldots,n\} \sum_{i\in I} a_i=1\}.$$

Then (φ,π) is a collection for J.

(5.3) Euclidean representation: Let k be an infinite field, considered as a commutative k-division algebra with equality ($\Omega=\{0,1,+,-,*,/\}\cup k$, where $\lambda\in k$ is interpreted as multiplication by λ, $P=\{=\}$). Let $\underline{n}\geq\underline{m}>0$, $J=(k^\times \times k^n)\times(k^\times\times k^m)\subset k^{n+m+2}$ $(k^\times=k\setminus\{0\})$. Think of inputs $\in J$ as pairs of polynomials A_1,A_2 such that $\deg A_1=n$, $\deg A_2=m$. Put

$$\varphi(A_1,A_2) = (Q_1,\ldots,Q_{t-1},A_t)$$

(see (4.1)) and let π be the partition of J into the fibres of the map

$$(A_1,A_2) \longmapsto (\deg Q_1,\ldots,\deg Q_{t-1},\deg A_t).$$

In other words, given nonnegative numbers $n_1,\ldots,n_t$ such that $n_i>0$ for $1<i<t$ and

$$\sum_1^t n_i = n, \quad \sum_2^t n_i = m,$$

let $D(n_1,\ldots,n_t)\subset J$ be the set of those (A_1,A_2) for which $\deg Q_i=n_i$ $(i<t)$, $\deg A_t=n_t$. Then the $D(n_1,\ldots,n_t)$ are just the π-classes. (φ,π) is a collection for J, which we call the euclidean representation.

We will skip a detailed semantics of computation trees and just state the following conclusion: An input $\underline{a}=(A,a_1,\ldots,a_n)$ fed into a computation tree B of the same type may or may not produce a leaf together with an output $\underline{b}=(A,b_1,\ldots,b_q)$ (at the root of B the variables are assigned the values $(a_1,a_2,\ldots,a_n,\infty,\infty,\ldots)$, then a directed path starting from the root together with an assignment to the variables for any vertex of the path is constructed). If it does, we say that B is defined on $\underline{a}$. If B is defined on a set J of inputs of the same length, we let $\varphi(\underline{a})=\underline{b}$, where $\underline{b}$ is the output produced by B on $\underline{a}$, and we let π be such that $\underline{a}$ and $\underline{a}'$ are in the same π-class iff the leaves that B produces at $\underline{a}$ and $\underline{a}'$ are in the same σ-class. Then (φ,π) is a collection for J. We say that <u>B computes (φ,π)</u>.

Now let a cost function

$$z: \Omega\cup P \longrightarrow \mathbf{R}^+$$

be given. By adding the costs of the various instructions encountered when going from the root of B to a leaf we may define the cost of any leaf of B. If B is defined on J, this gives us a function

$$t: J \longrightarrow \mathbf{R}^+,$$

the <u>cost of B on J</u> ($t(\underline{a})$ is the cost of the leaf of B produced by the input $\underline{a}$).

Finally, given J, a collection (φ,π) for J and a function

$$t: J \longrightarrow \mathbf{R}^+,$$

we say that <u>(φ,π) is computable in time t</u> if there is a computation tree B which computes (φ,π) and has cost $\leq t$ on J. We also say that (φ,π) is strictly computable in time t if in addition B is required to have $\sigma=\{\{v\}: v \text{ leaf of } B\}$. (I.e. B has to output all the information gained by performing tests.) Obviously, (φ,π) is computable in time t iff there is a partition π' of J finer than π such that (φ,π') is a collection, strictly computable in time t.

In order to eliminate the clumsy notion of a computation tree, we will now axiomatically characterize the correct statements of the form

(5.4) "(φ,π) is computable in time t".

To this end fix (Ω,P), z and J.

Axiom $(\mathrm{id},\{J\})$ is computable in time 0.

Rules of inference Let (φ,π) be a collection for J, computable in time t, $D\in\pi$ and

$$\varphi \upharpoonright D = (\varphi_0,\dots,\varphi_q).$$

(Thus for $\underline{a}\in D$ we have $\varphi_0(\underline{a})=A$, $\varphi_i(\underline{a})\in A$ for $i\geq 1$.)

(I) If $t\leq t'$ then (φ,π) is computable in time t'.

(II) If $\varphi'=\varphi$ on $J\setminus D$ and

$$\varphi' \upharpoonright D = (\varphi_0,\varphi_{i_1},\dots,\varphi_{i_p}),$$

where $1\leq i_1,\dots,i_p\leq q$, then (φ',π) is computable in time t.

(III) If $\varphi'=\varphi$ on $J\setminus D$ and

$$\varphi' \upharpoonright D = (\varphi_0,\varphi_1,\dots,\varphi_q,\omega(\varphi_1,\dots,\varphi_m)),$$

where $m\leq q$, $\omega\in\Omega$ m-ary and $(\varphi_1(\underline{a}),\dots,\varphi_m(\underline{a}))\in \mathrm{Def}\,\omega$ for any $\underline{a}\in D$, then (φ',π) is computable in time $t+z(\omega)1_D$ (where 1_D = indicator of D).

(IV) If

$$\pi' := (\pi\setminus\{D\})\cup\{\{\underline{a}\in D: (\varphi_1(\underline{a}),\dots,\varphi_m(\underline{a}))\in\rho\}, \{\underline{a}\in D: (\varphi_1(\underline{a}),\dots,\varphi_m(\underline{a}))\notin\rho\}\},$$

where $m\leq q$, $\rho\in P$ m-ary, then (φ,π') is computable in time $t+z(\rho)1_D$.

(V) If $\pi' = (\pi\setminus\{D,E\})\cup\{D\cup E\}$,

where $E\in\pi$, $E\neq D$ but length $(\varphi\upharpoonright E)=q$, then (φ,π') is computable in time t.

Theorem 5.1 The correct statements of the form

"(φ,π) is computable in time t"

are exactly those which can be deduced from the above axiom with the rules of inference I-V. (Similarly for "strictly computable" and rules I-IV.)

Proof "Deducible $\Longrightarrow$ correct": Straightforward. "Correct $\Longrightarrow$ deducible": Show that if (φ,π) is strictly computable in time t, then the corresponding statement may be deduced from the axiom by the rule I-IV, using induction on the size of a "strict" computation tree which computes (φ,π).

A convenient way to use Theorem 5.1 is by axiomatic induction: Given z and J, let $\mathfrak{A}$ be a statement about triples (φ,π,t) such that the above axiom and the rules I-V hold when all statements of the form (5.4) have been replaced by the corresponding statements $\mathfrak{A}(\varphi,\pi,t)$. Then we can conclude

(φ,π) computable in time t $\Longrightarrow \mathfrak{A}(\varphi,\pi,t)$.

As simple applications of axiomatic induction the following two propositions can be proved.

Proposition 5.2 Let (φ,π) be a collection for J, computable in time t and let $D_1\in\pi$. Moreover, let (φ_1,π_1) be a collection for J_1, computable in time t_1, and assume $\varphi(D_1)\subset J_1$. Put

$$\tilde{\varphi} = \begin{cases} \varphi & \text{on } J\setminus D_1 \\ \varphi_1\circ\varphi & \text{on } D_1 \end{cases}$$

$$\tilde{\pi} = (\pi\setminus\{D_1\})\cup\varphi^{-1}\pi_1$$

$$\tilde{t} = t+(t_1\circ\varphi)\cdot 1_{D_1}.$$

Then $(\tilde{\varphi},\tilde{\pi})$ is computable in time $\tilde{t}$.

Proposition 5.3 Let (φ_i,π_i) be collections for J, computable in time t_i $(i=1,2)$. Define φ by

$$\varphi(\underline{a}) = (A,b_1,\dots,b_{q_1},c_1,\dots,c_{q_2}),$$

where $\varphi_1(\underline{a})=(A,b_1,\dots,b_{q_1})$, $\varphi_2(\underline{a})=(A,c_1,\dots,c_{q_2})$, and let

$$\pi = \pi_1\wedge\pi_2.$$

Then (φ,π) is a collection for J, computable in time t_1+t_2.

Both propositions remain correct, when everywhere "computable" is replaced by "strictly computable".

Application Consider the collection (φ,π) of (5.3) (euclidean representation). If we take the cost function $z=1_{\{*,/\}}$, i.e. if we allow linear operations and tests for free and count the remaining multiplications/divisions, then Theorem 4.3 (or rather its proof) shows that

(5.5) (φ,π) is strictly computable in time $30n(H+6.5)$.

In fact, using Propositions 5.2 and 5.3, the proof of section 4 can easily be formalized for the present model.

6. Conversion of a rational fraction into a continued fraction: Lower bounds

Let k be an infinite field, considered as a k-field with equality ($\Omega=\{0,1,+,-,*,/\}\sqcup k$ with constants 0,1 and unary $\lambda\in k$, $P=\{=\}$). As before, let $z=1_{\{*,/\}}$.

Theorem 6.1 Assume that k is algebraically closed. Let $J\subset k^n$ be Zarisky open and (φ,π) be a collection for J, computable in time T. Let $D\in\pi$ and $\varphi\restriction D=(\varphi_1,\dots,\varphi_q)$ (disregarding $\varphi_o=k$). Then

1. D is a Zariski locally closed (see [12]) subset of k^n. $\varphi\restriction D$ is the restriction to D of a rational map $k^n \longrightarrow k^q$, whose domain of definition includes D. Graph $(\varphi\restriction D)$ is a locally closed subset of k^{n+q}.
2. $T\restriction D \geq \log \deg \operatorname{graph}(\varphi\restriction D)$.

Proof By axiomatic induction we first show: If (φ,π) is strictly computable in time T, then for any $\tilde{D}\in\pi$ with $\varphi\restriction\tilde{D}=(\tilde{\varphi}_1,\dots,\tilde{\varphi}_u)$ (say) there are rational functions $F_1,\dots,F_u$, $G_1,\dots,G_v$, $H_1,\dots,H_w$ on k^n such that

(6.1) $\tilde{D}=\{G_1=\dots=G_v=0,\ H_1=\dots=H_w\neq 0\}\cap J$

(6.2) The F_i are defined on $\tilde{D}$ and $\tilde{\varphi}_i=F_i\restriction\tilde{D}$

(6.3) $T\restriction\tilde{D}\geq L(F_1,\dots F_u,G_1,\dots,G_v,H_1,\dots,H_w)$.

(The domain of definition of a rational function F is the set of points in k^n for which the reduced denominator does not vanish. The condition G=0 is satisfied at a point $\underline{\alpha}$ iff $G(\underline{\alpha})$ is defined and equals 0. Similarly $H(\underline{\alpha})\neq 0$ is meant to imply that $H(\underline{\alpha})$ is defined.) The inductive proof is quite straightforward and we content ourselves with giving two instances of treating the rules of inference (V excluded). First, rule III with $\omega=/$: The case $\tilde{D}\neq D$ being clear let $\tilde{D}=D$. Then u=q and the rational functions $F_1,\dots,F_u,F_1/F_2,G_1,\dots,G_v,H_1,\dots,H_w$ will do. Second, rule IV: Take $F_1,\dots,F_u,G_1,\dots,G_v$, $F_1-F_2,H_1,\dots,H_w$ in case $\tilde{D}=D\cap\{\varphi_1=\varphi_2\}$ and $F_1,\dots,F_u,G_1,\dots,G_v,H_1,\dots,H_w,F_1-F_2$ in case $\tilde{D}=D\cap\{\varphi_1\neq\varphi_2\}$.

(6.1) and (6.2) imply the first assertion of the theorem. By (6.3), (1.3) and (1.2) (intersection with a linear space) we have

$$\begin{aligned} T\restriction\tilde{D} &\geq L(F_1,\dots,F_u,G_1,\dots,G_v,H_1,\dots,H_w)\\ &\geq \log\deg(F_1,\dots,F_u,G_1,\dots,G_v)\\ &\geq \log \sum_{\substack{C \text{ component of}\\ \operatorname{graph}(F_1,\dots,F_u)\cap(\{G_1=\dots=G_v=0\}\times k^u)}} \deg C \end{aligned}.$$

Now by (6.1) and (6.2)

$$\operatorname{graph}(\varphi\restriction\tilde{D})=\operatorname{graph}(F_1,\dots,F_u)\cap (\{G_1=\dots=G_v=0\}\times k^u)\cap((\{H_1\dots H_w\neq 0\}\cap J)\times k^u).$$

Thus the closure of any component of the graph of $\varphi\restriction\tilde{D}$ is also the closure of a component of $\operatorname{graph}(F_1,\dots,F_u)\cap(\{G_1=\dots=G_v=0\}\times k^u)$. Therefore

$$T\restriction\tilde{D}\geq \log \sum_{\substack{C \text{ component}\\ \text{of } \operatorname{graph}(\varphi\restriction\tilde{D})}} \deg C,$$

proving assertion 2 of the theorem.

We remark that there are finite partitions π of k^n into locally closed subsets, for which (id,π) is not strictly computable: Take $k=\mathbb{C}$, n=2 and $\pi=\{D,E,F\}$, where $D=\{y=x^2-1,x\neq 1\}$, $E=\{-y=x^2-1,x\neq -1\}$, $F=\mathbb{C}^2\setminus(D\cup E)$.

Now for any field k let

$$J_k(n,m)=\{k\}\times(k^\times\times k^n)\times(k^\times\times k^m).$$

This is an open subset of k^{n+m+2}. A typical input $(k,a_o,\dots,a_n,b_o,\dots,b_m)\in J_k(n,m)$ is interpreted as a pair of polynomials $A_1=\sum_o^n a_i x^{n-i}$, $A_2=\sum_o^m b_j x^{m-j}\in k[x]$ of degrees n and m respectively. Let (φ_k,π_k) be the euclidean representation for $J_k(n,m)$ (see (5.3)). Then π_k consists of the classes $D_k(n_1,\dots,n_t)$ $(t\geq 2, n_1\geq 0, n_2,\dots,n_{t-1}>0, n_t\geq 0, \sum_1^t n_i=n, \sum_2^t n_i=m)$, which by φ_k are mapped bijectively onto

$$J_k(n_1,\dots,n_t)=\{k\}\times\prod_{i=1}^t(k^\times\times k^{n_i}).$$

$J_k(n_1,\dots,n_t)$ is an open subset of k^{n+t}.

Lemma 6.2 Let k be algebraically closed.

1. $D_k(n_1,\dots,n_t)$ is a locally closed irreducible subvariety of k^{n+m+2}
2. $\varphi_k\restriction D_k(n_1,\dots,n_t)$ is an isomorphism of varieties

$$D_k(n_1,\dots,n_t)\longrightarrow J_k(n_1,\dots,n_t).$$

Its inverse is given by polynomials of degree $\leq t$.

3. Let $W_k \subset k^{n+m+2} \times k^{n+t}$ be the graph of $\varphi_k \upharpoonright D_k(n_1,\ldots,n_t)$. Then W_k is locally closed, irreducible and

$$\log \deg W_k \geq n(H(n_1,\ldots,n_t)-2).$$

4. Put $N=(n+m+2)+(n+t)$, $z=(n+m+2)$. There are polynomials $F_1,\ldots,F_z \in k[x_1,\ldots,x_N]$ of degree $\leq t$ such that

$$\overline{W}_k = \{F_1=\ldots=F_z=0\}.$$

Proof By (5.5) the euclidean representation is strictly computable. Thus by Theorem 6.1 $D_k(n_1,\ldots,n_t)$ is locally closed and $\varphi_k \upharpoonright D_k(n_1,\ldots,n_t)$ is a morphism into $J_k(n_1,\ldots,n_t)$. (3.4) interpreted as a function ψ on $J_k(n_1,\ldots,n_t)$ gives its inverse. So $\varphi_k \upharpoonright J_k(n_1,\ldots,n_t)$ is an isomorphism, $D_k(n_1,\ldots,n_t)$ is irreducible and ψ is defined by polynomials of degree $\leq t$. Since W_k is up to a permutation of the coordinates the same as graph ψ, (3.10) gives

$$\log \deg W_k \geq n(H-2).$$

Finally, let $f_1,\ldots,f_z \in k[x_{z+1},\ldots,x_N]$ be the polynomials defining ψ. Put $F_i = x_i - f_i$ for $1 \leq i \leq z$. Then by (3.4) $\deg F_i \leq t$ and $\overline{W}_k = \{F_1=\ldots=F_z=0\}$.

Theorem 6.3 Let k be algebraically closed and let the euclidean representation for $J_k(n,m)$ be computable in time T. Then any $D_k(n_1,\ldots,n_t)$ contains a dense open subset U such that

$$T \upharpoonright U \geq n(H(n_1,\ldots,n_t)-2).$$

Proof There is a refinement π' of π_k such that (φ_k,π') is strictly computable in time T. π' further subdivides $D_k(n_1,\ldots,n_t)$ into $D_1,\ldots,D_p$, say. By Theorem 6.1 the D_i are locally closed subsets of the irreducible variety $D_k(n_1,\ldots,n_t)$, so one of them (call it U) is a dense open subset of $D_k(n_1,\ldots,n_t)$. Then the graph of $\varphi_k \upharpoonright U$ and the graph of $\varphi_k \upharpoonright D_k(n_1,\ldots,n_t)$ have the same closure. Therefore the graph of $\varphi_k \upharpoonright U$ is irreducible and by Lemma 6.2, 3

$$\log \deg(\text{graph } \varphi_k \upharpoonright U) \geq n(H-2).$$

Theorem 6.1, 2 now yields $T \upharpoonright U \geq n(H-2)$.

Next we will discuss the euclidean representation for nonclosed fields. Let k be an infinite field, K its algebraic closure. A point of K^n that belongs to k^n is called rational. k^n has a Zariski topology (generated by the sets $\{f \neq 0\}$, where $f \in k[x_1,\ldots,x_n]$). This is also the topology induced by the Zariski topology in K^n (If $f \in K[\underline{x}]$, let $\{f_1,\ldots,f_z\}$ be its orbit under the action of $\mathrm{Gal}(K/k)$ on $K[\underline{x}]$. Put $g=\prod_1^z f_i$ if chark=0 and $g=(\prod_1^z f_i)^{p^e}$ with e sufficiently large if chark=p. Then $g \in k[\underline{x}]$ and $\{\underline{\xi} \in k^n: f(\underline{\xi})=0\}=\{\underline{\xi} \in k^n: g(\underline{\xi})=0\}$.) Since k is infinite, k^n is dense in K^n, hence it is an irreducible topological space. Since Euclid's algorithm is "field independent", we have

$$(6.4) \qquad D_k(n_1,\ldots,n_t) = D_K(n_1,\ldots,n_t) \cap J_k(n,m)$$

$$(6.5) \qquad \varphi_k \upharpoonright D_k(n_1,\ldots,n_t) = \varphi_K \upharpoonright D_k(n_1,\ldots,n_t)$$

(disregarding k and K in front of an input or output). Therefore $D_k(n_1,\ldots,n_t)$ is mapped onto $J_k(n_1,\ldots,n_t)$ under φ_K. Since k^{n+t} is dense in K^{n+t}, $J_k(n_1,\ldots,n_t)$ is dense in $J_K(n_1,\ldots,n_t)$ and therefore $D_k(n_1,\ldots,n_t)$ is dense in $D_K(n_1,\ldots,n_t)$, in particular irreducible. Also W_k is dense in W_K.

Corollary 6.4 Let k be an infinite field, $n \geq m \geq o$ and let B be a computation tree which computes the euclidean representation for $J_k(n,m)$ in time T. Assume that B computes also the euclidean representation for $J_K(n,m)$. Then any $D_k(n_1,\ldots,n_t)$ contains a dense open subset U such that

$$T \upharpoonright U \geq n(H(n_1,\ldots,n_t)-2).$$

Proof Let B compute (φ_K,π_K) in time $\tilde{T}$. Then

$$\tilde{T} \upharpoonright J_k(n,m) = T$$

and by Theorem 6.3 there is a dense open subset $\tilde{U}$ of $D_K(n_1,\ldots,n_t)$ such that

$$\tilde{T} \upharpoonright \tilde{U} \geq n(H-2).$$

Since $D_k(n_1,\ldots,n_t)$ is dense in $D_K(n_1,\ldots,n_t)$, $U := \tilde{U} \cap D_k(n_1,\ldots,n_t)$ is dense open in $D_k(n_1,\ldots,n_t)$. Moreover since $U \subset J_k(n,m)$ we have

$$T \upharpoonright U = \tilde{T} \upharpoonright U \geq n(H-2).$$

Next we try to free ourselves from the assumption that B compute the euclidean representation also over the algebraic closure of k. To this end we will have to estimate the degree of an unknown rational map that extends $\varphi_k \upharpoonright D_k(n_1,\ldots,n_t)$. A result of this nature is proved (for a similar purpose) in Strassen [24]. A more powerful method is contained in the proof of Theorem 1 of

Heintz-Sieveking [7]. The following lemma comes out of their approach.

Lemma 6.5 Let K be algebraically closed and $X\subset K^N$ be a Zariski closed set, all of whose components have the same dimension. Assume that there are polynomials $F_1,\dots,F_z \in K[x_1,\dots,x_N]$ of degree $\le d$ such that any component of X is also a component of $\{F_1=\dots=F_z=0\}$. Then if $X\subset Y\subset K^N$ and Y is closed, we have

$$\deg Y \ge \deg X \cdot d^{\dim X-\dim Y}.$$

Proof Induction on dimY: The start dimY=dimX being clear let dimY>dimX. Cleaning Y of superfluous components we may assume that any component of Y contains a component of X. Let $C_1,\dots,C_r$ be the components of Y of highest dimension. If all F_i would vanish on a C_ρ, then a component of X contained in C_ρ would not be a component of $\{F_1=\dots=F_z=0\}$. Thus for each ρ there is an $i_\rho\le z$ and a $c_\rho\in C_\rho$ such that $F_{i_\rho}(c_\rho)\neq 0$. The set

$$\{(\lambda_1,\dots,\lambda_r)\in K^r: \exists\sigma\le r \sum_{\sigma=1}^{r}\lambda_\rho F_{i_\rho}(c_\sigma)=0\}$$

is a union of r proper subspaces of K^r, therefore not all of K^r. So we can choose $\lambda_1,\dots,\lambda_r\in K$ such that $\sum_{\rho\le r}\lambda_\rho F_{i_\rho}$ does not vanish on any C_σ. Let

(6.6) $$\tilde{Y} = Y\cap\{\sum_{\rho\le r}\lambda_\rho F_{i_\rho}=0\}.$$

Then $\dim\tilde{Y}<\dim Y$ and $X\subset\tilde{Y}$. The induction hypothesis applied to $\tilde{Y}$ yields

$$\deg\tilde{Y} \ge \deg X \cdot d^{\dim X-\dim\tilde{Y}}$$

On the other hand, (6.6) and Bezouts inequality (1.2) gives

$$\deg\tilde{Y} \le \deg Y \cdot d.$$

The two last inequalities together complete the induction.

Theorem 6.6 Let k be infinite and $\varepsilon>0$. If the euclidean representation for $J_k(n,m)$ is computable in time T, then any $D_k(n_1,\dots,n_t)$ with $t\ge(\frac{1}{2}+\varepsilon)m$ contains a Zariski dense open subset U such that

$$T\restriction U \ge 2\varepsilon n\, H(n_1,\dots,n_1) - 5n.$$

Proof By axiomatic induction (over k) one easily shows: Let a collection (φ,π) for $J_k(n,m)$ be computable in time T. Then there is a collection $(\hat{\varphi},\hat{\pi})$ for $J_K(n,m)$, strictly computable in time $\hat{T}$, such that

(6.7) $\hat{\varphi}\restriction J_k(n,m) = \varphi$,

(6.8) for any π-class D there are $\hat{\pi}$-classes $\hat{D}_1,\dots,\hat{D}_p$ such that $D = \bigcup_1^p(\hat{D}_i\cap J_k(n,m))$,

(6.9) $T = \hat{T}\restriction J_k(n,m)$.

(Only rule III with ω=/ requires some care: One has to insert a test as to whether the denominator takes the value 0 or not.)

We apply this to the euclidean representation (φ_k,π_k) for $J_k(n,m)$. Unfortunately, $(\hat{\varphi},\hat{\pi})$ need not be the euclidean representation for $J_K(n,m)$. Let

(6.10) $$D_k(n_1,\dots,n_t) = \bigcup_1^p(\hat{D}_i\cap J_k(n,m)).$$

W.l.o.g. $\hat{D}_i\cap J_k(n,m)\neq\emptyset$ for all i. Let $\hat{E}_i$ be the closure of $\hat{D}_i$. Then (6.10) implies

$$D_k(n_1,\dots,n_t)\subset\bigcup_1^p\hat{E}_i.$$

Since $D_k(n_1,\dots,n_t)$ is irreducible, it is contained in some $\hat{E}_i$, say $D_k(n_1,\dots,n_t)\subset\hat{E}_1$. Since $\hat{D}_1$ is open in $\hat{E}_1$ and $\hat{D}_1\cap D_k(n_1,\dots,n_t)\neq\emptyset$, the irreducibility of $D_k(n_1,\dots,n_t)$ implies that

(6.11) $U := \hat{D}_1\cap D_k(n_1,\dots,n_t)$ is dense open in $D_k(n_1,\dots,n_t)$.

From (6.7) we see that

$$V := \text{graph}(\varphi_k\restriction U)\subset\text{graph}(\hat{\varphi}\restriction\hat{D}_1).$$

(6.11) implies that V is dense in W_k, thus also in W_K. Therefore

$$X := \overline{W}_K\subset\text{closure graph}(\hat{\varphi}\restriction\hat{D}_1) := Y.$$

Now we apply Lemma 6.5. We have dimX=n+t, dimY≤n+m+2, N=(n+m+2)+(n+t). By Lemma 6.2, 4 (with k replaced by K) we can take d=t, z=n+m+2. Thus

$$\log\deg Y \ge \log\deg X - (m+2-t)\log t$$
$$\ge n(H-2) - (m+2-t)\log t$$

by Lemma 6.2, 3. Now Theorem 6.1, 2, applied to $(\hat{\varphi},\hat{\pi})$ and $\hat{D}_1$ yields

$$\hat{T}\restriction\hat{D}_1 \ge n(H-2) - (m+2-t)\log t.$$

and therefore by (6.11) and (6.9)

$$T\restriction U \ge n(H-2) - (m+2-t)\log t.$$

Now

$$H \ge \sum_{i=2}^{t-2}\frac{1}{n}\log n \ge \frac{t-3}{n}\log n.$$

Thus for $t\ge(\frac{1}{2}+\varepsilon)m\ge m/2$ we have $m\log n\le 2nH+6\log n$ and therefore

$$T\upharpoonright U \geq n(H-2) - (m+2-(\tfrac{1}{2}+\varepsilon)m)\log n$$
$$= n(H-2) - (\tfrac{1}{2}-\varepsilon)\, m \log n - 2\log n$$
$$\geq n(H-2) - (1-2\varepsilon)nH - 5\log n$$
$$\geq 2\varepsilon nH - 5n.$$

In practice one frequently wants to compute the euclidean representation over a field $\mathbf{Z}_p$ (or several such fields, see [3]). Typically, p is not known in advance. So we are led to consider the type $\Omega=\{0,1,+,-,*,/\}$, $P=\{=\}$ and the following set of inputs:

(6.12) $$J'(n,m) = \bigcup_{p \text{ prime}} J_{\mathbf{Z}_p}(n,m).$$

A typical input $(\mathbf{Z}_p,a_o,\ldots,a_n,b_o,\ldots,b_n)$ is interpreted as a pair of polynomials $\sum_o^n a_i x^{n-i}$, $\sum_o^m b_j x^{m-j} \in \mathbf{Z}_p[x]$ of degrees n and m respectively. Let (φ',π') be the euclidean representation for $J'(n,m)$. If we put

(6.13) $$D'(n_1,\ldots,n_t) = \bigcup_{p \text{ prime}} D_{\mathbf{Z}_p}(n_1,\ldots,n_t),$$

we have

$$\varphi' = \bigcup_{p \text{ prime}} \varphi_{\mathbf{Z}_p}$$

$$\pi' = \{D'(n_1,\ldots,n_t)\colon t\geq 2,\ n_1\geq 0,\ n_2,\ldots,n_{t-1} > 0,$$
$$n_t\geq 0,\ \sum_1^t n_i=n,\ \sum_2^t n_i=m\}.$$

Given $n\geq m\geq 0$, Theorem 4.3 applies not only to infinite ground fields, but also to fields $\mathbf{Z}_p$ with p sufficiently large. Using a table look up procedure for small $\mathbf{Z}_p$, we obtain:

(6.14) The euclidean representation (φ',π') for $J'(n,m)$ is computable in time $$30n(H(n_1,\ldots,n_t)+6.5).$$

If we dismiss in the Knuth-Schönhage algorithm the symbolic multiplication of polynomials by interpolation in favor of a slower direct method, the algorithm works for any $\mathbf{Z}_p$. Thus (φ',π') is also strictly computable.

$J'(n,m)$ carries a Zariski topology, a basis being given by the sets

$$U_f := \{(\mathbf{Z}_p,\underline{a},\underline{b}) \in J'(n,m)\colon f(\underline{a},\underline{b})\neq 0 \text{ in } \mathbf{Z}_p\},$$

where $f \in \mathbf{Z}[x_1,\ldots,x_{n+m+2}]$. Since $U_f\neq\emptyset$ for $f\neq 0$, and $U_f \cap U_g = U_{fg}$, any two nonempty open sets intersect. Thus $J'(n,m)$ is irreducible. Similarly,

(6.15) $$J'(n_1,\ldots,n_t) := \bigcup_{p \text{ prime}} J_{\mathbf{Z}_p}(n_1,\ldots,n_t)$$

is irreducible in its Zariski topology.

<u>Lemma 6.7</u> 1. Let a collection (φ,π) for $J'(n,m)$ be strictly computable. Then any $D \in \pi$ is locally closed and

$$\varphi\upharpoonright D\colon D \longrightarrow \bigcup_{p \text{ prime}} \{\mathbf{Z}_p\} \times \mathbf{Z}_p^q$$

is Zariski continuous.

2. $$\varphi'\colon D'(n_1,\ldots,n_t) \longrightarrow J'(n_1,\ldots,n_t)$$

is a homeomorphism. In particular, $D'(n_1,\ldots,n_t)$ is irreducible.

<u>Proof</u> 1. Similar to the proof of Theorem 6.1, 1.

2. By 1.,

$$\varphi'\colon D'(n_1,\ldots,n_t) \longrightarrow \bigcup_p \{\mathbf{Z}_p\} \times \mathbf{Z}_p^{n+t}$$

is continuous. But φ' maps $D'(n_1,\ldots,n_t)$ into $J'(n_1,\ldots,n_t)$, whose Zariski topology is induced from the Zariski topology of $\bigcup_p \{\mathbf{Z}_p\}\times\mathbf{Z}_p^{n+t}$. Thus

$$\varphi'\colon D'(n_1,\ldots,n_t) \longrightarrow J'(n_1,\ldots,n_t)$$

is continuous. It is also bijective and its inverse is continuous (see (3.4)).

It follows easily from Lemma 6.7, 2, that all $D'(n_1,\ldots,n_t)$ are infinite and that any Zariski dense open subset U of $D'(n_1,\ldots,n_t)$ has asymptotic density one (along the decomposition (6.13)).

<u>Theorem 6.8</u> Let $n\geq m\geq 0$ and B be a computation tree that computes the euclidean representation for $J'(n,m)$ in time T. Assume that B also computes the euclidean representation for $J_{\mathbf{C}}(n,m)$. Then any $D'(n_1,\ldots,n_t)$ contains a dense open subset U' such that

$$T\upharpoonright U' \geq n(H(n_1,\ldots,n_t)-2).$$

<u>Proof</u> We apply B to the "combined" input set

$$J''(n,m) := J'(n,m) \cup J_{\mathbf{C}}(n,m).$$

$J''(n,m)$ also has a Zariski topology (defined by polynomials $\in \mathbb{Z}[x_1,\ldots,x_{n+m+2}]$). The Zariski topology on $J'(n,m)$ is induced by that of $J''(n,m)$, and the Zariski topology on $J_{\mathbf{C}}(n,m)$ is finer than the induced topology. Similarly for

$$J''(n_1,\ldots,n_t) = J'(n_1,\ldots,n_t) \cup J_{\mathbf{C}}(n_1,\ldots,n_t).$$

B computes the euclidean representation (φ'',π'') on $J''(n,m)$. Lemma 6.7 also holds for this situation, i.e.

$$\varphi'': D''(n_1,\ldots,n_t) \longrightarrow J''(n_1,\ldots,n_t)$$

is a homeomorphism. φ'' restricts to φ' on $D'(n_1,\ldots,n_t)$ and to $\varphi_{\mathbb{C}}$ on $J_{\mathbb{C}}(n_1,\ldots,n_t)$. Since obviously $J_{\mathbb{C}}(n_1,\ldots,n_t)$ is dense in $J''(n_1,\ldots,n_t)$, $D_{\mathbb{C}}(n_1,\ldots,n_t)$ is dense in $D''(n_1,\ldots,n_t)$. Similarly $D'(n_1,\ldots,n_t)$ is dense in $D''(n_1,\ldots,n_t)$.

Let T_1 be the cost of B on $J''(n,m)$. Then

(6.16) $\quad T_1 \restriction J'(n,m) \leq T.$

Each $D''(n_1,\ldots,n_t)$ is subdivided into locally closed sets D_i, on which T_1 is constant. Since $D''(n_1,\ldots,n_t)$ is irreducible, one of the D_i is dense open in $D''(n_1,\ldots,n_t)$, say D_1. Then $D_1 \cap D_{\mathbb{C}}(n_1,\ldots,n_t)$ is $\neq\emptyset$ and open in $D_{\mathbb{C}}(n_1,\ldots,n_t)$ with respect to the induced topology, thus also with respect to the Zariski topology. By Theorem 6.3 there is a dense open U in $D_{\mathbb{C}}(n_1,\ldots,n_t)$ such that

$$T_1 \restriction U \geq n(H-2).$$

We have $U \cap D_1 \neq \emptyset$ and therefore

(6.17) $\quad T_1 \restriction D_1 \geq n(H-2).$

Now $U' := D_1 \cap D'(n_1,\ldots,n_t)$ is dense open in $D'(n_1,\ldots,n_t)$ and by (6.16) and (6.17) we have

$$T \restriction U' \geq T_1 \restriction U' \geq n(H-2).$$

Theorem 6.9 Let $\varepsilon>0$, $n\geq m\geq 0$ and let the euclidean representation for $J'(n,m)$ be computable in time T. Then any $D'(n_1,\ldots,n_t)$ with $t\geq(\frac{1}{2}+\varepsilon)m$ contains a dense open subset U' such that

$$T \restriction U' \geq 2\varepsilon n\, H(n_1,\ldots,n_t) - 5n.$$

Proof Analogous to the proof of Theorem 6.8, with $\mathbb{C}$ being replaced by $\mathbb{Q}$ and using Theorem 6.6 instead of Theorem 6.3. One only has to show, that a computation tree, which computes the euclidean representation for $J'(n,m)$, also computes the euclidean representation for $J_{\mathbb{Q}}(n,m)$. This is achieved by looking at large p.

References

[1] Aho, A.V.-Hopcroft, J.E.-Ullman, J.C., The design sign and analysis of computer algorithms, Addison-Wesley, 1974.

[2] Borodin, A. and Munro, I., Computational complexity of algebraic and numeric problems, American Elsevier, 1975.

[3] Brown, W.S., On Euclid's algorithm and the computation of polynomial greatest common divisors, J. Assoc. Comput. Mach. 18(1971),478-504.

[4] Collins, G.E., Subresultants and reduced polynomial remainder sequences, J. Assoc. Comput. Mach. 14(1967), 128-142.

[5] Fano, R.M., Transmission of information, a statistical theory of communications, Wiley,1961.

[6] Hartshorne, R., Algebraic geometry, Springer, 1977.

[7] Heintz, J. and Sieveking, M., Lower bounds for polynomials with algebraic coefficients, Theor. Comput. Science 11(1980), 321-330.

[8] Knuth, D.E., The analysis of algorithms, Proc. Internat. Congress Math. (Nice, 1970), vol. 3, Gauthier-Villars, 1971, 269-274.

[9] Kung, H.T., On computing reciprocals of power series, Numer. Math. 22(1974), 341-348.

[10] Lehmer, D.H., Amer. Math. Monthly 45(1937), 227-233.

[11] Moenck, R., Fast computation of GCD's, Proc. Fifth Sympos. on Theory of Computing, ACM, New York, 1973, 142-151.

[12] Mumford, D., Introduction to algebraic geometry, Chapter I, Harvard University (mimeogr. notes)

[13] Perron, O., Die Lehre von den Kettenbrüchen, Teubner, 1913.

[14] Samuel, P., Méthodes d'algèbre abstraite en géométrie algebrique, Springer, 1967.

[15] Schnorr, C.P., An extension of Strassen's degree bound, preprint in: Prof. of the FCT Conf. in Berlin/Wendisch-Rietz, 1979, ed. L. Budach, Akademie-Verlag, 1979 (a), 404-416.

[16] Schönhage, A., Schnelle Berechnung von Kettenbruchentwicklungen, Acta Inform. 1(1971), 139-144.

[17] Schönhage, A., An elementary proof of Strassen's degree bound, Theor. Comput. Science 3(1976), 267-272.

[18] Shafarevich, I.R., Basic algebraic geometry, Part I, Springer, 1974.

[19] Sieveking, M., An algorithm for division of power series, Computing 10(1971), 153-156.

[20] Strassen, V., Gaussian elimination is not optimal, Numer. Math. 13(1969), 354-456.

[21] Strassen, V., Berechnung und Programm I, Acta Inform. 1(1972), 320-335.

[22] Strassen, V., Berechnung und Programm II, Acta Inform. 2(1973), 64-79.

[23] Strassen, V., Die Berechnungskomplexität von elementarsymmetrischen Funktionen und von Interpolationskoeffizienten, Numer. Math. 20(1973), 238-251.

[24] Strassen, V., Computational complexity over finite fields, SIAM J. Comput. 5(1976), 324-331.

[25] Strassen, V., Some results in algebraic complexity theory, Proc. Internat. Congress Math., Vancouver, 1974.

[26] Wall, H.S., Analytic theory of continued fractions, van Nostrand, 1948.

[27] van der Waerden, B.L., Einführung in die algebraische Geometrie, Springer, 1973.

Newton's Iteration and the Sparse Hensel Algorithm

Extended Abstract

Richard Zippel
MIT Laboratory for Computer Science
Cambridge, Mass. 02139

1. Introduction. This paper presents an organization of the p-adic lifting (or Hensel) algorithm that differs from the organization previously presented by Zassenhaus [Zas69] and currently used in algebraic manipulation circles [Mos73, Yun74, Wan75, Mus75]. Our organization is somewhat more general than the earlier one and admits the improvements that yielded the "sparse modular" algorithm [Zip79] more easily than the Zassenhaus algorithm. From a pedagogical point of view, the relationship between Newton's iteration and the p-adic algorithms is clearer in our formulation than with the Zassenhaus algorithm.

The problems for which p-adic polynomial techniques have been developed thus far are all based on the problem of lifting a factorization of a polynomial. For simplicity, begin with a monic polynomial F, over $\mathcal{R}$ a commutative ring with unit. We are trying to determine G and H, relatively prime elements of $\mathcal{R}[X]$ whose product is F. The first step is to pick $\mathfrak{p}$, a maximal, principal ideal of $\mathcal{R}$. Let $\mathcal{F} = \mathcal{R}/\mathfrak{p}$, and $\bar{F}$ be the image of F in $\mathcal{F}[X]$. Since $\mathcal{F}$ is "smaller" than $\mathcal{R}$, it is often easier to determine $\bar{G}$ and $\bar{H}$ (the images of G and H) than G and H. The lifting problem is to compute G and H from $\bar{G}$ and $\bar{H}$.

The approach taken in this paper is to convert this problem into a system of polynomial equations that must be solved in $\mathcal{R}$. The unknowns in this system correspond to the coefficients of X^i in G and H. The coefficients of $\bar{G}$ and $\bar{H}$ are a solution of the image problem in $\mathcal{F}$. From the p-adic point of view, these solutions are approximations to the true solutions, i.e., the coefficients of G and H. Using Newton's iteration we can "refine the approximate solutions" and deduce the exact coefficients of G and H.

The Zassenhaus of the lifting algorithm constructs equations involving G and H; the correction terms are polynomials in X that are added to G and H. Our organization deals exclusively with the coefficients of G and H, allowing finer control of the iteration process. This turns out to be the key differences that allows us to apply the probabilistic approaches previously used in the modular algorithm. (Historical note: Actually the sparse Hensel algorithm was discovered first.)

In the next section we present our development of the p-adic lifting algorithm. Section 3 demonstrates the reduction of several problems to systems of polynomial equations. In section 4 we present the modifications that allow the the algorithm to deal with sparse polynomials and overcome the "bad-zero" problem. The final section gives some conclusions and thoughts for future research.

2. The p-adic Solution of Systems of Equations. Throughout this paper we will use the following notation: $\mathcal{R}$ is a commutative ring with unit and $\mathfrak{p}$ a maximal, principal ideal of $\mathcal{R}$ which is generated by p. Since $\mathfrak{p}$ is maximal, $\mathcal{R}/\mathfrak{p}$ is a field which will be denoted by $\mathcal{F}$. Let $F(X)$ be a polynomial over $\mathcal{R}$ whose zeroes we wish to determine. There is a canonical map from $\mathcal{R}$ to $\mathcal{R}/\mathfrak{p}^k$, for any k. Thus F has a canonical image in $\mathcal{R}/\mathfrak{p}^k[X]$. By the zeroes of F in $\mathcal{R}/\mathfrak{p}^k$ we shall mean the zeroes of this canonical image polynomial.

Assume that ϖ is a root of F in the $\mathcal{R}_\mathfrak{p}$ and w_k the corresponding zeroes in $\mathcal{R}/\mathfrak{p}^k$. Using the p-adic metric, w_k forms a Cauchy sequence that converges to ϖ. Since $\mathfrak{p}$ is principal, ϖ can be written as power series in p:

$$\begin{aligned} \varpi &= w_{(0)} + w_{(1)}p + w_{(2)}p^2 + \cdots \\ w_k &= w_{(0)} + w_{(1)}p + \cdots + w_{(k)}p^k. \end{aligned} \tag{1}$$

Assume w_k is known, and better p-adic approximations to ϖ are desired (i.e. w_ℓ, $\ell > k$). By Taylor's theorem, $F(\varpi)$ can be expanded at w_k:

$$0 = F(\varpi) = F(w_k) + F'(w_k)(\varpi - w_k) + (\varpi - w_k)^2 R(x_k).$$

Using (1) and rearranging terms somewhat,

$$0 = F(w_k) + F'(w_k)[w_{(k+1)}p^{k+1} + \cdots + w_{2k+1}p^{2k+1}] + p^{2k+2}(\cdots).$$

The quantity in the brackets is $w_{2k+1} - w_k$. Thus,

$$w_{2k+1} - w_k \equiv F'(w_k)^{-1}F(w_k) \pmod{p^{2k+2}}$$

This is the usual quadratic version of "Newton's iteration." Since $F(w_k)$ is a multiple of p^{k+1}, this iteration can be rewritten as

$$w_{(k+1)} + w_{(k+2)}p + \cdots + w_{(2k+1)}p^k \equiv -\frac{[F(w_k)/p^{k+1}]}{F'(w_k)} \pmod{p^{k+1}}$$

The division inside the brackets takes place in $\mathcal{R}$.

A linear version of Newton's iteration may be derived by observing that $F'(w_k) \equiv F'(w_0) \pmod{p}$ (use Taylor's theorem again). Thus,

$$0 = F(w_k) + F'(w_0)w_{(k+1)}p^{k+1} + p^{k+2}(\cdots)$$

Solving for $w_{(k+1)}$ gives

$$w_{k+1} - w_k = w_{(k+1)} \equiv -\frac{[F(w_k)/p^{k+1}]}{F'(w_0)} \pmod{p}$$

Notice that with this iteration, $F'(w_0)^{-1} \pmod{p}$ needs to be computed only once. The k-th *correction term* $(w_{k+1} - w_k)$ is a linear function of the k-th *error term* $[F(w_k)/p^{k+1}]$.

The transition to several dimensions is straight-forward. Now let $F_1, \ldots, F_n$ $(= \vec{F})$ be polynomials over $\mathcal{R}$ in $X_1, \ldots, X_n$. We will assume that $(\varpi_1, \ldots, \varpi_n) = \vec{\varpi}$ is a zero of $\vec{F}$ in $\mathcal{R}_{\mathfrak{p}}$, and that the image of $\vec{\varpi}$ in $\mathcal{F}$ is $\vec{w}_0$ $(= (w_{01}, \ldots, w_{0n}))$. Again $\mathfrak{p}$ is assumed to be principal so

$$\varpi_i = \varpi_{(0i)} + \varpi_{(1i)}p + \varpi_{(2i)}p^2 + \cdots$$
$$w_{ki} = w_{(0i)} + w_{(1i)}p + w_{(2i)}p^2 + \cdots + w_{(ki)}p^k$$

The derivation of the multivariate iteration again begins with Taylor's theorem:

$$0 = F_j(\vec{w}_k) + \sum_{i=1}^{n} \frac{\partial F_j}{\partial X_i}(\vec{w}_k)(\varpi_i - w_{ki}) + \cdots \tag{2}$$

for $j = 1, \ldots, n$. The matrix

$$\begin{pmatrix} \frac{\partial F_1}{\partial X_1} & \frac{\partial F_1}{\partial X_2} & \cdots & \frac{\partial F_1}{\partial X_n} \\ \vdots & \vdots & & \vdots \\ \frac{\partial F_n}{\partial X_1} & \frac{\partial F_n}{\partial X_2} & \cdots & \frac{\partial F_n}{\partial X_n} \end{pmatrix}$$

is called the *Jacobian matrix* of $\vec{F}$. Equation (2) can be rewritten as

$$0 = \vec{F}(\vec{w}_k) + J(\vec{w}_k) \cdot (\vec{\varpi} - \vec{w}_k) + \cdots.$$

From this, the quadratic and and linear iterations may be derived:

$$\vec{w}_{(k+1)} + \vec{w}_{(k+2)}p + \cdots + \vec{w}_{(2k+1)}p^k \equiv -J(\vec{w}_k)^{-1} \cdot [\vec{F}(\vec{w}_k)/p^{k+1}] \pmod{p^{k+1}}$$

$$\vec{w}_{(k+1)} \equiv -J(\vec{w}_0)^{-1} \cdot [\vec{F}(\vec{w}_k)/p^{k+1}] \pmod{p} \tag{3}$$

Notice that the Jacobian matrix needs to be inverted only once in the linear iteration. This can be a distinct advantage over the quadratic algorithm, since it is often not necessary to compute many terms of $\vec{\varpi}$.

3. Reduction to Systems of Equations. This section will examine how the factorization problem can be converted to to a system of equations problem. It is not difficult to use identical techniques to deal with problems like the computation of greatest common divisors (GCDs) and square-free decomposition.

Let F be a polynomial over a ring $\mathcal{R}$ (which satisfies the usual conditions),

$$F(X) = F_0X^n + F_1X^{n-1} + \cdots + F_n$$

and assume it factors into two polynomials of degree r and s. Construct two polynomials of these degrees with formal parameters G_i and H_i:

$$G = G_0X^r + G_1X^{r-1} + \cdots + G_r,$$
$$H = H_0X^s + H_1X^{s-1} + \cdots + H_s.$$

by equating the coefficients of X^i in F and GH we have:

$$\begin{aligned} G_0H_0 &= F_0 \\ G_1 + H_1 &= F_1 \\ G_2 + G_1H_1 + H_2 &= F_2 \\ &\vdots \\ G_rH_s &= F_n \end{aligned}$$

This is a system of $n+1$ equations in the $r+s+2$ unknowns G_i, H_i. One solution to the system corresponds to the true factorization of F over $\mathcal{R}$, $F = G_TH_T$. Unfortunately, there are more unknowns than equations so the solution is not necessarily unique. Some of these additional solutions correspond associates of the factorization, *viz.* $F = (uG_T)(u^{-1}H_T)$. These associates can be eliminated by fixing one of the coefficients of G or H, usually the leading or trailing coefficient is easiest. The technique given by Wang [Wan78] is probably the best technique for determining the leading coefficient.

For simplicity, we will assume that F and its factors are monic. Since G_0 and H_0 are now assumed to be 1, there are only $r+s=n$ unknowns to be determined in the n remaining equations.

If more is known about the factorization, a more precise set of equations can be constructed. For instance, if G and H were known to be the same polynomial, the number of variables would be cut in half.

$$\begin{aligned} 2G_1 &= f_1 \\ 2G_2 + G_1^2 &= f_2 \\ &\vdots \\ G_r^2 &= f_n \end{aligned}$$

This approach can be used to great advantage when computing square-free decompositions as has been pointed out by Trager and Wang [Tra79].

This technique can also be applied to multivariate problems. Assume we know that

$$F(X,Y,Z) = (X^2 + W_1Y)(X^2 + W_2XY + W_3Y)$$

and we wish to determine the W_i. By equating the coefficients of the monomials in X and Y we have the following set of equations:

$$\begin{aligned} W_2 &= f_{31} \\ W_1 + W_3 &= f_{21} \\ W_1W_2 &= f_{12} \\ W_1W_3 &= f_{02} \end{aligned}$$

where f_{ij} refers to the coefficient of X^iY^j in F. Notice that the solution of this system may deduced immediately. In general, however, this will not be the case.

We have shown how a system of algebraic equations can be solved using Newton's iteration, and how a system of algebraic equations can be generated from a polynomial factoring problem. The following example illustrates these techniques.

Assume we wish to factor

$$P(X) = X^5 - 6X^4 + 14X^3 - 71X^2 - 29X + 7$$

over the integers. Since $P(X)$ is monic, its factors will also be monic. Factoring this polynomial in a number of finite fields [Ber70, Knu81], reveals that $P(X)$ cannot factor into smaller pieces than one quadratic and one cubic. The factorization modulo 2 is

$$P(X) = (X^3 + X + 1)(X^2 + X + 1).$$

The rational integers are a principal ideal domain, and (2) is a maximal ideal so Newton's iteration can be used. We will be lifting this factorization over $\mathbf{Z}/(p)$ to a factorization over $\mathbf{Z}/(p)^2$, $\mathbf{Z}/(p)^3$ and so on. Since we are lifting the product of a cubic and a quadratic, the resulting set equations is

$$\begin{aligned} F_1 &= G_1 + H_1 + 6 = 0, \\ F_2 &= G_2 + G_1H_1 + H_2 - 14 = 0, \\ F_3 &= G_3 + G_2H_1 + G_1H_2 + 71 = 0, \\ F_4 &= G_3H_1 + G_2H_2 + 29 = 0, \\ F_5 &= G_3H_2 - 7 = 0. \end{aligned}$$

From the factorization modulo 2, the starting point for the iteration is:

$$\begin{aligned} g_{01} &= 0, & h_{01} &= 1, \\ g_{02} &= 1, & h_{02} &= 1, \\ g_{03} &= 1. \end{aligned}$$

The Jacobian matrix for this system of equations is

$$J = \begin{pmatrix} 1 & 0 & 0 & 1 & 0 \\ h_{01} & 1 & 0 & g_{01} & 1 \\ h_{02} & h_{01} & 1 & g_{02} & g_{01} \\ 0 & h_{02} & h_{01} & g_{03} & g_{02} \\ 0 & 0 & h_{02} & 0 & g_{03} \end{pmatrix} = \begin{pmatrix} 1 & 0 & 0 & 1 & 0 \\ 1 & 1 & 0 & 0 & 1 \\ 1 & 1 & 1 & 1 & 0 \\ 0 & 1 & 1 & 1 & 1 \\ 0 & 0 & 1 & 0 & 1 \end{pmatrix},$$

whose inverse is

$$\begin{pmatrix} 1 & 1 & 1 & 0 & 1 \\ 0 & 1 & 1 & 1 & 0 \\ 1 & 1 & 0 & 1 & 0 \\ 0 & 1 & 1 & 0 & 1 \\ 1 & 1 & 0 & 1 & 1 \end{pmatrix}.$$

Using the linear iteration in (3), we begin by computing the first error term

$$\vec{F}(\vec{w}_0)/2 = \frac{1}{2}\begin{pmatrix} 6 \\ -12 \\ 60 \\ 24 \\ -6 \end{pmatrix} = \begin{pmatrix} 3 \\ -6 \\ 30 \\ 12 \\ -3 \end{pmatrix} = \begin{pmatrix} 1 \\ 0 \\ 0 \\ 0 \\ 1 \end{pmatrix}. \qquad \pmod 2$$

The first correction term is thus $<0,0,1,1,0>$. Continuing this process yields the following sequence of approximations to the p-adic zero of the polynomials

$$\begin{aligned} g_1 &= 0, & 0, && 0, && 0, && 0, && 0, && 0, \ldots \\ g_2 &= 1, & 1, && 5, && 13, && 13, && 13, && 13, \ldots \\ g_3 &= 1, & 3, && 7, && 7, && 7, && 7, && 7, \ldots \\ h_1 &= 1, & 3, && 11, && 27, && 59, && 123, && 251, \ldots \\ h_2 &= 1, & 1, && 1, && 1, && 1, && 1, && 1, \ldots \end{aligned}$$

Thus the cubic polynomial is $G = X^3 + 13X + 7$ while the quadratic is $H = X^2 - 5X + 1$. Notice that when converting p-adic integers to rational integers some care must be taken with sign of the rational integer. When $\mathcal{R}$ is a polynomial domain, this problem does not exist.

This iteration can be used if and only if the Jacobian matrix is invertible; that is, when the determinant of the Jacobian is invertible. The elements of the Jacobian are elements of $\mathcal{R}/\mathfrak{p}$, as is the determinant. Our requirement that $\mathfrak{p}$ be maximal ensures that $\mathcal{R}/\mathfrak{p}$ is a field and thus the Jacobian is invertible if it is not zero. The Jacobian is actually the Sylvester matrix for the resultant of G and H over $\mathcal{F}$. It can be zero only when G and H have a common factor over $\mathcal{F}$, which is assumed not to be the case.

For some problems, there will be more equations than unknowns and thus the Jacobian is not square. In this case, Newton's method is applied to a subset of the equations, a subset that yields an invertible Jacobian. The resulting solutions are then be verified by substitution into the full system. If the solutions fail to satisfy the full system, the system has no solutions.

4. The Sparse Hensel Algorithm. Thus far, our attention has centered on relatively simple rings $\mathcal{R}$. This section develops a technique for handling the ring $\mathcal{R} = \mathbf{Z}/(q)[X_1, \ldots, X_v]$, which often occurs in manipulation (q is a prime). Generally, we will assume that solutions to problems will be *sparse polynomials*, that is, many of the coefficients are zero. We have shown how to reduce problems to solving systems of equations. In this section we will assume that we are given a system of algebraic equations which are to be solved in $\mathbf{Z}/(q)[X_1, \ldots, X_v]$, and leave to the reader the construction of the equations from the original problem. The basic approach described can also be used to lift a solution in $\mathbf{Z}/(q)[X_1, \ldots, X_v]$ to a solution in $\mathbf{Z}[X_1, \ldots, X_v]$, but there are a number of details that complicate the problem. These details concern the factorization of irreducible polynomials over finite fields and are peripheral to the focus of this paper. Thus slightly,b different approaches are used for GCD computation, factoring and square-free decomposition.

For simplicity assume we are trying to obtain the zero

not invertible. If this occurs, it is necessary to return to the beginning and choose a new starting point. It is possible for all the elements of a finite field to be bad starting points for the iteration. For this reason, an effort is usually made to choose a relatively large finite field for the iteration if possible. If this is not possible the field may be enlarged by picking the starting point from an algebraic extension of the finite field. The correction terms must be elements of the original field, though, for the answer to lie in $\mathcal{R}$.

We will denote the zero of $F(W)$ by W_{XYZ}, the zero of $F(W)$ in $\mathbf{Z}/(13)[X,Y,Z]/(Z-z_0)$ as W_{XY}. Similarly W_X involves just X. For this example we will use $x_0 = 2$, $y_0 = 3$ and $z_0 = 7$ as the starting point. To generate the problem in $\mathbf{Z}/(13)[X]$ replace Y and Z by 3 and 7 respectively in $F(W)$

$$F(W) = W^3 + W - X^9 - 2X^7 - 6x^6 + 3X^5 + 5X^4 + 5X^3 - 6X^2 + 3.$$

Factoring this polynomial over $\mathbf{Z}/(13)[X]/(X-2)$ gives

$$F(W) = (W+6)(W^2 - 6w + 2).$$

Using Newton's iteration, starting at $W_{X0} = -6$ gives

$$W_X = -6 + 4(X-2) + 6(X-2)^2 + (X-2)^3 = X^3 + 5X + 2$$

Now we need to construct a system of equations that will be solved in $Z/(13)[Y]$. Since W is a cubic polynomial, there are four unknown coefficients to be determined. Replacing W by $AX^3 + BX^2 + CX + D$ in $F(W)$ and equating the coefficients of X with zero gives the following system of equations

$$\begin{aligned}
A^3 &= 1\\
3A^2B &= 0\\
3A^2C + 3AB &= 6Y^2\\
3AD^2 + 6ABC + B^3 &= 3YZ^3\\
6ABD + 3AC^2 + 3B^2C &= 12Y^4\\
6ACD + 3B^2D + 3BC^2 &= 12Y^3Z^3\\
3AD^2 + 6BCD + C^2 + A &= 8Y^6 + 3Y^2Z^6 + 1\\
3BD^2 + 3C^2D + B &= 12Y^5Z^3\\
3CD^2 + C &= 6Y^4Z^6 + 2Y^2\\
D^3 + D &= Y^3Z^9 + YZ^3
\end{aligned} \tag{4}$$

The zeroes of this system are the coefficients of X^i in W_{XYZ}. The initial values for this system are known from W_X: $A=1$, $B=0$, $C=5$, $D=2$.

Though Newton's method could again be applied at this point, we know a little more about the solution of this system than we knew about W_X. This is the point at which the probabilistic techniques of [Zip79] are applied. The coefficient of X^2 in W_{XYZ} is a polynomial in Y and Z. Since it is zero at the randomly chosen point $Y=3$, $Z=7$ we assume it is identically zero.

Using this assumption and replacing Z by 7 in (4) gives the following system of equations which can be solved for polynomials in $\mathbf{Z}/(13)[Y]$ by Newton's method.

$$\begin{aligned}
A^3 &= 1\\
3A^2C &= 6Y^2\\
3AD^2 &= 2Y\\
3AC^2 &= 12Y^4\\
6ACD &= -5Y^3\\
3AD^2 + C^2 + A &= -5Y^6 - 3Y + 1\\
3C^2D &= -5Y^5\\
3CD^2 + C &= -6Y^4 + 2Y^2\\
D^3 + D &= -5Y^3 - 5Y
\end{aligned}$$

The use of this simplified system of equations, which involves fewer unknowns, is the key idea of the sparse Hensel algorithm.

The solutions are

$$\begin{aligned}
A &= 1\\
C &= 5 - (Y-3) + 2(Y-2)^2 = 2Y^2\\
D &= 2 + 5(Y-3) = 5Y
\end{aligned}$$

Notice that the p-adic solutions are converted to polynomials, since the polynomials are likely to be sparse while the p-adic series are almost certainly not.

A system of equations must be constructed for the coefficients of the powers of Y in A, B, C and D. As before, we assume that the coefficients that are zero here are identically zero, thus only 3 coefficients need be determined.

$$\begin{aligned}
A &= R\\
C &= SY^2\\
D &= TY
\end{aligned}$$

The resulting system contains many equations but no more terms than (4). In general it could be solved using Newton's method, but it this particular case this is not necessary. The solution may be determined by inspection. This happens frequently for sufficiently sparse polynomials as was pointed out by Wang [Wan78]. Determining the values of particular variables can be used to great advantage at all stages, even if the complete solution is not determined in this manner.

$$\begin{aligned} R^3 &= 1 & 3R^2S &= 6 \\ R^2T &= Z^3 & RS^2 &= 4Z^3 \\ RST &= 2Z^3 & RT^2 &= Z^6 \\ S^2 &= 4 & R &= 1 \\ 3S^2T &= 4Z^3 & ST^2 &= 2Z^6 \\ S &= 2 & T^3 &= Z^9 \\ T &= Z^3 \end{aligned}$$

This algorithm assumes that certain coefficients are zero since their value at a randomly chosen point was zero. If a coefficient were non-zero contrary to the assumption, the iteration would not converge to a valid zero. Thus it is important to verify, as each new variable is introduced into the solution, that what has been computed is actually a solution. This can easily be done by substituting into the system of equations.

5. Conclusions. In this paper we have presented a different organization of the Hensel technique that is somewhat more flexible than the version commonly used in algebraic manipulation. Our version makes use of the probabilistic techniques developed in [Zip79] and for all practical purposes does not exhibit the exponential behavior of earlier version of the algorithm. Furthermore, it is applicable to any problem that can be phrased in terms of solving systems of polynomial equations. This includes the classical problems of computing GCDs and factorizations.

References

1. E. R. Berlekamp, "Factoring Polynomials over Large Finite Fields," *Math. of Comp.* **24,** 111 (1970), 713–735.
2. D. E. Knuth, *The Art of Computer Programming,* Vol. II, second edition, Addison-Wesley Publishing Company, Reading, Mass., (1981).
3. J. Moses and D. Y. Y. Yun, "The EZGCD algorithm," *Proceedings of ACM Nat. Conf.* (1973), 159–166.
4. D. R. Musser, "Multivariate Polynomial Factoring," *J. ACM* **22,** 2 (1975), 291–308.
5. B. M. Trager and P. S.-H. Wang, "On Square-free Decomposition," *SIAM Rev. of Comp.* **8,** 3 (1979), 300–305.
6. P. S.-H. Wang and L. P. Rothschild, "Factoring Multivariate Polynomials over the Integers," *Math. Comp.* **29,** (1975), 935–950.
7. P. S.-H. Wang, "An Improved Multivariate Polynomial Factoring Algorithm," *Math. Comp.* **32,** (1978), 1215–1231.
8. D. Y. Y. Yun, *The Hensel Lemma in Algebraic Manipulation,* Ph. D. thesis, Dept. of Mathematics, Massachusetts Institute of Technology, (1974).
9. H. J. Zassenhaus, "On Hensel Factorization I," *J. Number Theory* **1,** (1969), 291–311.
10. R. E. Zippel, "Probabilistic Algorithms for Sparse Polynomials," *Symbolic & Algebraic Computation (E. W. Ng, Ed.),* Springer-Verlag, Heidelberg, (1979).

Automatic Generation of Finite Difference Equations and Fourier Stability Analyses

Michael C. Wirth
Air Force Institute of Technology
Department of Mathematics
Wright-Patterson AFB, Ohio 45433

Abstract

Recently several software tools based on the algebraic manipulation system MACSYMA have been implemented which facilitate the design, analysis and construction of finite difference programs for the numerical solution of systems of partial differential equations. Two of them are described here. The FDIFF package converts scalar, non-linear partial differential equations into linear, finite difference approximations. It includes tools for discretization of the domain of the PDE's dependent variables, linearization of non-linear terms and conversion of derivative terms into finite difference expressions. A notation and algebra for building arbitrary finite difference operators is provided. The FSTAB package automatically performs local Fourier stability analyses on sets of finite difference equations by deriving amplification matrices.

1: Introduction

Computational physics typically involves the use of large programs written in a language such as FORTRAN to solve sets of partial differential equations (PDEs) numerically by finite difference or other approximations[6]. Often the limiting factor in solving such problems is not the speed of the computer used to perform the numerical calculations, but rather the prodigious manual effort required to develop the associated computer codes. Recent efforts have attempted to apply algebraic manipulation systems such as MACSYMA[3] to the automation of this code development task[2,5,7,8]. This paper discusses two specific software tools implemented in MACSYMA: FDIFF, a package for generating finite difference equations, and FSTAB, a package for analyzing the local stability properties of the difference equations. A simple example is also presented. Additional software tools and examples are provided in [8].

2: Generation of Finite Difference Equations

FDIFF is a package of MACSYMA functions which convert non-linear partial differential equations into linear, finite-difference approximations. This process involves: discretization of the domain of the PDE dependent variables, linearization of non-linear terms and conversion of derivative terms into finite difference expressions.

Before executing the discretization and linearization functions, it is expected that the user would have one or more PDEs such as

$$\frac{dV}{dT} = \frac{d^2 W}{dY^2} + \frac{d^2 W}{dX^2} + \frac{d^2 U}{dZ^2} + \frac{d^2 U}{dY^2} + \frac{d^2 U}{dX^2} \qquad (1)$$

The explicit dependencies of the dependent on the independent variables should have been established by executing a statement such as

```
DEPENDS([U,V],[X,Y,Z,T],W,[X,Y,T]);        (2)
```

which declares that U and V are functions of X, Y, Z, and T, but that W has no dependence on Z.

2.1: Discretization

The dependent variables in equation 1, which are defined over a continuous domain, can be converted into mesh variables, defined at discrete mesh points, with a call of the DISCRETIZE function of the form

```
DISCRETIZE(<expn>,<indvaru>,<indvarl>,
                    <indexu>,<indexl>).
```

The argument <expn> is an arbitrarily complex MACSYMA expression (e.g., the complete list of scalar PDEs). Arguments <indvaru> and <indvarl> are either single independent variables or lists of them which are associated with the superscripts and subscripts, respectively. Arguments <indexu> and <indexl> are the corresponding integer indices. If the variable EQN1 contained equation 1 as value, then the MACSYMA statement

```
DISCRETIZE(EQN1,T,[X,Y,Z],N,[I,J,K]);
```

would produce the expression

$$\frac{d}{dT}\left(V\Big|^{N}_{I,J,K}\right) = \frac{d^2}{dY^2}\left(W\Big|^{N}_{I,J}\right) + \frac{d^2}{dX^2}\left(W\Big|^{N}_{I,J}\right)$$

$$+ \frac{d^2}{dZ^2}\left(U\Big|^{N}_{I,J,K}\right) + \frac{d^2}{dY^2}\left(U\Big|^{N}_{I,J,K}\right) + \frac{d^2}{dX^2}\left(U\Big|^{N}_{I,J,K}\right)$$

DISCRETIZE replaces each dependent variable with an indexed functional form with indices determined by the logical intersection of the dependent variable's dependencies and the independent variables specified in the second and third arguments. Thus W does not appear with a K index in this equation. DISCRETIZE provides additional features for discretization in stages, remembering index names, generating default names, etc.

The indexed objects which are produced by the DISCRETIZE function have different internal and external representations. The external form is two-dimensional, consistent with normal notation for mesh values, such as

$$U\Big|^{N}_{I,J,K}$$

The vertical bar is used to distinguish this form from array references and exponents. The internal form, suitable for input as well, is a linear functional form using the noun (quoted) form of the special function, INDEXED

```
'INDEXED(U,[N],[I,J,K]).
```

2.2: Linearization

The function LINEARIZE handles polynomial nonlinearities. The approximation used is a simple first-order Taylor expansion about the N-th time level which leads to the following substitutions:

$$u\Big|^{N+1} * v\Big|^{N+1} \Longrightarrow u\Big|^{N+1} * v\Big|^{N} + u\Big|^{N} * v\Big|^{N+1} - u\Big|^{N} * v\Big|^{N}$$

$$\left(u\Big|^{N+1}\right)^{m} \Longrightarrow m*u\Big|^{N+1} * \left(u\Big|^{N}\right)^{m-1} - (m-1)*\left(u\Big|^{N}\right)^{m}$$

These substitutions are recursively applied for products of 3 unknowns, etc.

LINEARIZE is called by an expression of the form

```
LINEARIZE(<expn>,<index>);
```

where <expn> is an arbitrary MACSYMA expression containing indexed dependent variables and <index> is the base index value (e.g., N or N + 1/2), and returns the linearized approximation as the result.

2.3: Difference Operator Notation and Algebra

The FDIFF package converts derivatives of indexed expressions to finite differences using the pattern matching facilities provided in MACSYMA. The pattern matching rules used with FDIFF contain two parts: a pattern which matches the appropriate class of derivative expressions and a replacement expression which defines the resulting finite difference. Patterns for matching various derivatives can be built using the standard MACSYMA facilities. The replacement expressions are built using finite difference operator expressions and the special "<>" operator as described below.

In order to produce arbitrary finite difference expressions, the FDIFF package provides a notation and algebra for finite difference operators. The functional form, E(psi), is interpreted by the <> operator as a shift operator which translates its argument by one mesh spacing in the psi direction (where psi is an independent variable). Similarly

$$E(psi)^{m}$$

shifts its operand by an integral (or, in some cases, fractional) number, m, of mesh spacings. Arbitrary finite difference operators can be built as polynomial expressions in the shift operator, E, and any other, non-operator expressions. The basic difference operators are defined in FDIFF as two argument functions which return the appropriate expressions in E as shown in Table 1 (where DELTAVAR is a measure of mesh spacing).

Table 1: Basic FDIFF Difference Operators

Operator	Function Definition
Forward diff.	$DELF(PSI,M) := \dfrac{E(PSI)^{M} - 1}{M*DELTAVAR[PSI]}$
Central diff.	$DELC(PSI,M) := \dfrac{E(PSI)^{M/2} - E(PSI)^{-M/2}}{M*DELTAVAR[PSI]}$
Backward diff.	$DELB(PSI,M) := \dfrac{1 - E(PSI)^{-M}}{M*DELTAVAR[PSI]}$
Central average	$AVG(PSI,M) := \dfrac{E(PSI)^{M/2} + E(PSI)^{-M/2}}{2}$

Using the operator function definitions in Table 1, new operator functions can now be easily defined. For example,

$$ADELC(X,J) := E(T)*DELC(X,J); \qquad (3)$$

defines a centered difference operator which also advances its operand to the next time level.

$$DELF2(PSI,M) := DELF(PSI,M)^{\wedge}2; \qquad (4)$$

defines a forward, second difference of width 2*M in the PSI direction.

In Table 1, E was treated as a simple functional form, E(psi), and could be manipulated using the normal MACSYMA facilities for variables and functions, i.e., polynomial expressions in E could be expanded and simplified. The next step is to apply the difference operator expressions thus derived to suitable operands, where E is now interpreted as an operator, and produce a replacement part for a pattern matching rule.

The FDIFF package contains the binary, infix operator "<>" which is used in the form

{difference operator expression} <> {operand}

The "<>" operator serves two purposes. First, it acts as a "fence," allowing its left argument to be manipulated and simplified independently using the normal MACSYMA algebraic simplification facilities, but prevents MACSYMA from mixing products of E and the operand on the right hand side. Secondly, it interprets the difference operator expression on the left, applies it to the operand on the right and generates a MACSYMA expression which can be used as the replacement part of a pattern-replacement rule. This rule will perform the appropriate differencing operations when it is applied. That is, the "<>" operator does not do differencing, but rather generates code for a replacement expression which will accomplish the differencing when the rule is applied.

3: Local Fourier Stability Analysis

FSTAB is a package of MACSYMA functions which perform local Fourier stability analyses on sets of finite difference equations. The FSTABILITY function analyzes a set of non-linear finite difference equations and produces the amplification matrix for a Fourier perturbation to the finite difference solution[6]. It is called with three arguments

FSTABILITY(<eqns>,<advtimelvl>,<indices>)

<eqns> is a list of the finite difference equations for each of the scalar components of the solution vector. <advtimelvl> is the index of the advanced time level, e.g., N+1. <indices> is a list of the spatial indices used in the difference equations, e.g., [I,J,K]. Other global information is also used by the FSTABILITY function. The flag VERBOSE, if set to TRUE, causes FSTABILITY to print a message as it starts each major computational step. SMOOTHSOLN is a flag which will be described below. FSTABILITY also uses the array, CONTINUOUSVAR[INDEX], which contains the continuous variables associated with each index, e.g., CONTINUOUSVAR[I] = X. This bookkeeping array is normally created by the DISCRETIZE function (from the FDIFF package) when the partial differential equations are first discretized and is used by FSTABILITY in forming the $i*(\underline{k}.\underline{x})$ exponential phase factors.

FSTABILITY returns its results in three global variables, LHSVARS, RHSVARS and AMPMATRIX, and as a functional value in the form of a matrix equation

LHSVARS = AMPMATRIX.RHSVARS

LHSVARS is a list of perturbation coefficients at the advanced time level, RHSVARS is a list of perturbation coefficients at all other time levels, and AMPMATRIX is the amplification matrix.

FSTABILITY computes the amplification matrix for a Fourier perturbation to a list of non-linear difference equations by performing the following steps: Fourier perturbation terms (coefficient times exponential phase factor) are added to each dependent variable; the difference equations are Taylor expanded to first order in the perturbation coefficients; the original difference equations are subtracted out, removing zero order terms; if the solution varies smoothly (denoted by setting the flag SMOOTHSOLN to TRUE), the mesh indices are dropped at this point; the set of linear equations in the perturbation coefficients is solved for the coefficients in LHSVARS; and the right-hand-side is factored into the form AMPMATRIX.RHSVARS.

The eigenvalues of AMPMATRIX can be found directly by using two MACSYMA functions:

```
CP : CHARPOLY(AMPMATRIX,LAMBDA);
```

to compute and set CP to the characteristic polynomial of the amplification matrix in terms of LAMBDA, and

```
SOLVE(CP,LAMBDA);
```

which solves the characteristic polynomial for the eigenvalues, $LAMBDA_i$.

Unfortunately, this direct approach to solving for the eigenvalues of the amplification matrix can only be used for very simple problems. In typical cases, either the amplification matrix contains so many terms that the eigenvalue computation overtaxes the storage available in MACSYMA or the resulting expressions for the eigenvalues are so complex that they are difficult to analyze. As a result, it is usually necessary to simplify interactively the amplification matrix first as shown in the example below.

4: Example: A Simple Non-Linear Problem

A non-linear PDE is used in Section 8.6 of Richtmyer and Morton[6] for discussing the stability of finite difference equations. The same example is presented here as a series of input and output lines (C-lines and D-lines, respectively) from a MACSYMA computation. First, the differential equation is introduced and assigned as a value to the variable, PDE:

```
(C1) PDE : 'DIFF(U,T) = 'DIFF(U^5,X,2);
```

$$\text{(D1)} \qquad \frac{dU}{dT} = \frac{d^2}{dX^2}(U^5)$$

The dependence of U on X and T is specified:

```
(C2) DEPENDS(U,[X,T]);


(D2)                 [U(X, T)]
```

The PDE is discretized using N as the T index and I as the X index, and the result is assigned to DPDE:

```
(C3) DPDE : DISCRETIZE(PDE,T,X,N,I);


                                 2     5
                   d    |N     d      |N
(D3)               -- (U| ) = --- (U|  )
                   dT   |I       2    |I
                               dX
```

The difference rules to be used are now defined. DT is a simple forward difference in T:

```
(C4) DEFINERULE(DT,'DIFF(ANY,T),DELF(T,1)<>ANY)$
```

where ANY is a pattern variable which matches any expression. DXX is a centered second difference in X, split into an implicit portion at the advanced time level, multiplied by THETA, plus an explicit portion at the current time level, multiplied by (1-THETA):

```
(C5) DEFINERULE(DXX,'DIFF(ANY,X,2),
        THETA*E(T)*DELC(X,1)^2<>ANY +
        (1-THETA)*DELC(X,1)^2<>ANY)$
```

The rules DT and DXX are applied to DPDE, producing a finite difference equation which is assigned to FDE:

```
(C6) FDE : APPLY1(DPDE,DT,DXX);


                                   5                  5
       |N+1     |N               |N+1               |N+1
      U|       U|     THETA U|          2 THETA U|
       |I       |I               |I+1               |I
(D6)  -----  - ---  = ------------ - --------------
       DT       DT          2                  2
                          DX                 DX

              5               5        5
           |N+1            |N       |N
   THETA U|       THETA U|       U|
           |I-1            |I+1     |I+1
 + ------------ - ------------ + ------
         2              2           2
       DX             DX          DX

              5         5             5        5
           |N        |N            |N       |N
 2 THETA U|      2 U|     THETA U|       U|
           |I        |I            |I-1     |I-1
 + ------------ - ------ - ------------ + ------
         2           2          2           2
       DX          DX         DX          DX
```

This equation is non-linear in the unknown, U, at the advanced time level, N+1. Next it is linearized about the N time level and assigned to LFDE:

```
(C7) LFDE : LINEARIZE(FDE,N);
```

```
        |N+1     |N
       U|       U|
        |I       |I
(D7)   -----  - ---  =
        DT       DT

                    4                5
          |N      |N+1         |N
THETA (5 U|      U|     - 4 U|      )
          |I+1    |I+1         |I+1
------------------------------------
                   2
                 DX

                    4              5
             |N     |N+1       |N
  2 THETA (5 U|    U|    - 4 U|  )
             |I     |I         |I
- --------------------------------
                    2
                  DX

                  4               5
           |N      |N+1        |N
  THETA (5 U|     U|     - 4 U|      )
           |I-1    |I-1        |I-1
+ -----------------------------------
                   2
                 DX

           5          5                  5         5
         |N         |N                 |N        |N
  THETA U|         U|       2 THETA U|       2 U|
         |I+1       |I+1               |I        |I
- ------------ + ------ + ------------ - ------
        2          2            2           2
      DX         DX           DX          DX

           5          5
         |N         |N
  THETA U|         U|
         |I-1       |I-1
- ------------ + ------
        2          2
      DX         DX
```

Note that the result is linear in U at the advanced time level, but non-linear in U, in general. This causes the stability of the finite difference scheme to be affected by the values of the solution. Next products are rationally expanded, like terms are combined and the result is reassigned to LFDE:

```
(C8) LFDE : RATEXPAND(LFDE)$
```

If we turn on the VERBOSE flag,

```
(C9) VERBOSE:TRUE$
```

then FSTABILITY will print a message as it starts each major computational step. FSTABILITY is called with three arguments: LFDE, the difference equation; N+1, the highest time level; and I, the spatial index.

```
(C10) FSTABILITY(LFDE,N+1,I)$
Add perturbation terms.
Taylor expand in the perturbations.
Subtract zero-order terms.
Drop indices on dependent variables.
Simplify the phase factors.
Solve the linear equations.
```

The answer is saved in three global variables in the form LHSVARS = AMPMATRIX.RHSVARS, where AMPMATRIX is a 1 x 1 matrix, and LHSVARS and RHSVARS are lists (vectors) of length 1.

```
(C11) LHSVARS;

                          |N+1
(D11)                 [dU|    ]
                          |K
                            X

(C12) RHSVARS;

                          |N
(D12)                 [dU|   ]
                          |K
                            X
```

Since AMPMATRIX is a 1 x 1 matrix, its single element is its eigenvalue.

```
(C13) LAMBDA:AMPMATRIX[1,1];

                                      2 %I DX K
                              4                X
(D13) ((5 DT THETA - 5 DT) U   %E

                                            %I DX K
                               4      2            X
   + ((10 DT - 10 DT THETA) U   - DX ) %E

                             4
   + (5 DT THETA - 5 DT) U  )

                            2 %I DX K
                   4                 X
/(5 DT THETA U  %E

                                            %I DX K
                        4      2                   X
   + (- 10 DT THETA U  - DX ) %E

                      4
   + 5 DT THETA U  )
```

Now comes the rub. This form of LAMBDA is sufficiently complicated to obscure the conditions on DT, DX and THETA which insure that the absolute value of LAMBDA is less than 1 + O(DT) for all values of K_X. And MACSYMA does not provide the needed insight either.

What we need to do is to manipulate the value of LAMBDA, rewriting it in various forms until we find one with a simple enough structure to be understood. This type of manipulation is usually easy for humans to do with pencil and paper, but frustratingly difficult to do in automated algebra systems. MACSYMA and related systems lack the goal-directed behavior of humans who can easily explain to one another, "collect terms of similar structure to see if anything cancels," or "define a new variable to simplify the common subexpressions." This issue was noted by Moses in a review article on algebraic simplification[4]. It also motivated Fateman to build an interactive expression editor for MACSYMA, MAC-ED, to at least improve the man-machine interface[1].

After introducing new variables, Z and S,

$$Z = e^{(i\ DX\ K_X)},$$

$$S = \frac{5\ DT\ U^4}{DX^2} \qquad (5)$$

and a number of interactive manipulation steps, LAMBDA is reduced to

```
                          2
                   (Z - 1)
(D24)    1 - -------------------
                        2   Z
             THETA (Z - 1)  - -
                              S
```

For local stability, we require that the absolute value of LAMBDA be bounded by 1 + O(DT), for all values of the wavenumber, K_X, and consequently all phase angles of Z. By examination of expression D24 for LAMBDA, we can see that there is no restriction on S for values of THETA greater than 1/2. For values of THETA between 0 and 1/2, we obtain a worst-case condition for LAMBDA by considering the shortest wavelength perturbation supported by the mesh, i.e., where K_X = pi/DX and Z = -1. Since LAMBDA is automatically less than +1 for Z = -1, we are interested in the -1 limit on LAMBDA. We would like to solve the inequality

-1 < LAMBDA

for S. But MACSYMA has no facilities for solving inequalities, so we solve the equality instead and obtain the inequality by inspection. Solving LAMBDA = -1 for S with Z = -1

```
(C25) SOLVE(LAMBDA=-1,S),Z=-1;

Solution:
                         1
(E25)           S = - -----------
                      4 THETA - 2

(D25)                 [E25]
```

we obtain one solution, E25. Pulling the lead minus sign into the denominator and recalling the definition of S from equation 5, we can manually construct the inequality which must be satisfied for local stability when THETA is between 0 and 1/2.

```
(C26) DENOM(PART(E25,2));


(D26)              4 THETA - 2


(C27) 5*DT*U^4/DX^2 < 1/(-%);


                      4
               5 DT U            1
(D27)          ------- < -----------
                   2      2 - 4 THETA
                 DX
```

Note that this condition depends on the solution value, U, as we expected for a non-linear problem.

The final inequality, D27, was obtained by a fair amount of human interaction. For example, in line C27 we performed a manual and purely syntactic manipulation. Thus there is no guarantee that the resulting expression for LAMBDA is semantically correct. This step was required because of a lack of suitable expression editing facilities and the inability to handle inequalities in MACSYMA.

5: Acknowledgments

This work was done with the aid of MACSYMA, which was developed at the MIT Laboratory for Computer Science and supported by the National Aeronautics and Space Administration under grant NSG 1323, by the Office of Naval Research under grant N00014-77-C-0641, by the U. S. Department of Energy under grant ET-78-C-02-4687, and by the U. S. Air Force under grant F49620-79-C-020. The work described in this paper was assisted by Brendan McNamara, who helped define the requirements for the package, by George Carrette, who provided programming assistance with the two-dimensional display package in MACSYMA, and by Ellen Golden, who as usual provided help in many areas of MACSYMA use and administration.

6: Postscript

Several MACSYMA facilities proved to be especially useful in constructing the FDIFF and FSTAB packages. These included:

1. Pattern matching for implementing the finite difference rules.

2. Hash-coded arrays and array-associated functions for keeping various tables, indexed by arbitrary expressions.

3. Syntax extensions for adding new operators such as "<>".

On the other hand, limitations in the current MACSYMA system constrained the implementation and use of the new software packages:

1. The lack of an operator algebra necessitated the use of the "<>" operator artifice.

2. Inadequate memory space for large, complicated sets of PDEs restricts the use of the packages.

3. The lack of a user-level interface to the two-dimensional output routines required that rather arcane LISP code be written to display indexed variables.

4. The lack of a goal-directed simplifier requires that amplification matrix eigenvalues be simplified with manual intervention.

5. The lack of a general purpose inequality manipulation package also necessitates manual processing.

REFERENCES

[1] Fateman, R. J., "MAC-ED, An Interactive Expression Editor for MACSYMA," Proc. of the 1979 MACSYMA Users' Conference, (MIT Laboratory for Computer Science, Washington, D.C., June 20-22, 1979).

[2] Lanam, D. H., A Package for Generating and Executing FORTRAN Programs with MACSYMA, MS Project Report, Univ. of California, Berkeley (November 1980).

[3] MACSYMA Reference Manual, Laboratory for Computer Science, MIT, Cambridge, Version Nine (December 1977).

[4] Moses, J. "Algebraic Simplification: A Guide for the Perplexed," Communications of ACM, 14, 527-537 (August 1971).

[5] Ng, E. W., Symbolic-Numeric Interface: A Review, Lecture Notes in Computer Science No. 72, (Springer-Verlag, Heidelberg, New York, 1979).

[6] Richtmyer, R. D. and K. W. Morton, Difference Methods for Initial-Value Problems, (John Wiley & Sons, New York, 1967), Second edition.

[7] Wirth, M. C., "Symbolic Vector and Dyadic Analysis," SIAM Journal On Computing, 8(3), 306-319 (August 1979).

[8] Wirth, M. C., On the Automation of Computational Physics, PhD Thesis, Univ. of California, Davis; also available as: Lawrence Livermore National Laboratory, Livermore, Rept. UCRL-52996 (October 1980).

An Algorithmic Classification of Geometries in General Relativity

Jan E. Åman, Anders Karlhede

Institute of Theoretical Physics, University of Stockholm,
Vanadisvägen 9, S–113 46 Stockholm, Sweden

ABSTRACT

The complicated coordinate transformations in general relativity make coordinate invariant classification schemes extremely important. A computer program, written in SHEEP, performing an algorithmic classification of the curvature tensor and a number of its derivatives is presented. The output is a complete description of the geometry. The problem to decide whether or not two solutions of Einstein's equations describe the same gravitational field can be solved if the (non-) existence of a solution to a set of algebraic equations can be established. The classification procedure has been carried through for a number of fields, and solutions previously believed to describe physically different situations have been shown to be equivalent. We examplify with a physically interesting class of geometries.

PHYSICAL BACKGROUND

General relativity describes gravity geometrically, as curvature of space-time. The fundamental object is the metric $g_{\mu\nu}(x^{\rho})$ which gives the distance ds between arbitrary neighbouring points with coordinates x^{μ} and $x^{\mu} + dx^{\mu}$

$$ds^2 = g_{\mu\nu}(x^{\rho})dx^{\mu}dx^{\nu}$$

The gravitational field is thus described by the functions $g_{\mu\nu}(x^{\rho})$.

The theory is invariant under arbitrary (reasonably regular) coordinate transformations. There are no preferred coordinate systems, any continuous numbering of the points in space-time is allowed. The form of the metric, however, is highly dependent on the choice of coordinate system. The complexity of the coordinate transformations makes this a serious problem. It is difficult to interpret the metric: truely physical effects are confused with effects introduced by the special choice of coordinates. In particular, it is non-trivial to decide whether or not two metrics $g_{\mu\nu}(x^{\rho})$ and $\tilde{g}_{\mu\nu}(\tilde{x}^{\rho})$ describe the same gravitational field (i.e. if there is a coordinate transformation $\tilde{x}^{\mu}(x^{\nu})$ transforming $\tilde{g}_{\mu\nu}$ into $g_{\mu\nu}$). This is the so called equivalence problem. These problems are of fundamental origin since they are a result of the invariances of the theory.

To illustrate the situation we look at a simple two-dimensional example: a sphere with two different coordinate systems.

$$ds^2 = d\theta^2 + \sin^2\theta\, d\varphi^2$$

$$ds^2 = \left[\frac{x^2}{(x^2+y^2)(1-(x^2+y^2))} + \frac{y^2}{x^2+y^2}\right]dx^2 + \frac{2\,xy}{1-(x^2+y^2)}\,dx\,dy + \left[\frac{x^2}{x^2+y^2} + \frac{y^2}{(x^2+y^2)(1-(x^2+y^2))}\right]dy^2$$

The equivalence can here be established by calculating the curvature scalars. Spaces of constant curvature are equivalent if and only if these scalars are equal.

For a general space the following theorem holds (see e.g. [1]). The local geometry of an n-dimensional Riemannian manifold is completely described by the curvature tensor R_{ijkl} and its 1/2n(n+1) lowest covariant derivatives $R_{ijkl;m}$, ...(In the following we restrict ourselves to n=4.) This provides an alternative description of the geometry to the one given by the metric. By referring the tensors to a field of tangent space frames with a constant frame metric (Lorentz frames for instance) the components of the tensors will be invariant under coordinate transformations. They will change under Lorentz transformations, but a canonical frame can always be determined by requiring the tensors to have a special form.

Brans based a procedure for obtaining an invariant description of a gravitational field on this theorem [2]. The method seemed to require enormous calculations and be of limited usefulness. In [3] we modified Brans' work and obtained an efficient scheme for a complete coordinate invariant description of space-times.

The first step in the invariantization is to calculate the curvature tensor and transform it to a standard frame. This is the Petrov classification [4] which is widely used in general relativity. A computer program was developed for this by d'Inverno and Russell-Clark and successfully applied to the Harrison metrics [5]. The Petrov classification is incomplete: it does not distinguish all inequivalent metrics. By including a finite number of the covariant derivatives of the curvature tensor the classification is made complete.

We have found it convenient to use the Newman-Penrose spinor formalism [6]. The equivalents of the curvature tensor R_{ijkl} are the curvature spinors Ψ_{abcd}, $\Phi_{abc'd'}$ and Λ. A tangent space frame corresponds to a dyad and a Lorentz transformation is replaced by a SL(2,C) rotation of the dyad. Only the Weyl spinor Ψ_{abcd} and its derivatives will appear since we restrict ourselves to vacuum space-times where $\Phi_{abc'd'} = \Lambda = 0$. The non-vacuum computer program is not yet sufficiently tested.

CLASSIFICATION PROCEDURE

In short the classification of a vacuum metric consists of the following steps (for more details consult [3,7]).

1) Write the metric in a spinor dyad form

$$ds^2 = 2\omega_{oo'}\omega_{11'} - 2\omega_{o1'}\omega_{1o'}$$

where

$$\omega_{ab'} = \sigma_{\mu ab'} dx^\mu$$

and read off $\sigma_{\mu ab'}$ which is the input to the computer.

2) The spin coefficients and the Weyl spinor are calculated by the computer and the Petrov type determined as in [5].

3) Transform to a standard dyad where the Weyl spinor (and later on its covariant derivatives) has a special form. In Petrov types I, II and III this determines the dyad completely. For types D and N a two parameter subgroup H of SL(2,C) leaves the Weyl spinor invariant.

4) If the dyad was changed in the previous step, recalculate the spin coefficients and the Weyl spinor in the new dyad. This is more straightforward than transforming them.

5) The computer determines the number n_o of functionally independent functions among Ψ_{abcd} by investigating the rank of the Jacobian formed from the real and imaginary parts of Ψ_{abcd} as functions of the coordinates.

6) The computer calculates the symmetrized part of the covariant derivative of the Weyl spinor. (Only the symmetrized parts are algebraically independent [8].)

7) Determine for Petrov types D and N the group H_1 ($H_1 \subseteq H_o$) of dyad transformations leaving the Weyl spinor and its derivative invariant. If $H_1 \neq H_o$, transform to a standard dyad and recalculate all quantities.

8) The computer determines the number $n_o + n_1$ of functionally independent functions among the components of the Weyl spinor and its derivative.

9) If $n_1 = 0$ and $\dim(H_1) = \dim(H_0)$, the procedure stops. This not being the case, steps 6) - 8) are repeated with successively higher symmetrized covariant derivatives of the Weyl spinor until $n_{p+1} = 0$ and $\dim(H_{p+1}) = \dim(H_p)$ for the p+1:st derivative.

The procedure stops at the p+1:st derivative, where $p+1 \leq 7$ since there are at most four functionally independent functions and $\dim(H_0) \leq 2$. It should be stressed that a situation where high derivatives is needed is very unusual. In our applications on the Harrison metrics and vacuum D metrics we have not needed higher derivatives than the second.

The number of Killing vectors in the geometry is $4 - (n_0 + n_1 + \ldots + n_p) + \dim(H_p)$ and H_p is the isotropy supgroup (see [1] or [3]).

Two metrics $g_{\mu\nu}(x^\rho)$ and $\tilde{g}_{\mu\nu}(\tilde{x}^\rho)$ are equivalent if and only if the algebraic equations $\Psi(x) = \Psi(\tilde{x})$, $D\Psi(x) = D\tilde{\Psi}(\tilde{x})$, ... $D^k\Psi(x) = D^k\tilde{\Psi}(\tilde{x})$ (where $D^k\Psi$ stands for the symmetrized components of the k:th covariant derivative of the Weyl spinor) have a solution $\tilde{x}^\mu = \tilde{x}^\mu(x^\nu)$. It is a non-algorithmic procedure to decide this. In those cases we have considered so far it has been possible to establish (non-)equivalence.

THE COMPUTER PROGRAM

A computer program carrying out the classification procedure has been developed by one of us (J.E.Å.) [7,9,10]. At present only the vacuum case is completed but the generalization to non-vacuum solutions is under consideration. The program is written in the algebraic language SHEEP [11] (which at present runs on a PDP-10, but will soon be available on other computers as well) with a number of additions, some of them discussed below. Among currently available languages SHEEP seems to be the most suitable for the following reasons:

- It is fast, 10-20 times faster than REDUCE for this kind of calculations.

- It is interactive.

- It prints formulae in a nice multiple line format and it is easy to instruct SHEEP how to write a new identifier in such a format.

The original language SHEEP does, however, have some limitations which have been handled by a number of additions. These additions are written in Standard LISP, using functions special for SHEEP.

One extension allows SHEEP to use complex quantities and coordinates exploiting relations as $f^*{}_{,z} = (f_{,z^*})^*$, where * means complex conjugation and comma partial differentiation. The sphere mentioned above can be described in complex coordinates z and z^* as:

$$ds^2 = \frac{dz\,dz^*}{(1 + \frac{z z^*}{4})^2}$$

Although a metric in general relativity can always be written in a real form a complex form may be more suitable and is often used.

One of the important advantages of SHEEP is its handling of tensors. However, for our purposes we need to calculate components of spinors rather than tensors. This made an extensive addition to SHEEP necessary.

As an example consider $\sigma_{\mu ab'}$. Here μ is a coordinate index taking the values 0, 1, 2 and 3 while the spinor indices a and b' take values 0 and 1 only. Furthermore it has an hermitean symmetry $\sigma_{\mu ab'} = \sigma_{\mu ba'}{}^*$ for real coordinates. When the component $\sigma_{310'}$ is requested during a calculation the $\sigma_{301'}$ component is picked up and complex conjugated. This mechanism resembles the way SHEEP handles antisymmetric tensors.

The savings achieved by the use of spinors and in particular symmetrized spinors rather than tensors are crucial. If we have a worst case (we doubt it exists) the 7:th covariant derivative of the curvature tensor $R_{ijkl;m;n;p;q;r;s;u}$ has to be calculated. This object has 86016 real components, related by Bianchi identities which are technically complicated to make use of. Calculation of that many algebraic expressions seems almost unrealistic. When using symmetrized parts of covariant derivatives the number of components we have to calculate in order to obtain the 7:th covariant derivative of the spinors Ψ, Φ and Λ will only be 187, 169 of them beeing complex.

One of the subproblems mentioned above is the determination of the number of functionally independent functions by calculating the rank of the

Jacobian (the functional determinant). The functions to be considered are the real and imaginary parts of Ψ, Φ, Λ and their covariant derivatives. These functions are studied successively to see if they are functionally independent of the previously tested functions. (There are at most four independent functions since this is the dimension of space-time.) Suppose we have k functionally independent functions: f1, f2, ... fk. A new function g is independent of theese if, for

$$k=0: \quad g_{,k} \neq 0$$

$$k=1: \quad 2 \sum_{\substack{k=0\\p=0\\k<p}}^{3} \epsilon^{lmkp} f1_{,k}\, g_{,p} \neq 0$$

$$k=2: \quad 2 \sum_{m=0} \sum_{\substack{k=0\\p=0\\k<p}}^{3} \epsilon^{lmkp} f1_{,k}\, g_{,m} \neq 0$$

and analogously for k=3. When a new independent function is found the result is stored as the vector $J1_k$, the antisymmetric tensor $J2^{lm}$ or the vector $J3^k$ respectively.

AN EXAMPLE

We illustrate our method by classifying the Kerr-NUT geometry. This includes the Kerr rotating black hole (1=0), the Schwarzschild non-rotating black hole (1=a=0) and the NUT geometry (a=0). In the following, user input is underlined. Immediately after each input line comes the responce from SHEEP (some unimportant output removed) and thereafter an indented comment discussing what was done.

.R FRAME

SHEEP 81-04-28

> Asks the computer to run the frame version of SHEEP.

*(SYSIN DSK: (CLASSI.SPI) (KIIA.NUD))

KIIA.NUD
Petrov class D, Kinnersley Case II.A (Kerr-NUT)
W Kinnersley, J. Math. Phys. 10 (1969) 1195.

"FINISHED LOADING"

> Loads the classification program as well as an already prepared file containing the necessary information about the Kerr-NUT metric.

*(PRINPUT)

Variables from 0 to 3 : u r x y

a is a constant

1 is a constant

m +il is a constant

m -il is a constant

$p = il - iacos(x) + r$

$p_{,r} = 1$

$p_{,x} = iasin(x)$

$S = 1/2(m+il)p^{-1}rp^{*-1} + 1/2(m-il)p^{-1}rp^{*-1} +$

$+1/2p^{-1}1^{2}p^{*-1} - 1/2p^{-1}a^{2}p^{*-1} - 1/2p^{-1}r^{2}p^{*-1}$

$S_{,r} = 1/2(m+il)p^{-1}p^{*-1} + 1/2(m-il)p^{-1}p^{*-1} - Sp^{*-1} -$

$-1/2p^{-1} - p^{-1}S - 1/2p^{*-1}$

$S_{,x} = iSap^{*-1}sin(x) - ip^{-1}Sasin(x)$

LIESUL

1 FOR r SUBSTITUTE $1/2p + 1/2p^{*}$

Used when making: ZDU LIE

RIESUL

1 FOR S

SUBSTITUTE $1/2(m+il)p^{-1}rp^{*-1} + 1/2(m-il)p^{-1}rp^{*-1} +$

$+1/2p^{-1}1^{2}p^{*-1} - 1/2p^{-1}a^{2}p^{*-1} - 1/2p^{-1}r^{2}p^{*-1}$

2 FOR $sin^{2}(x)$

SUBSTITUTE $1 - pa^{-2}p^{*} - 21a^{-1}cos(x) + 1^{2}a^{-2} + a^{-2}r^{2}$

3 FOR cos(x)

SUBSTITUTE $1/2ipa^{-1} - 1/2ia^{-1}p^{*} + 1a^{-1}$

4 FOR r SUBSTITUTE $1/2p + 1/2p^{*}$

5 FOR 1 SUBSTITUTE -1/2i(m+il) +1/2i(m-il)

Used when making: RIE

D2PSISUL

1 FOR m -il

SUBSTITUTE $-2iacos(x) + (m+il) - p + p^{*}$

Used when making: D2PSI

This function presents you a list of the coordinates, variables and their non-zero derivatives and lists of substitutions to be used during the calculation.

***(WMAKE DS2F)**

$$ds^2 = 2(-S^{1/2}du+S^{-1/2}dr+$$

$$+(-pS^{1/2}a^{-1}p^{*}+S^{1/2}a+S^{1/2}1^{2}a^{-1}+S^{1/2}a^{-1}r^{2})dy)(S^{1/2}du+$$

$$+(pS^{1/2}a^{-1}p^{*}-S^{1/2}a-S^{1/2}1^{2}a^{-1}-S^{1/2}a^{-1}r^{2})dy) -$$

$$-2((2)^{-1/2}ip^{-1/2}ap^{*-1/2}\sin(x)du+(2)^{-1/2}p^{1/2}p^{*1/2}dx+$$

$$+(-(2)^{-1/2}ip^{-1/2}1^{2}p^{*-1/2}\sin(x)-$$

$$-(2)^{-1/2}ip^{-1/2}a^{2}p^{*-1/2}\sin(x)-$$

$$-(2)^{-1/2}ip^{-1/2}r^{2}p^{*-1/2}\sin(x))dy)(-$$

$$-(2)^{-1/2}ip^{-1/2}ap^{*-1/2}\sin(x)du+(2)^{-1/2}p^{1/2}p^{*1/2}dx+$$

$$+((2)^{-1/2}ip^{-1/2}1^{2}p^{*-1/2}\sin(x)+$$

$$+(2)^{-1/2}ip^{-1/2}a^{2}p^{*-1/2}\sin(x)+$$

$$+(2)^{-1/2}ip^{-1/2}r^{2}p^{*-1/2}\sin(x))dy)$$

This is the Kerr-NUT metric written in null tetrad or spinor dyad form i.e. it is constructed as $ds^2 = 2\,\omega_{00'},\ \omega_{11'},\ -2\,\omega_{01'},\ \omega_{10'}$.

***(CLASSIFY0)**

Vacuum solution

Since this geometry is found to represent an empty space we only have to concider Ψ for the rest of this calculation.

***(WMAKE PSI)**

$$PSI_2 = -(m+il)p^{-3}$$

***(PETROV)**

Please check if PSI2 is really nozero !
If so, Petrov type is D

$$PSI_2 = -(m+il)p^{-3}$$

D

The Petrov type is determined to be D (unless m+il=0). For Petrov type D we use a standard dyad where Ψ is of the form $\Psi_0 = \Psi_1 = \Psi_3 = \Psi_4 = 0$. This determines the dyad up to the two-parameter group of transformations

$$H_0 = \left\{ \begin{pmatrix} 0 & 1 \\ 1 & 0 \end{pmatrix}, \begin{pmatrix} a & 0 \\ 0 & 1/a \end{pmatrix} \right\} \quad a \in C$$

***(FUNTST PSI)**

New function, probably independent:

$$f1 = Re(PSI_2)$$

Please check if J1 really nozero:

$$J1_1 = -3(m+il)p^{-4} - 3(m-il)p^{*-4}$$

$$J1_2 = -3i(m+il)p^{-4}a\sin(x) + 3i(m-il)ap^{*-4}\sin(x)$$

New function, probably independent:

$$f2 = Im(PSI_2)$$

Please check if J2 really nozero:

$$J2^{03} = 36(m+il)(m-il)p^{-4}ap^{*-4}\sin(x)$$

$$n_0 = 2$$

Determines the number of functionally independent functions of Ψ, treating real and imaginary parts separately.

***(WMAKE DPSI)**

$$DPSI_{20'} = 3(m+il)p^{-4}S^{1/2}$$

$$DPSI_{21'} = -3(2)^{-1/2}i(m+il)p^{-9/2}ap^{*-1/2}\sin(x)$$

$$DPSI_{30'} = -3(2)^{-1/2}i(m+il)p^{-9/2}ap^{*-1/2}\sin(x)$$

$$DPSI_{31'} = 3(m+il)p^{-4}S^{1/2}$$

Computes the symmetrized part of the first covariant derivative of the Weyl spinor ($DPSI_{20'} = (D\Psi)_{20'} = \Psi_{(0001;1)0'}$). The dyad is now completely determined. We define our standard dyad for Petrov Type D by requiring $|(D\Psi)_{20'}| = |(D\Psi)_{31'}|$ and $(D\Psi)_{21'} = (D\Psi)_{30'}$. If these conditions were not fulfilled we would have performed a dyad transformation.

***(FUNTST DPSI)**

$$n_1 = 0$$

No additional functionally independent function is found.

*(WMAKE D2PSI)

$D^2PSI_{20'} = -12(m+il)p^{-5}S$

$D^2PSI_{21'} = 12(2)^{-1/2}i(m+il)p^{-11/2}S^{1/2}a\,p^{*-1/2}\sin(x)$

$D^2PSI_{22'} = 6(m+il)p^{-6}a^2p^{*-1}\sin^2(x)$

$D^2PSI_{30'} = 24(2)^{-1/2}i(m+il)p^{-11/2}S^{1/2}a\,p^{*-1/2}\sin(x)$

$D^2PSI_{31'} = -3(m+il)p^{-4}Sp^{*-1} - 3/2(m+il)p^{-4}p^{*-1} -$

$-12(m+il)p^{-5}S - 3/2(m+il)p^{-5}a^2p^{*-2}\sin^2(x) +$

$+6(m+il)p^{-6}a^2p^{*-1}\sin^2(x) + 3/2(m+il)^2p^{-5}p^{*-1}$

$D^2PSI_{32'} = 24(2)^{-1/2}i(m+il)p^{-11/2}S^{1/2}a\,p^{*-1/2}\sin(x)$

$D^2PSI_{40'} = 6(m+il)p^{-6}a^2p^{*-1}\sin^2(x)$

$D^2PSI_{41'} = 12(2)^{-1/2}i(m+il)p^{-11/2}S^{1/2}a\,p^{*-1/2}\sin(x)$

$D^2PSI_{42'} = -12(m+il)p^{-5}S$

The symmetrized part of the second covariant derivative is calculated.

*(FUNTST D2PSI)

$n_2 = 0$

No additional functionally independent function is found in the second covariant derivative. Since the continuous transformation freedom of the dyad is not reduced (the dyad is completely determined already by $D\Psi$) the algorithmic classification stops here.

RESULTS

When classifying a number of the Harrison metrics we found that three of these earlier belived to be distinct [5] are in fact equivalent, i.e. there exist coordinate transformations between them [7]. The vacuum Petrov type D solutions are all known [12] and include many of the physically most important solutions. We have now classified all of them, except one, and hope to complete this work in the near future [13].

We believe the program will allow clasification of many of the known metrics given in e.g. "Exact Solutions of Einstein's Fields Equations" [14].

It is a pleasure to thank Professor Bertel Laurent and Dr. Inge Frick for numerous fruitful discussions.

REFERENCES

1. E. Cartan, "Lecons sur la Geometrie des Espaces de Riemann" (Gauthier-Villars, Paris) (1946).
2. C.H. Brans, J. Math. Phys. 6 (1965) 94.
3. A. Karlhede, Gen. Rel. Grav., 12 (1980) 693.
4. A.Z. Petrov, Doklady Akad. Nauk, SSSR 105 (1955) 905.
5. R.A. d'Inverno and R.A. Russel-Clark, J. Math. Phys. 12 (1971) 1258.
6. E. Newman and R. Penrose, J. Math. Phys. 3 (1962) 566.
7. J.E. Åman and A. Karlhede, Phys. Lett. 80A (1980) 229.
8. R. Penrose, Ann. Phys. 10 (1960) 171.
9. A. Karlhede and J.E. Åman, Eurosam 1979, Lecture Notes in Computer Science, Springer Verlag, 72 (1979) 42.
10. A. Karlhede and J.E. Åman, Abstracts of contributed papers, 9th International Conference on General Relativity and Gravitation (1980) 104.
11. I. Frick, "The Computer Algebra System SHEEP, what it can and cannot do in General Relativity." Preprint, Inst. of Theor. Physics, Univ. of Stockholm (1977).
12. W. Kinnersley, J. Math. Phys. 10 (1969) 1195.
13. A. Karlhede and J.E. Åman, "A Classification of the Vacuum D Metrics." Preprint, University of Stockholm (1981).
14. D. Kramer, H. Stephani, M.A.H. MacCallum and E. Herlt, "Exact Solutions of Einstein's Field Equations" (VEB Deutcher Varlag der Wissenschaften, Berlin and Cambridge University Press) (1980).

Formulation of Design Rules for NMR Imaging Coil by using Symbolic Manipulation

John F. Schenck
M. A. Hussain

General Electric Corporate Research
and Development Center
Schenectady, New York 12301

INTRODUCTION

A common problem in electrical technology is to design a current carrying coil that will produce a given magnetic field. For over a hundred years an equation, the Biot-Savart law, has been available that defines precisely the magnetic field at any point as a line integral along the path of the electric currents that are the sources of the field. In principle then, the design problem is straightforward - it is merely necessary to invert the Biot-Savart law and find a path whose line integral has the given values at the specified field points. However, the actual solution is not trivial and there is a continuing need for improved computational methods for relating magnetic fields to their sources.

One reason for this continuing need is that the Biot-Savart law gives the field at a single point but the design problem to be solved is almost always given as a specification of the field over an extended region of space. Historically, a common example of such a specification has been the problem of producing a uniform magnetic field. In this case the requirement is simply to design a set of coils such that the field varies less than a prescribed amount over a prescribed region of space. Here we are not interested so much in the precise value of the field at a point, instead we are interested in controlling the way in which the field varies within the vicinity of a given point. The natural mathematical approach to this problem is to introduce some form of series expansion of the field, wherein the expansion coefficients are functions of the configuration of the coil.

Two forms of series expansion will be considered here - the ordinary Taylor series and the expansion in spherical harmonics. Other possible expansions - such as Bessel function methods for cylindrical coordinate problems - usually involve integrals over some eigenparameter rather than discrete sums and are not directly competitive with the methods discussed here.

In either of these two types of expansions the leading term is constant over the field points and all the other terms are given functions of position. The problem of designing for a uniform field then is to specify coils that produce a large value for the leading term and either zero, or at least acceptably small, values for all other coefficients. In practice it is never possible to make the coefficients of all the unwanted orders vanish. However, the very high order terms usually are insignificant in the region of interest and the design is accomplished by eliminating as much as possible the contributions of selected low order terms.

Taylor series expansions in a single spatial variable, such as x, are very familiar and contain, in addition to a constant, terms proportional to x, x^2, x^3 and so on. In three dimensions there are more terms of each order - for example there are three first order terms, x, y and z, six second order terms, x^2, y^2, z^2, xy, xz and yz. In general the Taylor series for a function of three spatial variables will have (n+1)(n+2)/2 terms of order n.

It can be shown that, in a region where there are no sources of magnetic field, each cartesian component of the magnetic field satisfies Laplace's equation. As a result each such component can also be expanded in a series using the spherical harmonics as the basis functions

(Reference 1, p. 1271). The expansion in spherical harmonics is formally equivalent to the Taylor series expansion, however, it has several practical advantages. First of all the spherical harmonic expansion has only (2n+1) terms of order n. Therefore, although the two expansions are identical for n=0 and n=1, for n=2 and greater the spherical harmonic series has fewer terms. Because each term in the spherical harmonic expression satisfies Laplace's equation, it represents a possible magnetic field configuration. Therefore, it is reasonable to attempt to find coil designs that control, to a high degree of accuracy, only a single term in the expansion. For n values greater than two the Taylor series terms do not satisfy Laplace's equation. The result is that the terms in the spherical harmonic expansion of the field can be controlled in an "orthogonal" fashion that the Taylor series terms cannot. Finally, we have found during the present study, that the coefficients of the spherical harmonic expansion are much simpler functions of the source coordinates than are the coefficients for the Taylor series.

In this paper we describe how symbolic manipulation can be used to determine the spherical harmonic expansion. We intend, in a subsequent paper, to analyze various coil designs in terms of these expansions.

NMR FOR MEDICAL DIAGNOSIS

We were led to an examination of the spherical harmonic expansion of magnetic fields during the design of a medical imaging system based on the phenomenon of nuclear magnetic resonance. NMR permits an analysis of the properties of magnetic nuclei in atoms and of their environments. The material to be studied is placed in an intense and highly uniform magnetic field. Radio frequency energy is fed into the nuclei and they are thereby stimulated to emit a radio frequency signal that is detected and analyzed. For many years the study of test-tube sized samples has been a standard technique in analytical chemistry. Recently it has been demonstrated that NMR can be performed on human patients and can provide a non-invasive method of producing cross sectional images of the internal anatomy and measurements of intracellular biochemical processes[2-8].

To design an effective NMR medical imaging system it is necessary to scale up from test-tube sized samples to human whole-body dimensions. In addition imaging requires that the applied magnetic field have superimposed on it a set of very uniform gradient fields of the form $\partial B_z/\partial x$, $\partial B_z/\partial y$ and $\partial B_z/\partial z$. The uniformity of the main field must be maintained to within 1 part in 10,000 or better over the entire cross section of the patient. Also the gradient fields must maintain their linearity over the same region without significant contamination from higher order derivatives. These two requirements represent a major challenge to the design and fabrication of large magnet systems. NMR promises to become a major new modality of medical diagnosis and the success of the method will require an efficient and effective solution to the field uniformity and gradient field design problem.

HISTORICAL BACKGROUND

The spherical harmonics are widely used in theoretical physics - in the quantum theory of angular momentum for example. Their use for the representation of magnetic fields dates from the last century[10]. The first success in the design of the coils to produce uniform fields was the invention of the Helmholtz coil more than a century ago[9]. A coil design capable of producing a highly uniform gradient field was proposed by Maxwell (Reference 10, Article 715). McKeehan[11] suggested a number of methods for designing cylindrically symmetric systems using the harmonic functions. The theory was taken to its most successful level in a classic series of papers by Milan Wayne Garrett[12-14]. Using the theory of zonal harmonics he thoroughly analyzed the fields produced by cylindrically symmetric current loops. Garrett's work has been the basis for the design of most of the magnet systems used in NMR imaging[15]. No definitive study of the spherical harmonic analysis of systems that lack cylindrical symmetry has been published, although several authors[16-20] have shown how this approach can be used in the design of shim coil systems.

In the present paper we use the computer techniques of symbolic manipulation to make a start toward extending the method of Garrett to a general treatment of the spherical harmonic representation of magnetic fields without the restriction to cylindrical symmetry. We believe that these methods will be of major utility in the design of gradient coils and correction coils for NMR imaging systems.

FORMULATION OF THE PROBLEM

Many different notations have been used for the spherical harmonics. We will follow that used by MacMillan[21] and define two sets of functions $r^n C_{mn}(\theta,\phi)$ and $r^n S_{mn}(\theta,\phi)$ as the solid spherical harmonics to be used as the basis functions for the expansions. Extended treatment of these and related functions are available (Reference 1, pp. 1264-1284 and Reference 22, pp. 139-168). It is necessary to distinguish carefully between the coordinates of source and field points. We will use x,y and z as the cartesian coordinates of the <u>source</u> points and u,v and w as the cartesian coordinates of the <u>field</u> point. The spherical coordinates of the <u>field points</u> are taken as r, θ and ϕ according to the conventional definitions.

The functions $C_{mn}(\theta,\phi)$ and $S_{mn}(\theta,\phi)$ are referred to as the surface spherical harmonics and depend only on angular variables as shown in Equation 1. The $P_n^m(\theta)$ are the associated Legendre functions.

EQUATION 1

$$C_{mn} = \cos m\phi \; P_n^m(\theta)$$

$$S_{mn} = \sin m\phi \; P_n^m(\theta)$$

The functions for m = 0 (the case considered by Garrett) have cylindrical symmetry around the w axis and are called <u>zonal harmonics</u>. The functions with m = n depend on u and v, but not on w and are called <u>sectorial harmonics</u>. The full set of functions, with no restrictions of m or n, are referred to as the <u>tesseral harmonics</u>.

The solid harmonics are products of r^n with a surface harmonic C_{mn} or S_{mn}. The solid harmonics can be used to expand any function f(u,v,w) in a region of space where Laplace's equation, $\nabla^2 f = 0$, is satisfied. The solid spherical harmonics can be expressed simply in either cartesian coordinates (Reference 1, p. 1271) or in spherical coordinates.

The system geometry is shown in Figure 1.

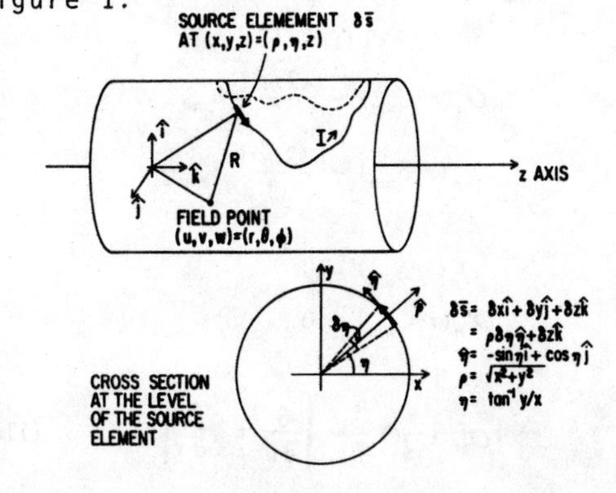

FIGURE 1. System Geometry.

In NMR only B_z, the z component of the magnetic field, is relevant and in whole-body imaging systems it is convenient to locate the correction and gradient coils on the surface of a cylinder. Using the Maxwell equations, it can be shown that, in a region of space that contains no sources, each cartesian component of B, i.e., B_x, B_y and B_z, independently satisfies Laplace's equation. Therefore, in such a region, we may formally write B_z in the form of Equation 2.

EQUATION 2

$$B_z = \frac{\mu_0 I}{4\pi r_0} \sum_{n=0}^{\infty} \sum_{m=0}^{n} \left[A_{mn} \left(\frac{r}{r_0}\right)^n C_{mn}(\theta,\phi) + B_{mn} \left(\frac{r}{r_0}\right)^n S_{mn}(\theta,\phi) \right]$$

Within the region of convergence (Reference 12, p. 1093 and pp. 1103-1104), the coefficients A_{mn} and B_{mn} completely specify the field of interest, B_z.

The quantity r_o is introduced as an arbitrary parameter with the dimensions of a length. Its use makes the parameters A_{am} and B_{mn} dimensionless. At an appropriate stage, r_o be set equal to some fundamental length in the system. In our case after the derivation are completed we will set r_o equal to a, the cylindrical radius of the coils. Equation 3 is the Biot-Savart law which gives the field at each field point as a line integral along the current elements that constitute the sources. Here r_o is the permeability of free space ($4\pi \times 10^{-7}$ henry/meter) and I is the current in amperes.

EQUATION 3

$$\bar{B} = \frac{\mu_0 I}{4\pi} \oint \frac{\delta \bar{s} \times \bar{R}}{R^3}$$

$$\delta \bar{B} = \frac{\mu_0 I}{4\pi} \frac{\delta \bar{s} \times \bar{R}}{R^3}$$

As indicated in Equation 4 the vector $\bar{R}$ is a position vector from the source to the field point. The distance between the source and field points is given by the scalar R (Equation 4).

EQUATION 4

$$\bar{R} = (u-x)\hat{i} + (v-y)\hat{j} + (w-z)\hat{k}$$

$$R^2 = (u-x)^2 + (v-y)^2 + (w-z)^2$$

To characterize the variation of a field in the vicinty of a given point we would like to utilize the Biot-Savart law to determine the coefficients

A_{mn} and B_{mn}. However, these coefficients obviously depend on the path chosen for the current sources and therefore must be recalculated everytime the coil configuration is changed. To overcome this problem we define a differential field (Equation 3a). The differential field is a function only of the source point (x,y,z), the field point (u,v,w) and the orientation of the current element $I\ \delta\bar{s}$. The differential path vector, $\delta\bar{s}$, is specified for both cartesian and cylindrical coordinates in Equation 5.

EQUATION 5

$$\delta\bar{s} = (\delta x\hat{i} + \delta y\hat{j} + \delta z\hat{k})\ ,\ \text{(cartesian coordinates)}$$

$$\delta\bar{s} = \rho\delta\eta\hat{\eta} + \delta\rho\hat{\rho} + \delta z\hat{k}\ ,\ \text{(cylindrical coordinates)}$$

There is a degree of ambiguity introduced by using the differential form of the Biot-Savart law[23]. This is because any vector function whose curl is equal to zero may be added to the right side of Equation 3a without changing any physically measureable quality[23]. We follow the custom of neglecting these possible additions as they do not appear to simplify the problem of coil design.

Consider the expression for B_z for a coil constrained to the surface of a cylinder. On such a surface $\rho = 0$. Also, because of the vector cross product in the Biot-Savart law the z-component of $I\ \delta\bar{s}$ makes no contribution to the field component B_z. Therefore the differential field B_z, has the simple form shown in Equation 6.

EQUATION 6

$$\delta\beta_z = \frac{\mu_0 I}{4\pi} f_\eta d\eta$$

$$f_\eta(x,y,z,u,v,w) = \frac{x^2 + y^2 - (ux+vy)}{[(u-x)^2 + (v-y)^2 + (w-z)^2]^{3/2}}$$

We now use Equation 7 to define coefficients a_{mn} and b_{mn} that are not dependent on the current path.

EQUATION 7

$$f_\eta(x,y,z,u,v,w) = \frac{1}{r_0}\sum_{n=0}^{\infty}\sum_{m=0}^{n}\left[a_{mn}\left(\frac{r}{r_0}\right)^n C_{mn} + b_{mn}\left(\frac{r}{r_0}\right)^n S_{mn}\right]$$

From the coefficients a_{mn} and b_{mn} and equation 8 we may calculate the coefficients A_{mn} and B_{mn} for any coil wound on the surface of a cylinder.

EQUATION 8

$$A_{mn} = \oint a_{mn}\,\delta\eta \qquad B_{mn} = \oint b_{mn}\,\delta\eta$$

The coefficients a_{mn} and b_{mn} can be determined once we notice that all the expansion functions, r^nC_{mn} and r^nS_{mn}, except r^oC_{oo} are zero at the zero of the field coordinates ($r = 0$ or, equivalently, ($u = v = w = 0$). Thus, by setting r Equal to zero in equation 7 and evaluating f_η from Equation 6 we find the coefficient a_{oo} immediately as shown in Equation 9.

EQUATION 9

$$a_{00} = f_\eta(x,y,z,0,0,0,) = [f_\eta]_{r=0} = \frac{r_0(x^2+y^2)}{(x^2+y^2+z^2)^{3/2}}$$

To evaluate the higher order coefficients it is necessary to differentiate both sides of equation 7 repeatedly. By using the integral representations of the solid spherical harmonics (Reference 1, p. 1270), we have derived the differential formulas listed in Table I. From these formulas we were led to define differential operators O_m and O'_m shown in Equation 10 that have the important property (Equation 11) that if they are applied to all of the solid spherical harmonics of order n only the functions with m=n will give a non-zero result. It is sometimes useful to note that O_m and O'_m are real and imaginary parts of a complex operator (Equation 11a).

EQUATION 10

$$O_m = \sum_{s=0}^{t}(-1)^s\frac{m!}{(2s)!(m-2s)!}\frac{\partial^{m-2s}}{\partial u^{m-2s}}\frac{\partial^{2s}}{\partial v^{2s}}$$

$$O'_m = \sum_{s=0}^{t'}(-1)^s\frac{m!}{(2s+1)!(m-2s-1)!}\frac{\partial^{m-2s-1}}{\partial u^{m-2s-1}}\frac{\partial^{2s+1}}{\partial v^{2s+1}}$$

where $t = \frac{m}{2}$ and $t' = \frac{m-2}{2}$ for m even

and $t = \frac{m-1}{2}$ and $t' = \frac{m-1}{2}$ for m odd

EQUATION 11

$$O_n(r^nC_{mn}) = \frac{(2m)!}{2}\delta_{mn}$$

$$O'_n(r^nS_{mn}) = \frac{(2m)!}{2}\delta_{mn}$$

$$O_n(r^nS_{mn}) = 0$$

$$O'_n(r^nC_{mn}) = 0$$

$$O_m + jO'_m = \left(\frac{\partial}{\partial u} + j\frac{\partial}{\partial v}\right)^m \qquad \text{(11a)}$$

TABLE I. Cartesian Derivatives for the Solid Spherical Harmonics.

$$\frac{\partial}{\partial w} r^n C_{mn} = (n+m) r^{n-1} C_{m,n-1}$$

$$\frac{\partial}{\partial w} r^n S_{mn} = (n+m) r^{n-1} S_{m,n-1}$$

$$\left.\begin{aligned} \frac{\partial}{\partial u} r^n C_{0n} &= -r^{n-1} C_{1,n-1} \\ \frac{\partial}{\partial v} r^n C_{0n} &= -r^{n-1} S_{1,n-1} \end{aligned}\right\} m=0$$

$$\left.\begin{aligned} \frac{\partial}{\partial u} r^n C_{mn} &= -\frac{1}{2} r^{n-1} C_{m+1,n-1} + \frac{(n+m)(n+m-1)}{2} r^{n-1} C_{m-1,n-1} \\ \frac{\partial}{\partial u} r^n S_{mn} &= -\frac{1}{2} r^{n-1} S_{m+1,n-1} + \frac{(n+m)(n+m-1)}{2} r^{n-1} S_{m-1,n-1} \\ \frac{\partial}{\partial v} r^n C_{mn} &= -\frac{1}{2} r^{n-1} S_{m+1,n-1} - \frac{(n+m)(n+m-1)}{2} r^{n-1} S_{m-1,n-1} \\ \frac{\partial}{\partial v} r^n S_{mn} &= \frac{1}{2} r^{n-1} C_{m+1,n-1} + \frac{(n+m)(n+m-1)}{2} r^{n-1} C_{m-1,n-1} \end{aligned}\right\} m \neq 0$$

TABLE II. The Differential Operators O_m and O'_m.

	O_m	O'_m
$m=0$	1	0
$m=1$	$\frac{\partial}{\partial u}$	$\frac{\partial}{\partial v}$
$m=2$	$\frac{\partial^2}{\partial u^2} - \frac{\partial^2}{\partial v^2}$	$2\frac{\partial^2}{\partial u \partial v}$
$m=3$	$\frac{\partial^3}{\partial u^3} - 3\frac{\partial}{\partial u}\frac{\partial^2}{\partial v^2}$	$3\frac{\partial}{\partial v}\frac{\partial^2}{\partial u^2} - \frac{\partial^3}{\partial v^3}$
$m=4$	$\frac{\partial^4}{\partial u^4} - 6\frac{\partial^2}{\partial u^2}\frac{\partial^2}{\partial v^2} + \frac{\partial^4}{\partial v^4}$	$4\frac{\partial^3}{\partial u^3}\frac{\partial}{\partial v} - 4\frac{\partial}{\partial u}\frac{\partial}{\partial v^3}$

Table II gives an explicit listing of these operators through m=4.

To evaluate the coefficient a_{mn} both sides of Equation 7 are differentiated ℓ times with respect to w, where where ℓ =n-m. This produces a function wherein the desired coefficient, A_{mn} multiplies the solid spherical harmonic $r^m C_{mm}(\theta,\phi)$. The operator O_m is then applied. The result is an expansion in solid spherical harmonics of the function $O_m \partial^\ell f/\partial w^\ell$, in which the constant term, proportional to $r^o C_{oo}$, is multiplied by a_{mn}. This expansion is evaluated at r=0 and the coefficient a_{mn} is thereby determined. The coefficients b_{mn} are determined in a similar fashion using the operator O'_m in place of O_m. The results are shown explicitly in Equation 12.

EQUATION 12

$$\left.\begin{aligned} a_{on} &= \frac{r_0^{n+1}}{n!}\left[\frac{\partial^n f_\eta}{\partial w^n}\right]_{r=0} \\ b_{on} &= 0 \end{aligned}\right\} m=0$$

$$\left.\begin{aligned} a_{mn} &= \frac{2r_0^{n+1}}{(n+m)!}\left[O_m \frac{\partial^l}{\partial w^l} f_\eta(x,y,z,u,v,w)\right]_{r=0} \\ b_{mn} &= \frac{2r_0^{n+1}}{(n+m)!}\left[O'_m \frac{\partial^l}{\partial w^l} f_\eta\right]_{r=0} \end{aligned}\right\} \begin{aligned} m &\neq 0 \\ l &= n-m \end{aligned}$$

The function of f (x,y,z,u,v,o) has the property that $\partial f/\partial z = -\partial f/\partial w$. If we use this property and differentiate equation 12 with respect to z we find a useful identity (Equation 13).

EQUATION 13

$$\frac{\partial a_{mn}}{\partial z} = -\frac{(n+m+1)}{r_0} a_{m,n+1}$$

This relationship leads to an alternative method of calculating a_{mn} and b_{mn} that involves less computation than Equation 12. As shown in Equation 14, we can first apply the operators O_m and O'_m to f_η and set r=0 to determina a_{mm} and b_{mm}. Then a_{mn} and b_{mn} may be found by n-m differentiations with respect to z. Because we are able t0 set u=v=w=0 before doing any z differentiation, the differentiation with respect to z in Equation 14 is much simpler than the corresponding differentiation with respect to w in Equation 12.

EQUATION 14

$$\left.\begin{aligned} a_{on} &= \frac{(-1)^n r_0^{n+1}}{n!} \frac{\partial^n}{\partial z^n} [f_\eta]_{r=0} \\ b_{on} &= 0 \end{aligned}\right\} m = 0$$

$$\left.\begin{aligned} a_{mn} &= (-1)^{m+n} \frac{2r_0^{n+1}}{(n+m)!} \frac{\partial^{n-m}}{\partial z^{n-m}} [O_m f_\eta]_{r=0} \\ b_{mn} &= (-1)^{m+n} \frac{2r_0^{n+1}}{(n+m)!} \frac{\partial^{n-m}}{\partial z^{n-m}} [O'_m f_\eta]_{r=0} \end{aligned}\right\} m \neq 0$$

ROLE OF SYMBOLIC MANIPULATION

The calculation of a_{mn} and b_{mn} by either Equations 12 or 14 involves repeated differentiation f from Equation 6. If m and n are not very small the labor of calculating the derivatives by hand is intolerable and the likelihood of error becomes almost certain. However, this procedure is well within the capabilities of MACSYMA for all values of m and n likely to be of interest. Furthermore, although the intermediate steps in the calculations are exceptionally cumbersome, a dramatic simplification results at the final steps. The procedure is to carry out the symbolic differentiations of Equation 6 that are called for by Equations 11 or 14, and then to set u = v = w =0 to find the coefficients as functions of x,y and z. the substitutions

$$x = \rho \cos \eta$$
$$y = \rho \sin \eta$$

are then made and a trigonometric reduction is carried out. The a_{mn} then have the simple, form indicated in Equation 15 where g_{mn} is a polynominal in z and of order n+2-m. The coefficients b_{mn} are given by the same expression with sin $m\eta$ replacing cos $m\eta$.

EQUATION 15

$$a_{mn} = r_0^{n+1} \rho^m \frac{\cos m\eta}{(z^2+\rho^2)^{\frac{2n+3}{2}}} g'_{mn}(z,\rho)$$

It is useful to introduce a dimensionless variable $Z=z/\rho$ which leads to

EQUATION 16

$$a_{mn} = \frac{r_0^{n+1}}{\rho^{n+1}} \frac{\cos m\eta}{(1+Z^2)^{\frac{2n+3}{2}}} g_{mn}(Z)$$

Finally, we may take ρ equal to a, the radius of the cylinder on which the coils are wound, and we may also set the arbitrary parameter r_o equal to a. This produces the final expression for the coefficients which are given in Equation 17.

EQUATION 17

$$a_{mn} = \frac{\cos m\eta}{(1+Z^2)^{\frac{2n+3}{2}}} g_{mn}(Z)$$

$$b_{mn} = \frac{\sin m\eta}{(1+Z^2)^{\frac{2n+3}{2}}} g_{mn}(Z)$$

The polynominals g_{mn} (Z) are tabulated in Table III.

CONCLUSION

To solve certain engineering problems encountered in the design of NMR medical imaging systems we have extended the zonal harmonic method of Garrett to include a wider range of current source configurations. In particular we have considered cases where the origin of the expansion is not necessarily on the z axis of a circular coil set. This permits the treatment of fields that do not have cylindrical symmetry.

The approach has been to define the expansion coefficients, a_{mn} and b_{mn}, for the differential form of the Biot-Savart law. They are more generally useful than the coefficients A_{mn} and B_{mn} which are defined only for complete current paths, and which must be recomputed if the current path is changed. We have used the complete set of solid tesseral harmonics, rather than just the solid zonal harmonics. Consequently the present expansions include (2n+1) terms of order n rather than just one. Algorithms have been developed for the calculation of the coefficients.

By use of symbolic differentiation we have determined the expansion coefficients for the B_z field produced by a current element that is located on the surface of a cylinder. Although the algorithm leads to extremely cumbersome expressions in the intermediate stages of the calculation, the final expressions are simple, elegant and easy to use. It is possible that a mathematician skilled in the properties of spherical harmonics could have found these expressions by conventional methods of calculation. However, we are convinced that we would not have found them to use in our engineering efforts without the availability of symbolic manipulation.

Now that expressions for the a_{mn} and b_{mn} are explicitly available, design rules for imaging coils may be formulated in terms of two questions. First, what coefficients A_{mn} and B_{mn}, in Equation 2 are necessary to produce the desired field? And second, what coil configuration on the cylindrical surface will produce the desired coefficients and suppress the unwanted or contaminating terms. To determine these paths for the current windings we use Equations 17 and 8. For example, for the coil to produce the x gradient field we wish to produce a field of the A_{11} form. Symmetry considerations restrict the contaminating term to those of the form A_{mn}, where m and n are odd. The design procedures is to determine the acceptable amplitudes of the contaminating terms and to find a path, using Equations 17 and 8, that produce an efficient A_{11} term and keeps the contaminating terms within the prescribed limits.

The work reported here does not complete the treatment of the spherical harmonic approach to coil design. Several possible extensions will now be mentioned. By analogy with Garrett's approach the functions of Equation 15 could be integrated as functions of ρ and z to permit the treatment of extended coils in addition to the filamentary circuit elements treated in the present paper. It is likely that the functions of Equation 15 could be expressed in a still simpler form by using spherical coordinates in place of cylindrical coordinates for the source position. The a_{mn} and b_{mn} do not satisfy Laplace's equation in the source coordinates and cannot therefore be expressed as spherical harmonics. However, by analogy to Equation 12 of Reference 14, it is likely that they can be expressed as functions closely related to the spherical harmonics. and if so, it should provide useful recursion relations for the a_{mn} and b_{mn}.

The algorithms presented here could be used to develop expansion coefficients for several other important fields. The fields B_x and B_y can be treated in the same way we have treated B_z above. Also expansion coefficients can now be computed for source elements that have radial components and are not constrained to a cylindrical surface.

It would be particularly helpful to find an analytic expression, analogous to Equation 3a, for the differential contribution of an arbitrary current element to the <u>magnetic scalar potential</u> $\delta\Omega$, at a given field point. We do not know if such a function exists (see Reference 10, Articles 417-422), but if it does it should be expandable in solid spherical harmonics and the expansions for all three field components B_x, B_y and B_z would follow immediately from Table I and from the relations $\delta B_x = -\partial\delta\Omega/\partial x$ and so on. An alternative to this approach may be provided by expanding the differential contribution to the magnetic vector potential.

In general magnetic fields are produced either by current carrying coils or by magnetized materials. To treat the fields of magnetized materials, in a fashion analogous to treatment of current-carrying coils outlined above, it would be useful to calculate the coefficients a_{mn} and b_{mn} for a magnetic dipole at an arbitrary source location.

Finally, it may be possible to use the spherical harmonic expansion to produce a general mathematical approach to the inverse problem. That is, given the problem of producing a field where A_{mn} and B_{mn} are prescribed and knowing the functional form of a_{mn} and b_{mn}, develop a general method for computing coil designs.

ACKNOWLEDGEMENTS

It is a pleasure to acknowledge helpful suggestions and discussions with H. R. Hart, W. A. Edelstein, R. W. Redington, D. E. Priest and W. E. Lorenson and assistance in the preparation of the manuscript from Maria Smith, Kathy Nichols and Arlene Agresta.

REFERENCES

1. P. M. Morse and H. Feshbach, Methods of Theoretical Physics, McGraw-Hill, New York, 1953

2. R. Damadian, "Tumor Detection by Nuclear Magnetic Resonance", Science 171, pp. 1151-1153, 1971.

3. P. C. Lauterbur, "Image Formation by Induced Local Interactions: Examples Employing Nuclear Magnetic Resonance", Nature 242, pp. 190-191, 1973.

4. W. S. Hinshaw, P. A. Bottomley and G. N. Holland, "Radiographic Thin-Section Image of the Human Wrist by Nuclear Magnetic Resonance", Nature 270, pp. 722-723, 1977.

5. W. A. Edelstein, J. M. S. Hutchison, G. Johnson and T. Redpath, "Spin Warp NMR Imaging and Applications to Human Whole-Body Imaging," Physics in Medicine and Biol. 25, pp. 751-756, 1980.

6. G. N. Holland, R. C. Hawkes, and W. S. Moore, "Nuclear Magnetic Resonance (NMR) Tomography of the Brain: Coronal and Sagittal Sections", J. Computer Assisted Tomography, 4, pp. 429-433, 1980.

7. B. Chance, S. Eleff and J. S. Leigh, Jr., "Noninvasive, Nondestructive Approaches to Cell Bioenergetics", Proc. Natl. Acad. Sci. 77, pp. 7430-7434, 1980.

8. B. D. Ross, G. K. Radda, D. G. Gadian, G. Rocker, M. Esiri and J. Falconer-Smith, "Examination of a Case of Suspected McArdle's Disease by ^{31}P Nuclear Magnetic Resonance", N. Engl. J. Med. 304, pp. 1338-1342, 1981

9. L. W. McKeehan, "Gaugain-Helmholtz (?) Coils for Uniform Magnetic Fields", Nature 133, pp. 832-833, 1934.

10. J. C. Maxwell, A Treatise on Electricity and Magnetism, Dover, New York, 1954.

11. L. W. McKeehan "Combinations of Circular Currents for Producing Uniform Magnetic Fields", Rev. Sci. Instruments 7, pp. 150-153, 1936.

12. M. W. Garrett, "Axially Symmetric Systems for Generating and Measuring Magnetic Fields, Part I" J. Appl. Phys. 22, pp. 1091-1107, 1951.

13. M. W. Garrett, "Computer Programs Using Zonal Harmonics for Magnetic Properties of Current Systems with Special Reference to the IBM 7090" Oak Ridge National Laboratory, ORNL-3318, 1962.

14. M. W. Garrett, "Thick Cylindrical Coil Systems for Strong Magnetic Fields with Field or Gradient Homogeneities of the 6th to 20th Order", J. Appl. Phys. 38, pp. 2563-2586, 1967.

15. D. I. Hoult and R. E. Richards, "Critical Factors in the Design of Magnetic Resonance Spectrometers," Proc. R. Soc. Lond. 344, pp. 311-340, 1975.

16. W. A. Anderson, "Electrical Current Shims for Correcting Magnetic Fields", Rev. Sci. Instruments 32, pp. 241-250, 1961.

17. W. Franzen, "Generation of Uniform Magnetic Fields by Means of Air-Core Coils", Rev. Sci. Instruments 33, pp. 933-938, 1962.

18. M. J. E. Golay, "Nuclear Magnetic Resonance

Apparatus", US Patent 3,569,823, 1971.

19. M. J. E. Golay, "Homogenizing Coils for NMR Apparatus", US Patent 3,622,869, 1971.

20. M. D. Sauzade and S. K. Kan, "High Resolution Nuclear Magnetic Resonance Spectroscopy in High Magnetic Fields", in Advances in Electronics and Electron Physics v. 34 (edited by L. Marton), pp. 1-93, Academic Press, New York, 1973.

21. W. D. MacMillan, The Theory of the Potential, Dover, New York, 1958.

22. W. R. Smythe, Static and Dynamic Electricity (3rd edition), McGraw-Hill, New York 1968.

23. J. A. Stratton, Electromagnetic Theory, McGraw-Hill, New York, 1941, p.232.

TABLE III. **Polynomials for the Z Dependence of the Numerators of the Expansion Coefficients.**

$m = 0$	
$g_{00} = 1$	$g_{05} = \frac{21}{8} Z(8Z^4 - 20Z^2 + 5)$
$g_{01} = 3Z$	$g_{06} = \frac{7}{16}(64Z^6 - 240Z^4 + 120Z^2 - 5)$
$g_{02} = \frac{3}{2}(2Z - 1)(2Z + 1)$	$g_{07} = \frac{9}{16} Z(64Z^6 - 336Z^4 + 280Z^2 - 35)$
$g_{03} = \frac{5}{2} Z(4Z^2 - 3)$	$g_{08} = \frac{45}{128}(128Z^8 - 896Z^6 + 1120Z^4 - 280Z^2 + 7)$
$g_{04} = \frac{15}{8}(8Z^4 - 12Z^2 + 1)$	
$m = 1$	
$g_{11} = -(Z^2 - 2)$	$g_{12} = -Z(Z - 2)(Z + 2)$
$g_{13} = -\frac{1}{4}(4Z^4 - 27Z^2 + 4)$	$g_{14} = -\frac{1}{4} Z(4Z^4 - 41Z^2 + 18)$
$g_{15} = -\frac{1}{8}(8Z^6 - 116Z^4 + 101Z^2 - 6)$	$g_{16} = -\frac{1}{8} Z(8Z^6 - 156Z^4 + 225Z^2 - 40)$
$m = 2$	
$g_{22} = -\frac{1}{4}(2Z^2 - 3)$	$g_{23} = -\frac{1}{4} Z(2Z^2 - 5)$
$g_{24} = -\frac{1}{24}(12Z^4 - 46Z^2 + 5)$	$g_{25} = -\frac{1}{8} Z(4Z^4 - 22Z^2 + 7)$
$g_{26} = -\frac{1}{64}(32Z^6 - 240Z^4 + 150Z^2 - 7)$	
$m = 3$	
$g_{33} = -\frac{1}{24}(3Z^2 - 4)$	$g_{34} = -\frac{1}{8} Z(Z^2 - 2)$
$g_{35} = -\frac{1}{64}(8Z^4 - 23Z^2 + 2)$	$g_{36} = -\frac{1}{192} Z(24Z^4 - 95Z^2 + 24)$
$m = 4,5,6$	
$g_{44} = -\frac{1}{192}(4Z^2 - 5)$	$g_{45} = -\frac{1}{192} Z(4Z^2 - 7)$
$g_{46} = -\frac{1}{1920}(40Z^4 - 96Z^2 + 7)$	$g_{55} = -\frac{1}{1920}(5Z^2 - 6)$
$g_{56} = -\frac{1}{1920} Z(5Z^2 - 8)$	$g_{66} = -\frac{1}{23040}(6Z^2 - 7)$

Computation for Conductance Distributions of Percolation Lattice Cells

Rabbe Fogelholm
Dept.of Computer Science,
University of Exeter,
Exeter EX4 4QL, UK.

Abstract: For a network, built from links whose conductances are given by some discrete statistical distribution, including a finite zero-conductance probability, and where links are assumed independent, the distribution of conductances for the network as a whole is of interest in the study of percolation on lattices. This quantity is computed by different methods for a set of test networks. It is found that the computation is more efficiently done by manipulating the networks themselves in a suitable representation rather than by computing with symbolic expressions for their conductance. In particular, with ordinary computer algebra systems there were severe limitations due to expression growth in this study.

1. Introduction

In recent years, the percolation transition that occurs in lattice models of disordered conductors has generated considerable interest [1] as a simple model of phase changes in matter.

The basic formulation of the model comprises a lattice into which conducting links are inserted with probability p between nearest-neighbour pairs of lattice points (this scheme is known as "bond percolation", there is also "site percolation" in which lattice sites support conduction with probability p). For small values of p, clusters of connected lattice points are of limited size, and the system behaves as an insulator. At some critical probability p_c the cluster size diverges and conducting channels of unlimited length appear. For $p>p_c$ there is an "infinite cluster" giving the system a conductor-like behaviour.

This "percolation transition" is similar to thermodynamic critical points in physical systems. Macroscopic properties of the system close to the critical point are well described by power laws; e.g. the conductivity σ for $p>p_c$ is given by $\sigma = \sigma_o(p-p_c)^t$ where the critical exponent t is believed to depend only on the dimensionality of the lattice.

The percolation model was first formulated in 1957 by Broadbent and Hammersley [2] who considered the flow of fluids through porous media. The significance of percolation as a very simple model that exhibits critical behaviour has emerged in the 1970s: Kirkpatrick (1973)[3] studied the conductivity of simulated resistor lattices and estimated the critical exponent in 2 and 3 dimensions. In 1974 Toulouse [4] conjectured that the critical dimensionality, at and above which the critical exponents take integral and dimensionality-independent values, is 6 for percolation systems. In 1975 Young and Stinchcombe [5] used coordinate-space renormalization to estimate the correlation length critical exponent in 2 dimensions. Numerous recent references on the subject are given in the review article by Stauffer [1].

The use of computers has been essential to several aspects of the work within this field. Theoretical results can be tested against measurements on physical systems such as conductor-insulator mixtures or granular thin films, and against computer simulations of large finite lattices. Computer simulation has the advantage of truly representing the microscopic details of the model, but the effects of limited sample size are severe as the coarseness of the connected structures diverges when the critical point is approached [6].

Ordinary symbolic computing, as provided by the general-purpose systems now available, doesn't seem to have been used within the field. There are however various examples of computer manipulation of non-numeric mathematical objects and the use of unconventional data structures: Sykes and Glen (1976) [7] computed the so-called perimeter polynomials for small clusters on various 2-dimensional lattices. Reynolds et al (1980) [8] computed the probability of connected wall-to-wall paths in small lattice cells as polynomials in p. Fogelholm (1980) [9] utilized transformations of networks, represented as cyclic list structures, to obtain an efficient method for the conductance of large network samples close to p_c.

The present work is an investigation of different methods for computing conductance distributions of "network cells" (see the next section for details). Although this computation enters only as one piece of a renormalization calculation of critical exponents it is of interest as a test case - the study of lattice models is a field of growing interest and there is good reason to explore the potential of symbolic computing as a tool of this trade.

2. Statement of the problem

The technique of coordinate-space renormalization [5,10,11] has been applied with a fair amount of success in the percolation problem. The idea is that, close to the critical point, the correlation length (= a typical diameter of finite clusters) is large and the system can be equally well described in terms of a less fine lattice. Hence, the original problem is mapped onto an isomorphic problem on a coarser lattice by subdividing the lattice into cells and forming cell averages of the quantity of interest. By calculating the probability of nearest-neighbour bonds in the renormalized lattice, and the change of scale of the averaged quantity under this mapping one can derive its critical exponent.

Recent studies of the bond percolation conductivity by this technique [10, 12-15] actualize the following computational problem:
Given some particular "network cell" (see Figure 1 for an example) in which the conductance of each bond, or link, is given by a discrete statistical distribution, what is the corresponding distribution for the entire cell?

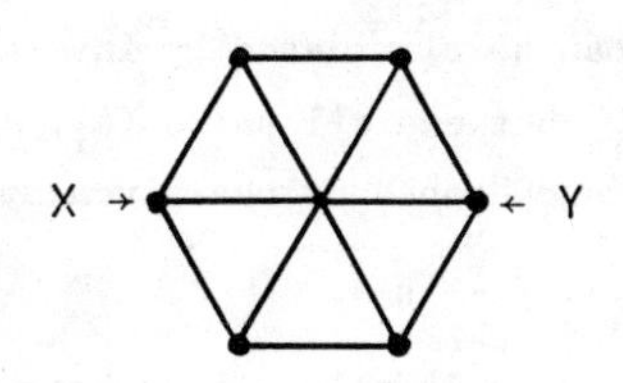

Figure 1. Example of 2-dimensional "network cell". The statistical distribution of the conductance between terminals X and Y is sought.

This distribution contains information on the renormalization behaviour of the associated lattice; in particular the weight of the zero-conductance case reveals how the bond probability is rescaled in the transformation.

The conductance distribution f of a single bond that has unity conductance with probability p and zero conductance otherwise is

$$f(s) = q\,\delta(s) + p\,\delta(s-1)\ , \quad q = 1-p \ , \qquad (1)$$

where s is conductance and δ is the Dirac delta function. For a composite network this generalizes to

$$f(s) = W_o\,\delta(s) + \sum_{i=1}^{R} W_i\,\delta(s-s_i)\ , \qquad (2)$$

where the summation is over the set of possible conductance values s_i, and where W_i are homogeneous polynomials in p and q.

For small cells the distributions can be calculated by hand, but the complexity of the calculation rises very steeply with the number of bonds in the cell and it is therefore of interest to consider how to compute these distributions by machine, especially as there are in principle advantages in using larger cells [16].

3. Comparison of different methods

The first method (M1) is a straightforward investigation of the entire set of possible states of the network cell, where the geometry of the links and nodes is fixed a priori. For a network of L links there are 2^L ways of assigning zero/unity conductance to individual links. The probability W of a state with K conducting links is $W = p^k q^{L-k}$ and its conductance, as seen between the terminal nodes, is obtained by successively eliminating the non-terminal nodes in a way that leaves the terminal-to-terminal conductance unchanged. The general rule for such eliminations is:

Assume the node A_o is connected to neighbouring nodes $A_1, A_2, \ldots, A_M$ via conductances $s_1, s_2, \ldots, s_M$. The node A_o can be eliminated by inserting replacement links s_{ij} between all pairs (A_i, A_j) of neighbours, with conductance values given by

$$s_{ij} = s_i s_j / \sum_{k=1}^{M} s_k . \tag{3}$$

This may introduce parallel links between certain node pairs; these may be merged immediately since the conductance of parallel conductors is additive. For M=2 and M=3 this scheme boils down to the usual rules for series-connected conductances and the Y - Δ transformation for conductances. The order in which internal nodes are eliminated is not significant for the final result, but simple arguments [17] suggest that the computational effort is minimized by eliminating lowest-valence nodes first (the valence of a node is the number of links emanating from it).

A LISP [18] program was written for method M1. Network nodes were represented by GENSYM atoms whose values were association lists of neighbour nodes paired with link conductances. Conductance values were computed using rational arithmetic, with GCD cancellations in every stage. No difficulties with large denominators were encountered.

The computational cost of eliminating a node goes as the square of its valence since it involves operations on the set of pairs of neighbours. We can thus estimate the run-time T as

$$T = \text{const} \cdot 2^L \cdot \sum_{i=1}^{N} V_i^2 \tag{4}$$

where L is the number of links, N the number of nodes, and V_i the valence of the i:th node. As can be seen from Figure 2 this estimate holds quite well over several orders of magnitude in the run-time.

The terminal-to-terminal conductance of a network with L links is obviously a function of L variables $s_1, s_2, \ldots, s_L$ that represent the conductances of individual links. In method M1, network graphs were repeatedly simplified down to a 2-node form whilst building up a numeric result. We can however equally well build a symbolic conductance expression in L variables; this needs to be done only once and any simplifications of this expression will be of advantage in subsequent repeated evaluations.

A REDUCE [19] program was written for this approach (M2). The network was represented by a symbolic adjacency matrix, its elements being rational expressions formed from the set of kernels S(1), S(2), ..., S(L), representing the conductances of links in the initial network. The final conductance expression, after the elimination of all internal nodes, was then evaluated for all possible assignments of zero/unity conductances to the symbols S(1), ..., S(L).

Some caution was required at this point: if zero conductance is simply represented by the number zero there is the problem of 0/0-type subexpressions that arise for many instances of the network. In REDUCE this could be circumvented by first representing zeros by some symbol EPSILON, simplifying, and finally substituting 0 for EPSILON.

In practical use this approach turned out to be much less successful than the M1 method in important respects: run-times were longer by a factor 30-90, and the generated conductance expression exceeded available memory for all but the two smallest networks. This latter difficulty could not be resolved by means of the switches for user control over the evaluation process - it appears that the expressions generated by this problem grow very quickly with the problem size when converted to the canonical form utilized by REDUCE.

Without too much hope for improvement, the same algorithm was re-coded for SHEEP [20], which is a fairly small LISP-embedded algebra package with special facilities for the manipulation of tensors (this attempt is referred to as M3). The advantages with SHEEP would be a larger free-storage area and tighter control over the evaluation (the simplifier has to be called explicitly when needed) but on the other hand the SHEEP simplifier does not provide general GCD cancellations. In practical tests only the very smallest of the test networks could be treated successfully due to problems with expression growth and the evaluation of 0/0-type subexpressions.

Eventually a LISP program (M4) was written for the conductance expression approach. Since standard algebraic simplification procedures seemed to do more harm than good no attempts were made to look for simplifications while building new expressions. Instead, attention was paid to compactness of the internal representation and sharing of subexpressions. With detailed knowledge of the internal structure of expressions it was easy to write an efficient evaluator; in the absence of simplifications every zero-denominator subexpression can be interpreted as zero for this problem. Furthermore, with shared subexpressions it becomes economical to build an association list of pointers to subexpressions paired with their current value.

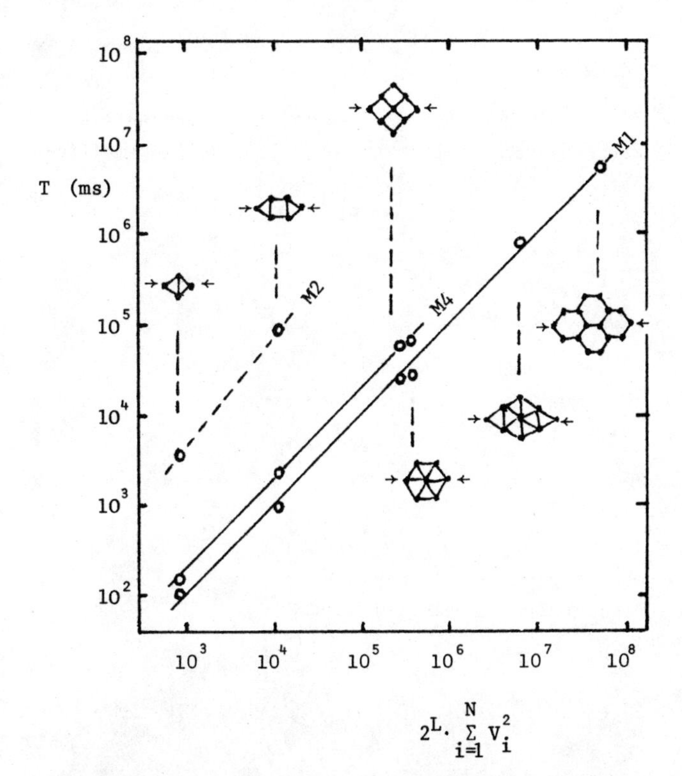

Figure 2. Actual run-time T versus the estimate of Eq.(4) for a set of test networks, using methods M1, M2, and M4. The unbroken lines are drawn with unity slope.

Nevertheless, this approach too failed to improve on the initial straightforward method of generating the networks and computing their conductance one by one (M1). Run-times were higher by almost a factor 2 (see Figure 2) and space usage was never a problem in the first place.

4. Discussion

Whether or not a conductance expression is derived for the network cell, the number of the states that have to be studied can be brought down by taking advantage of symmetry. However, since the savings in run-time are less than a factor 4 for 2-dimensional networks this possibility was not investigated in the present work.

It remains an open question if a radically better method can be devised for this particular problem - it may be that there is some non-standard simplification method that can turn the conductance expression into a more efficient "program" for conductance computations than the original network graph itself.

Apart from the fact that run-times increase very steeply with network size, the space occupied by the result will also become a problem eventually. However, this can be dealt with by various forms of data reduction (like, e.g. collecting a histogram instead of exact conductances), depending on how the result is to be used.

Applications of this work are at a preliminary stage so far. A re-investigation of a calculation by Stinchcombe and Watson (1976) [10] has been made, using method M1, with the resultant distribution used as input to REDUCE for the final stages of the analysis. In ref. 10, the cell of Figure 1 was considered with the restriction that conducting paths should be no more than 3 steps long, thereby reducing the number of configurations to 91. A full analysis of all 4096 states yielded a fixed-point probability p^* in better agreement with p_c (p^* is the value of p where the probability that the cell conducts equals p; among other things it serves as an estimate of p_c). The estimate for the conductivity critical exponent was however only slightly improved - this study thus confirms that the essential physics of the problem is well contained in the restricted-length paths used in ref. 10.

In summary then, it has been shown that symbolic computation has some relevance to this test problem in lattice models, although it seems that general purpose algebra packages quickly run into severe memory problems. From the point of view of the design of computer algebra systems, the need for extensibility in various directions is illustrated; non-standard data structures for the representation of graphs and non-standard procedures for manipulating them are of interest. While it is obviously important that computer algebra systems (my experience is with REDUCE, but this comment may be of more general relevance) provide a clean and robust face to the user who has no need to consider the inner workings of the system, it is also important that the user who is compelled to do some work at LISP level is given appropriate support in the form of accessibility of below-top-level routines, and documentation at a lower level than found in the typical User's Manual. The value of low-level work is likely to be further enhanced by any possible developments of LISP towards better performance in symbolic computing applications [21].

Acknowledgments

Stimulating discussions with Professor J.A.Campbell, and his critical reading of the manuscript, has been of great value for this work. It is also with great pleasure that I acknowledge the hospitality which I have experienced during my stay at Exeter University.

This work was financed by the Swedish Natural Science Research Council (grant no. F-PD 4620-100), except for computing facilities which were provided by the British Science Research Council (grant no. GR/B/0151.7).

References

1. D. Stauffer, Phys.Reports 54, 1 (1979)
2. S.R. Broadbent and J.M. Hammersley Proc.Camb.Philos.Soc. 53, 629 (1957)
3. S. Kirkpatrick, Rev.Mod.Phys. 45, 574 (1973)
4. G. Toulouse, Nuovo Cim. 23 B, 234 (1974)
5. A.P. Young and R.B. Stinchcombe, J. Phys. C 8, L535 (1975)
6. J.P. Straley, AIP Conf.Proc. 40, p.118 (1978)
7. M.F. Sykes and M. Glen, J.Phys. A 9, 87 (1976)
8. P.J. Reynolds, H.E. Stanley,and W. Klein, Phys.Rev. B 21, 1223 (1980)
9. R. Fogelholm, J. Phys. C 13 L571 (1980)
10. R.B. Stinchcombe and B.P. Watson, J.Phys. C 9, 3221 (1976)
11. P.J. Reynolds, W. Klein, and H.E. Stanley, J.Phys. C 10, L167 (1977)
12. J. Bernasconi, Phys.Rev. B 18, 2185 (1978)
13. J.P.Straley, J.Phys. C 12, 3711 (1979)
14. C.J. Lobb and D.J. Frank, J.Phys. C 12, L827 (1979)
15. B. Shapiro, J. Phys. C 13, 3387 (1980)
16. P.J. Reynolds, H.E. Stanley, and W. Klein, J.Phys. A 11, L199 (1978)
17. R. Fogelholm, Report TRITA-FYS-5075, Royal Institute of Technology (1980)
18. I.B. Frick, Univ. of Utah Symbolic Computation Group, Technical Report TR-2 (1978)
19. A.C. Hearn, REDUCE-2 User's manual, Univ. of Utah Computational Physics Group report UCP-19 (1973)
20. I.B.Frick, The computer algebra system SHEEP, what it can and cannot do in general relativity, preprint, Inst. Theor. Phys., Univ. of Stockholm (1979)
21. J.A. Campbell and J.P. Fitch, Proc. 1980 LISP Conference, (The LISP Company, Redwood Estates, California, 1980)

Breuer's Grow Factor Algorithm in Computer Algebra

J.A. van Hulzen
Twente University of Technology
Department of Applied Mathematics
P.O. Box 217, 7500 AE Enschede
The Netherlands

Abstract:
Computer algebra can be of importance in solving electrical network synthesis problems. It enables to generate automatically FORTRAN-programs needed to solve the underlying systems of nonlinear multivariate polynomial equations. Such systems are obtained by equating the numerical target function and a symbolically computed equivalent. Breuer's grow factor algorithm is extended not only to eventually improve the result of required determinant calculations, but also to optimize FORTRAN-code describing the system and the associated Jacobian. It is indicated how this common subexpression search approach can be extended.

1. Introduction

A common subexpression search (css) can be relevant for simplification [20], it can assist in improving the readibility of output [7,8,15] and consequently also in preparing optimized code for numerical evaluation [14,15].

If such a css has to be considered as part of general purpose simplification facilities is doubtful. When applying algebra systems additional information is not seldom provided by the structure of the problem under consideration. Certainly when assuming that code-preparation -often related to certain classes of problems- is part of a generation mechanism of programs for numerical purposes. A well known example is the production of Jacobians [6,23,15,11]. Especially here the role of simplification (or a css) is quite specific. All methods make use of the fact that partial derivatives of identical subexpressions are identical. So one needs only to transform a given, fixed set of expressions into a form on which the remaining steps of the computation, like differentiation and code generation, can be performed most efficiently (citing Moses [13]).

Such automatic program facilities ought to operate on a simple and comprehensive description of the problem as input. So it might be possible that the input format requires some algebraic "preprocessing" to construct the set of expressions needed to prepare the arithmetic code. This suggests that an additional presumption must be that a css can be applied on any internal representation delivered by this preprocessing: nested, factored, expanded or what so ever. This arithmetic is extensive, and the structure of the problem is supposed to require repetitions of identical sequences of arithmetic instructions, thus making code optimization, via a css, worthwhile the effort; certainly when dealing with "production" programs.
This in contrast to Knuth's conclusion that for an average FORTRAN-program, a compiler with facilities to optimize arithmetic is really overdone [12]. This eventually explains why little is done in this area, except, for instance, Breuer's restrictive proposals [3].
Another reason might be that the usual dag-model in flow graph analysis is a questionable instrument. This in view of considerations concerning commutativity and associativity [2,5]. A third aspect is, of course, the lack of information about error-sensitivity of the code under consideration. Although a css itself is not necessarily a good instrument to structure arithmetic as to minimize chop off and round off error propagation, it can contribute to it, when combined with facilities, similar to proposals of Stoutemyer [19].
The complexity of the structure of a css depends on the class of expressions on which it is supposed to operate. But with the "kernel-mechanism" of Reduce in mind [9] we limit this class to the set of multivariate polynomials over Z. However, to start with, we assume to have an even more restrictive class, which easily allows to introduce a notion of optimization via Breuer's algorithm in NETFORM [16] and in relation to that for mixed symbolic-numeric methods for solving network synthesis problems.

2. The role of Breuer's algorithm in NETFORM

NETFORM is a REDUCE program designed for the symbolic analysis of electrical networks. It operates on a simple graph-description of a

©1981 ACM 0-89791-047-8/81-0800-0100 $00.75

specific network as input. This input is used to construct the associated incidence matrix, describing the structure equations in accordance with Kirchhoff's laws.
This structural information allows to obtain transfer-functions symbolically, essentially via determinant calculations [17]. However, this matrix representation of the network under consideration is not always the most adequate description of it. This in terms of the determinant calculations. When, for example, considering a subcycle of the network, a serie of resistances in it can be replaced by one voltage meter. Discovering such specific physical properties of a network is similar to a Breuer-like search for detecting common subexpressions (cse's). Such a search can lead to a sparser and somewhat larger matrix. This allows, in turn, a more compact and comprehensible description of the required determinants. It often implies an overall reduction of CPU-times.
Once having tools to obtain symbolic expressions for characteristic entities of an electrical network, the interesting question arises whether such facilities can be of practical value for solving network synthesis problems. Is a realistic alternative possible for the often cumbersome numerical approaches [4]? Assume a target function H_n with constant numerical coefficients is available. Its symbolic equivalent H_s, also a rational function, can be produced with NETFORM. Hence coefficient matching or so called frequency matching might allow to determine the different components of the realisation under consideration. This matching results in a system of polynomial equations, where the indeterminates symbolize network components.
The numerator and denominator of H_s are in fact determinants, computed via a nested minor and thus given in a compact nested form. Attempts to solve nonlinear systems numerically imply the need for Jacobians when Newton's method is used. A realistic wish is to obtain the partial derivatives in a structure-preserving way. This leads to the conclusion that, before actually differentiating, a css on the expressions as such may deliver considerable savings in computing time and will essentially improve the comprehensibility of the produced FORTRAN programs. An extended version of Breuer's algorithm, currently being implemented, is a most promising and simple tool to achieve this [10,22].
A detailed and more technical description of this approach to solving network synthesis problems is given by Smit [18]. Our aim is to illustrate the profitability of problem dependent simplification approaches. Hence, implicity, we advocate for subroutine library facilities, certainly for reasonable portable systems, like Reduce.

3. A notion of optimization

An initial set E_0 of N_0 multivariate polynomials in $Z[V_0]$ is assumed, where V_0 is a set of M_0 indeterminates. Our css is an iterative process. During an iteration step $i(\geq 1)$ the application of a profitcriterium delivers either a decision to terminate the process or $k(\geq 1)$ cse's. The instances of these cse's are replaced by new indeterminates. The descriptions of the cse's are added to the set E_{i-1}, thus producing a set E_i of $N_i = N_{i-1} + k$ multivariate polynomials in $Z[V_i]$, where V_i $(\supset V_{i-1})$ is a set of $M_i = M_{i-1} + k$ indeterminates.
As indicated in section 2 our present interest is quite specific. E_0 is either a set of linear expressions, associated with an incidence matrix, or a set of nested expressions, describing a transfer-function and its partial derivatives, with respect to the networkcomponents. This allows to consider E_i, $i \geq 0$, as a set of sums or products over V_i of a certain length and as given in
Definition 3.1: Let $s_i(\alpha,\beta)$ $[p_i(\alpha,\beta)]$ denote a sum [product] over $V_i = \{x_1, x_2, \ldots, x_{M_i}\}$ of length $\alpha + \beta \geq 0$ such that $s_i(\alpha,\beta) = ps_i(\alpha) + cs_i(\beta)$ $[p_i(\alpha,\beta) = pp_i(\alpha).cp_i(\beta)]$ where

$ps_i(\alpha)$ $[pp_i(\alpha)]$ denotes a primitive sum [product] of length $\alpha \geq 0$ given by

- $ps_i(0) = 0$ $[pp_i(0) = 1]$
- $ps_i(\alpha) = \sum_{k=1}^{\alpha} x_{m_k}$ $[pp_i(\alpha) = \prod_{k=1}^{\alpha} x_{m_k}]$ where
  $0 < \alpha \leq M_i$, $1 \leq m_k \leq M_i$, $m_n \neq m_\ell$ if $1 \leq n < \ell \leq \alpha$,

and

$cs_i(\beta)$ $[cp_i(\beta)]$ denotes a composite sum [product] of length $\beta \geq 0$ given by

- $cs_i(\beta) = 0$ $[cp_i(0) = 1]$,
- $cs_i(\beta) = \sum_{k=1}^{\beta} p_i(\nu_k, \mu_k)$ $[cp_i(\beta) = \prod_{k=1}^{\beta} s_i(\nu_k,\mu_k)]$,
  where $p_i(\nu_n,\mu_n) \neq p_i(\nu_\ell\ \mu_\ell)$ $[s_i(\nu_n,\mu_m) \neq s_i(\nu_\ell\ \mu_\ell)]$ if $1 \leq n < \ell \leq \beta$ □

Definition 3.1 allows an arbitrary often nesting of subexpressions. Therefore we assume the number of primitive sums [products] in an expression e to be finite and denoted by $N_s(e)$ $[N_p(e)]$.
Empty primitives are included for reasons which will become apparent.
The category of NETFORM-related problems allows the exclusion of a repeated occurrence of indeterminates in primitives and of identical (sub)sums [products] as factor [term] in composites. So, according to definition 3.1, all numerical coefficients are in Z_2 and exponentiation is excluded.
The initial set E_0, for instance consisting of N_0 sums $e_k = s_0(\alpha_k, \beta_k)$, can be considered as the definition of N_0 arithmetic assignment statements prescribing how numerical values have to assigned to the e_k (the output), assuming the x_i's (forming a set of inputnames) are replaced by a set of inputvalues. So we consider E_0 as a representation of a block of straightline code [1]. Then its arithmetic complexity, taken over all permissible inputs, can be defined as $AC_i = n_i^+ + \phi n_i^*$, where n_i^ω stands for the number of binary ω-operations and ϕ for the ratio of the cost of a binary multiplication and addition

(including assignments). This representation is not necessarily an optimal way to obtain the output, i.e. some primitive equivalence preserving transformations, such as elimination of redundant computations [1], thus attempting to minimize AC_0 via a css, might be profitable. So we base the profitcriterium on reducing the arithmetic complexity, as stated in:

Definition 3.2: A transformation $T:E_i \to E_{i+1}$, $i \geq 0$, is called an optimization iff $AC_i > AC_{i+1}$. The set $E_\ell = T^\ell(E_0)$, $\ell \geq 0$, is called an optimal description of E_0 with respect to T iff $AC_0 > AC_1 > \ldots > AC_\ell = AC_{\ell+1}$. □

T is supposed to contain a css, i.e. the profit-criterium to obtain the cse's. In case E_0 is a set of linear expressions (now called primitive sums), associated with an incidence-matrix, this view on a css has only a formal meaning. It allows to use Breuer's grow factor algorithm and his profit-criterium. Once the transferfunction is produced an extended version of Breuer's algorithm can be used to really produce optimized FORTRAN-code. Definition 3.1 suggests that the nested expressions to be elaborated can be decomposed in sets of primitive (sub) sums and -products.

4. Breuer's grow factor algorithm: an informal review

Breuer's heuristic algorithm is iterative. It allows to optimize a set E_0 of N_0 primitive sums $e_k = s_0(\alpha_k,0)$ or products $e_k = p_0(\alpha_k,0)$ by operating on an (N_0,M_0) - incidence matrix B_0, defining the primitives, and weights, associated with its rows and columns, to obtain intuitively optimal subexpression choises and leading to an eventually optimized version $E_\ell = T^\ell(E_0)$ of E_0, defined by an (N_ℓ, M_ℓ) - matrix B_ℓ.
$B_i = \| b_{k,j} \|$ reflects "the state of the art" after step $i \geq 0$: $b_{k,j} = 1$ if x_j occurs in e_k, otherwise $b_{k,j} = 0$. The intermediate length α_k of e_k is used as rowweight. The intermediate weight of the j-th column, its number of ones, reflects the "density" of x_j in E_i.
A cse s is a primitive which is contained in at least 2 elements of E_i, its parents, and which has at least a length 2. If Js_i is the set of (column) indices of the indeterminates forming s and if Ps_i is the set of (row)indices of its parents then $ws_i = \{|Js_i| -1\}\{|Ps_i|-1\} > 0$ is called the profitcriterium, reflecting the reduction in binary arithmetic operations when replacing s by a new indeterminate and adding s as new expression to the next version of E_0, implying that after each iteration the weights have to be reset. To deliver E_ℓ in its final form some renaming can flipping can be required [1].
So the transformation T, introduced in definition 3.2, can now be considered as a composite function, such that $E_\ell = T(E_{\ell-1}) = R^{-1}T_\beta\, R(E_{\ell-1}) = R^{-1}\, T_\beta{}^\ell R\,(E_0)$.
The value of ℓ depends on a termination condition, say profit. Assume $profit_\ell$ = FALSE and $profit_i$ = TRUE, i=0, 1, ..., ℓ-1. Then

$$(4.1)\begin{cases} R \quad : E_0 \to (D_0,\ profit_0) \\ T_\beta : (D_i, profit_i) \to (D_{i+1}, profit_{i+1}), i=0,..,\ell-1 \\ R^{-1} : (D_\ell, profit_\ell) \to E_\ell \end{cases}$$

Here $D_i = (B_i, \underline{\alpha_i}, \underline{\delta_i})$, where $\underline{\alpha_i}$, $\underline{\delta_i}$ are vectors of row-and columnweights, respectively. R^{-1} can eventually be replaced by, say F, to produce FORTRAN-code from D_ℓ, directly.
T_β denotes a Breuer-step, operating on D_i to select a pair $CSE_i = (Js_i, Ps_i)$, such that ws_i is intuitively maximal. If $ws_i > 0$ then D_{i+1} is constructed with (D_i, CSE_i). If $ws_i = 0$ then the process is terminated, i.e. $profit_i \leftarrow ws_i > 0$.
Before attempting to "discover" a cse redundancy in B_i, if existing, is eliminated. Rows (columns) are redundant if their weights are less then 2, and thus can be disregarded when searching cse's. If a row is redundant the column sharing its one can become redundant too, i.e. repeated removal of redundancy can imply that no further profit can be gained.
To find an s we select columns of maximal weight from a non-redundant (sub)matrix SB_i of B_i. Then we compute the weightfactors of these columns, being the sums of the rowweights of their nonzero elements. These weightfactors reflect the 1-density associated with a column of maximal weight, thus delivering a "preview" of the structure of potential parents. It allows an intuitively optimal choise to extend Js_i, initially empty, by adding the index of the most left column with maximal weightfactor. Ps_i is simply the set of row indices of the nonzero elements of this column. This process is continued as long as the profit is not decreasing, implying that Js_i is growing and Ps_i is eventually shrinking. Finally if $ws_i > 0$ then B_i is updated resulting in B_{i+1}. So T_β can be considered as a composite function $U_\beta C_\beta E_\beta$, such that

$$(4.2)\begin{cases} E_\beta : (D_i, profit_i) \to (D_i, profit_i, SB_i) \\ C_\beta : (D_i, profit_i, SB_i) \to (D_i, profit_i,\ CSE_i) \\ U_\beta : (D_i, profit_i, CSE_i) \to (D_{i+1}, profit_{i+1}) \end{cases}$$

A detailed description is given in [22].
An example:

Given E_0 =

$$\{e_1 = x_3 + x_5 \quad x_7,\\ e_2 = x_2 + x_3 + x_4 + x_6 + x_7,\\ e_3 = x_1 + x_2 + x_3 + x_4 + x_6,\\ e_4 = x_1 + x_5,\\ e_5 = x_1 + x_3 + x_4,\\ e_6 = x_1 + x_4 + x_5 + x_6\}$$

We find E_3 =

$$\{s_1 = x_4 + x_6\\ s_2 = s_1 + x_2 + x_3,\\ s_3 = x_1 + x_5,\\ e_1 = x_3 + x_5 + x_7,\\ e_2 = s_2 + x_7,\\ e_3 = x_1 + s_2,\\ e_4 = s_3,\\ e_5 = x_1 + s_1,\\ e_6 = s_1 + s_3\}$$

5. An extended Breuer algorithm

In a more general situation, when E_0 consists of nested expressions for instance, the function $R: E_0 \to (D_0,\ profit_0)$, see (4.1), ought to produce a decomposition D_0 of E_0 resulting in two incidence matrices and additional information like weights and "history".
According to definition 3.1 E_i, $i \geq 0$, consists of N_i, eventually nested, sums. With each of the $Ns(E_i) = \sum_{n=1}^{N_i} Ns(e_n)$ (sub)sums $s_{ik}(\alpha_k, \beta_k)$ we

can associate a vector $\underline{\rho\sigma_{ik}}=(\rho_{k,1}, \ldots, \rho_{k,M_i})$, representing its primitive part $ps_{ik}(\alpha_k)$, its primitive part weight $\alpha\sigma_{ik} = \alpha_k \geq 0$, its number of composite sumterms $\beta\sigma_{ik} = \beta_k \geq 0$ and its father $\psi\sigma_{ik} \in \{0,1,\ldots,Np(E_i)\}$; $\psi\sigma_{ik} = 0$ if $s_{ik}(\alpha_k, \beta_k)$ is a "root", i.e. a top-level sum. Similar characterisations of the $Np(E_i)$ (sub) products $p_{ik}(\alpha_k, \beta_k)$ can be introduced. This leads to

Definition 5.1 A description D_i of E_i, $i \geq 0$, is a 4-tuple $(D\sigma_i, H\sigma_i, D\pi_i, H\pi_i)$ such that

- $D\sigma_i = (B\sigma_i, \underline{\alpha\sigma_i}, \underline{\delta\sigma_i})$, where
  - $B\sigma_i$ is an $(Ns(E_i), M_i)$ incidence matrix with $\underline{\rho\sigma_{ik}}$ as its k-th row.
  - $\underline{\alpha\sigma_i}^T = (\alpha\sigma_{i,1}, \ldots, \alpha\sigma_{i,Ns(E_i)})$.
  - $\underline{\delta\sigma_i} = (\delta\sigma_{i,1}, \ldots, \delta\sigma_{i,M_i})$ with $\delta\sigma_{ik} = \sum_{j=1}^{Ns(E_i)} \rho\sigma_{j,k}$, called the weight of the k-th column of $B\sigma_i$.
- $H\sigma_i = (\underline{\psi\sigma_i}, \underline{\beta\sigma_i})$, where
  - $\underline{\psi\sigma_i}^T = (\psi\sigma_{i,1}, \ldots, \psi\sigma_{i,Ns(E_i)})$.
  - $\underline{\beta\sigma_i}^T = (\beta\sigma_{i,1}, \ldots, \beta\sigma_{i,Ns(E_i)})$.
- $D\pi_i = (B\pi_i, \underline{\alpha\pi_i}, \underline{\sigma\pi_i})$ and $H\pi_i = (\underline{\psi\pi_i}, \underline{\beta\pi_i})$ are similar to $D\sigma_i$ and $H\sigma_i$. □

The transformation T_β: $(D_i, profit_i) \rightarrow (D_{i+1}, profit_{i+1})$, as introduced in section 4, can now be considered as a composite function $TE_\beta = M_\Pi E_\Sigma M_\Sigma E_\Pi$.
The function E_Π is supposed to apply T_β on the pair $(D\pi_i, \pi p_i)$, where πp_i denotes the profit, associated with the primitive product scheme, defined by $D\pi_i$. If profit is gained, it might be possible that the common subproduct can be added, as new indeterminate, to the sumscheme $D\sigma_i$, before this scheme itself is subject to a Breuer search. The decision criterium is

$$(5.1)\quad dc_{ik} = \{\ \beta\pi_{ik}=0 \wedge \alpha\pi_{ik}=1 \wedge \psi\pi_{ik} > 0\ \}.$$

This condition reflects that the parent, defined by $(\underline{\rho\pi_{ik}}, \psi\pi_{ik}, \beta\pi_{ik}, \alpha\pi_{ik})$ is purely primitive $(\beta\pi_{ik}=0)$, identical to the cse $(\alpha\pi_{ik}=1)$ and contained as composite sumterm in the sum defined by $(\underline{\rho\sigma_{i\ell}}, \psi\sigma_{i\ell}, \beta\sigma_{i\ell}, \alpha\sigma_{i\ell})$ where $\ell=\psi\pi_{ik}>0$.
So if πp_{i+1}= TRUE then $D\pi_i$ is updated as part of T_β. The criterium (5.1) can be used to split the set of parents in two disjoint subsets, one of incorporable parents (dc_{ik}= TRUE), say $p\pi$, and another being its complement (dc_{ik}= FALSE).
If $|p\pi|>1$ then $(D\sigma_i, H\sigma_i)$ is modified, using the function M_Σ, as follows:
A new column is added to $D\sigma_i$ by extending all $\underline{\rho\pi_{ik}}$ such that $\rho_{\ell,M_{i+1}}=1$ for all $\ell \in p\pi$ and $\rho_{\ell,M_i}=0$ otherwise, by setting $\beta\sigma_{i,\ell} \leftarrow \beta\sigma_{i,\ell} - 1$ $\alpha\sigma_{i,\ell} \leftarrow \alpha\sigma_{i,\ell} + 1$ for all $\ell \in p\pi$ and by extending $\underline{\delta\sigma_{ik}}$ with $\delta\sigma_{i,M_i+1} \leftarrow |p\pi|$.
This leads to an intermediate description $ID_i = (ID\sigma_i, IH\sigma_i, ID\pi_i, IH\pi_i)$ of D_i, where $(ID\pi_i, p\pi_{i+1}) = T_\beta(D\pi_i, p\pi_i)$ and $IH\pi_i= H\pi_i$. This mechanism motivates the incorporation of initially empty primitive sums and products.

E_Σ and M_Π are similar to E_Π and M_Σ if the role of primitive sums and products is interchanged.
Apparently the terminating condition, reflected by $profit_i$ depends on two Breuer searches and two modifications. So let $profit_i = \sigma p_i \vee \pi p_i$.
This implies that the optimization attempts are only terminated if none of both schemes delivers profit. In view of the modification steps, it might happen that these partial profits have to be reset if one scheme is extended due to profit gained with the other scheme. However in practice we merge both schemes [10], which is allowed in our particular situation. To conclude with we give the overall structure of the extended Breuer algorithm;

$$R\ :\ E_0 \rightarrow (D_0, \sigma p_0, \pi p_0)$$
$$T_\beta\ :\ (D_i, \sigma p_i, \pi p_i) \rightarrow (D_{i+1}, \sigma p_{i+1}, \pi p_{i+1}),\quad i=0, 1, \ldots, \ell-1$$
$$R^{-1}\ :\ (D_\ell, \sigma p_\ell, \pi p_\ell) \rightarrow E_\ell$$

where

$$T_\beta\ (D_i, \sigma p_i, \pi p_i) = M_\Pi\ E_\Sigma\ M_\Sigma\ E_\Pi\ (D_i, \sigma p_i, \pi p_i)$$

with

$$E_\Pi\ :\ (D_i, \sigma p_i, \pi p_i) \rightarrow (D\sigma_i, H\sigma_i, \sigma p_i, ID\pi_i, H\pi_i, \pi p_{i+1})$$
$$M_\Sigma\ :\ (D\sigma_i, H\sigma_i, \sigma p_i, ID\pi_i, H\pi_i, \pi p_{i+1}) \rightarrow (ID\sigma_i, IH\sigma_i, \sigma p_i, ID\pi_i, H\pi_i, \pi p_{i+1})$$
$$E_\Sigma\ :\ (ID\sigma_i, IH\sigma_i, \sigma p_i, ID\pi_i, H\pi_i, \pi p_{i+1}) \rightarrow (D\sigma_{i+1}, H\sigma_{i+1}, \sigma p_{i+1}, ID\pi_i, H\pi_i, \pi p_{i+1})$$
$$M_\Pi\ :\ (D\sigma_{i+1}, H\sigma_{i+1}, \sigma p_{i+1}, ID\pi_i, H\pi_i, \pi p_{i+1}) \rightarrow (D_{i+1}, \sigma p_{i+1}, \pi p_{i+1})$$

An example:

Given E_0 =	We find E_4 =
$\{e_1= a+b+cd+(g+f)(a+bcd),$	$\{p_1=cd, s_1=a+b, p_2=bp_1,$
$e_2= a+f+cd(g+f)(a+bcd),$	$s_2=a+p_2, s_3=f+g, p_3=s_2s_3,$
$e_3= a+gcd(g+cd)(a+b)\}$	$e_1=s_1+p_1+p_3,$
	$e_2=a+f+p_1p_3,$
	$e_3=a+gp_1(g+p_1)s_1\}$

6. Further generalisations?

The optimization method, proposed in the previous sections, operates on a restricted class of multivariate polynomials as input. This is fine for network-problems, but in general of no practical value. Allowing a repeated use of indeterminates in primitives and of identical (sub)sums [products] as factor [term] in composites, by dropping some of the restrictions of definition 3.1, introduces multivariate polynomials in $Z[V_i]$ in any preferable representation. But it implies that both $B\sigma_i$ and $B\pi_i$ become integer matrices, defining the constant coefficients in linear expressions (primitive sums) and the exponents in the monomials (primitive products), which are not necessarily multilinear, respectively. We also need to add a coefficient vector to $B\pi_i$ and an extra vector to $B\sigma_i$, defining possible powering of primitive sums. What are the consequences? The definition of AC_i must be revisited, implying that the profit criteria are more complicated. The Breuer searches however remain applicable, be it in a different setting.
A css on sums is now a search for submatrices of $B\sigma_i$ with rank 1, or stating this more precisely a search for identical primitive parts of subexpresions of the primitives, described by $B\sigma_i$.
The product scheme requires only minor changes in the extended Breuer algorithm.
However, being more concrete is beyond the scope

of this short note.
When considering the profit to be gained as an integer-valued parameter Hearn's problem of substitution [8] can be approached in a flexible way. Such a general algorithm might also be of practical value when trying to solve systems of multivariate polynomial equations, according to the William-Moses approach [7,21].

References

[1] A.V. Aho and J.D. Ullmann, The Theory of Parsing, Translating and Compiling, vol. II: Compiling, Prentice Hall (1973).

[2] A.V. Aho, S.C. Johnson and J.D. Ullmann, Code generation for expressions with common subexpression, J. ACM vol. 24, no.1 (Jan. 1977), pp. 146-160.

[3] M.A. Breuer, Generation of optimal code for expressions via factorization, Comm. ACM. vol. 12, no. 6 (June 1969), pp. 333-340.

[4] D.A. Calahan, Computer design of linear frequency selective networks, Proceedings of the IEEE, vol. 53 no. 11 (November 1965), pp. 1701-1706.

[5] P.J. Downey, R. Sethi and R.E. Tarjan, Variations on the common subexpression problem, J. ACM Vol. 27 no.4 (Oct. 1980), pp. 758-771.

[6] H. Eisenpress and A. Bomberault, Efficient symbolic differentiation using PL/I-FORMAC, IBM New York, SC Report no. 320-2956 (September 1968).

[7] M.L. Griss, The output of large expressions in Reduce, Utah Computational Physics Group, Operational Note no. 14 (May 1974).

[8] A.C. Hearn, The problem of substitution, Proceedings of the 1968 Summer Institute on Symbolic Mathematical Computation (R.G. Tobey, editor), IBM Boston Programming Center, Cambridge Mass (June 1969), pp. 3-20.

[9] A.C. Hearn, Reduce 2: A system and language for algebraic manipulation, Proceedings of the Second Symposium on Symbolic and Algebraic Manipulation (S.R. Petrick, editor), ACM New York (1971), pp. 128-133.

[10] B.J.A. Hulshof, Code optimization in NETFORM, BSC Thesis (in Dutch), Twente University of Technology, in preparation.

[11] G. Kedem, Automatic differentiation of computerprograms, ACM TOMS Vol. 6 no. 2 (June 1980), pp. 150-165.

[12] D.E. Knuth, An empirical study of Fortran programs, Software Practice and Experience Vol.1 (1971), pp. 105-133.

[13] J. Moses, Algebraic simplification: A guide for the perplexed, Comm. ACM Vol. 14 no. 8 (August 1971), pp. 527-537.

[14] E. Ng, Symbolic-Numeric Interface: a review, Symbolic and Algebraic Computation, (E. Ng, editor) Springer LNCS no. 72 (1979), pp. 330-345.

[15] E. Ng and B. Char, Gradient and Jacobian computation for numerical applications, Proceeding 1979 MACSYMA Users' Conference (V.E. Lewis, editor), M.I.T., Cambridge Mass (June 1979), pp. 604-621.

[16] J. Smit, NETFORM manual (second edition), Twente University of Technology (Jan. 1981).

[17] J. Smit, A cancellation free algorithm, with factoring and truncation capabilities, for the efficient solution of large sparse systems of equations, These proceedings.

[18] J. Smit, Sparse Kirchhoff equations, an effective support tool for the numeric and symbolic solution of large sparse tableau based systems, THT/EL Report 1231-AM 0581 (Jan. 1981).

[19] D.R. Stoutemyer, Automatic error analysis using computer algebraic manipulation, ACM TOMS Vol. 3 no. 1 (March 1977), pp. 25-43.

[20] R.G. Tobey, Experience with FORMAC algorithm design, Comm. ACM Vol.9 no. 8 (August 1966), pp. 589-597.

[21] J.A. van Hulzen, A note on solving systems of polynomial equations with floating-point coefficients, Symbolic and Algebraic Computation (E. Ng, editor), Springer LNCS no. 72 (1979), pp. 346-357.

[22] J.A. van Hulzen, Breuer's grow factor algorithm in computer algebra, Memorandum No. 332, Department of Applied Mathematics, Twente University of Technology (april 1981).

[23] D.D. Warner, A partial derivative generator, Bell Labs, Computing Science Technical Report # 28 (April 1975).

An Implementation of Kovacic's Algorithm for Solving Second Order Linear Homogeneous Differential Equations

B. David Saunders[1]
Department of Mathematical Sciences
Rensselaer Polytechnic Institute
Troy, NY 12181

1. Introduction.

Kovacic [3] has given an algorithm for the closed form solution of differential equations of the form ay" + by' + cy = 0, where a, b, and c are rational functions with complex coefficients of the independent variable x. The algorithm provides a Liouvillian solution (i.e. one that can be expressed in terms of integrals, exponentials and algebraic functions) or reports that no such solution exists.

In this note a version of Kovacic's algorithm is described. This version has been implemented in MACSYMA and tested successfully on examples in Boyce and DiPrima [1], Kamke [2], and Kovacic [3]. Modifications to the algorithm have been made to minimize the amount of code needed and to avoid the complete factorization of a polynomial called for. In Section 2 these issues are discussed and in Section 3 the author's current version of the algorithm is described.

---

[1] Partially supported by National Science Foundation grant MCS-7909158.

2. Issues.

Some notation: For any function u of x we write u' for du/ux, and C denotes the complex numbers, C[x] the polynomials over C, and C(x) the rational functions over C.

By a standard change of variables,

(1) $au'' + bu' + cu = 0, \quad a \neq 0,$
$a, b, c \in C(x)$

is equivalent to

(2) $y'' = ry, \quad r = (2b'a-2ba'+b^2-4ac)/4a^2$

in that (1) has a Liouvillian solution if and only if (2) has a Liouvillian solution ($u = \exp(\int -b/2a\, dx)\cdot y$).

Using the Galois theory of differential fields, Kovacic shows that equation (2) has a Liouvillian solution if and only if it has a solution of the form $y = e^{\int \omega\, dx}$ where ω is algebraic of degree 1, 2, 4, 6, or 12 over C(x). In turn, equation (2) has a solution of the form $y = e^{\int \omega\, dx}$ if and only if the Ricatti equation

(3) $\omega' + \omega^2 = r$

has a solution ω which is algebraic of degree n = 1, 2, 4, 6, or 12 over C(x). The algorithm then consists in finding such ω. This is done largely by analysis of the singularities of r. If no such ω exists it can be asserted that equation (1) has no Liouvillian solution whatsoever.

The goal of the algorithm then is to determine the coefficients of the minimal polynomial of ω, $m_\omega(z) = z^n + \sum_{i=0}^{n-1} m_i z^i$,

$m_i \varepsilon C(x)$. Kovacic determines for each possible degree n in turn candidate polynomials m_ω and conditions on them in such a way that if the conditions are met then ω is a solution to the Ricatti equation and if the conditions are not met for any of the candidate m_ω then no ω of degree n is a solution.

Computationally, three features of the algorithm deserve special attention.

1. Candidates and conditions must be generated and the condidates tested against the conditions for each possible degree separately, (though degrees 4, 6, 12 can be treated simultaneously). In the version described in Section 3, we have taken advantage of similarities in the candidate generation process to minimize the amount of code.
2. It is assumed that the poles of r are known (analysis of singularities dominates the process of determining the candidate solutions). This implies a complete splitting of the denominator of r, which can be difficult in itself and requires also that computations be carried out in an algebraic extension field (of the constant field generated by the coefficients of r) over which the denominator splits. But the solution - if found - generally has coefficients in a simpler extension involving only certain square roots. The version of the algorithm described here avoids explicit realization of the individual odd order poles but does require determination of the even order poles. There is reason to think that one could also avoid splitting the even order square free factors of the denominator, but this has not been achieved as yet. It should be remarked that this matter only begins to cause computational concern when the denominator of r has a factor t^k, k even, with t irreducible of degree 2 or greater, a rare occurrence.
3. The number of candidates which must be checked is in the worst case $(n + 1)^{k+1}$ where k is the number of even order poles of r. Thus the algorithm requires exponential time, though in practice this is unlikely to be of much concern in that k is generally small and aspects of the production of candidates dominate the computing time. For example k = 1 seems to be the upper bound for textbook problems (cf. [1], [2]).

3. The Algorithm.

Kovacic describes his algorithm by treating three cases. Case 1 (n=1): search for a rational function ω satisfying the Ricatti equation $\omega' + \omega^2 = r$, case 2 (n=2): search for a solution ω quadratic over C(x), and case 3 (n=4, 6, or 12): search for a higher degree function. In each case one first computes a number d and rational function θ, then a polynomial P of degree d whose coefficients are determined by d linear equations expressed in terms of θ. Each case has 3 steps, (1) determination of parts for d and θ (by analysis of the poles of r), (2) assembly of trial d's and θ's from the parts, (3) determination from each trial d and θ a system of linear equations for the coefficients of P. If any of these systems is solvable the desired function ω may be expressed in terms of the corresponding θ and P. Basically, the coefficient of z^{n-1} in the minimal polynomial $m_\omega(z)$ of ω is $-P'/P - \theta$ and the other coefficients are expressible in terms of P, θ, and the input r via a recursive formula. For example if n=1 we have $\omega = P'/P + \theta$ (then $e^{\int \omega\, dx} = Pe^{\int \theta\, dx}$). If n=2, ω satisfies $\omega^2 - \phi\omega + \phi'/2 + \phi^2/2 - r = 0$, where $\phi = P'/P + \theta$.

The algorithm outlined below unifies step 1 and step 2 across the three cases. Thus only one description (piece of code) for these steps (with n as a parameter) is needed. Step 3 remains in 3 parts, one for each case, and is as described by Kovacic. Hence step 3 is not developed in detail below. Step 3 can be given a

unified treatment also, and it may be best to do this. However, such a unification has not been implemented as yet.

Kovacic's algorithm (modified)

[Given $r \in C(x)$, solve $\omega' + \omega^2 = r$ for ω.]

Step 0: [Preliminaries.]

(a) Let $r = s/t$, with gcd $(s,t) = 1$, $s, t \in C[x]$. Compute the square free factorization of t:

$$t = t_1 t_2^2 t_3^3 \cdots t_m^m.$$

Let $o(\infty) = \deg(t) - \deg(s)$ [order of ∞ as a zero of r].

(b) [Necessary conditions.]

Form a set L of possible degrees over $C(x)$ of a solution ω. [L is a subset of {1, 2, 4, 6, 12}.]

$1 \in L$ only if

$t_i = 1$ for all odd $i \geq 3$ and $o(\infty)$ is even or $o(\infty) > 2$.

$2 \in L$ only if

$t_2 \neq 1$ or $t_i \neq 1$ for some odd $i \geq 3$.

4, 6, $12 \in L$ only if

$t_i = 1$ for all $i > 2$ (i.e., $m \leq 2$) and $o(\infty) \geq 2$.

Step 1: [Form parts for d and θ.]

(a) [Fixed parts.]

$d_{fix} = \frac{1}{4}(\min(o(\infty), 2) - \deg(t) - 3\deg(t_1))$.

$\theta_{fix} = \frac{1}{4}(t'/t + 3t_1'/t_1)$.

(b) [Poles of order 2.]

Find the roots $c_1, \ldots, c_{k_2}$ of t_2.

For i = 1 to k_2 let $d_i = \sqrt{1 + 4b}$, $\theta_i = d_i/(x - c_i)$.

(c) [High order poles.]

If $1 \in L$ then find the roots $c_{k_2+1}, \ldots, c_k$ of $t_4, t_6, \ldots, t_m$.

For $i = k_2 + 1$ to k let $d_i = b/a$, $\theta_i = 2[\sqrt{r}]_{c_i} + d_i/(x\ c_i)$.

[Here a, b, $[\sqrt{r}]_c$ are as described by Kovacic [3, pg. 21, 34, 45]. They are computed from the Laurent series expansion of r at c. However the parts d_i and θ_i are different. This reflects the computation of d_{fix} and θ_{fix} above and the different scheme for assembling parts in step 2. Also, parts d_0 and θ_0 corresponding to the 'zero' at infinity are computed as in (b) or (c) if $o(\infty) = 2$ or $o(\infty) < 2$ respectively (but replace $d_0/(x - c_0)$ by 0).]

Step 2: [Form trial d's, θ's.]

For n in L (taken in increasing order) do:

If n = 1 then m = k else m = k_2.

For all sequences $s = (s_0, s_1, \ldots, s_m)$ where each $s_i \in \{-\frac{1}{2}n, -\frac{1}{2}n+1, \ldots, \frac{1}{2}n\}$, [Start with $s = (-\frac{1}{2}n, \ldots, -\frac{1}{2}n)$, view s as a m-digit number base n + 1 which is incremented by 1 at each pass until $s = (\frac{1}{2}n, \ldots, \frac{1}{2}n)$] do:

Let $d_s = n \cdot d_{fix} - \sum_{i=0}^{m} s_i d_i$.

If d_s is an integer ≥ 0 then let $\theta_s = n \cdot \theta_{fix} + \sum_{i=0}^{m} s_i \theta_i$ and apply step $3_n(d_s, \theta_s)$.

If step 3_n is successful then stop - solution is found.

[It is expressed in terms of θ_s and a polynomial P of degree d_s found in step 3_n by solving a system of linear equations.]

End 'for all sequences s do'.

End 'for n in L do'.

[If this point is reached, all possibilities for a solution have been tried, so...]

Stop - no solution exists.

The d_s and θ_s computed in step 2 are Kovacic's d's and θ's, except that in case 1 he uses 'ω' rather than 'θ' (see [3, pg. 23, 35, 46]). Thus this version may be verified by calculations to show that the same expressions are indeed obtained, and then reference to Kovacic's proof of the algorithm.

It is a tribute to the expressive power of MACSYMA that the code for this algorithm is about the same length as Kovacic's (mathematician's) description and contains little in the way of extra 'programmer's' details. One exception to this is the vector s used to keep track of the trial d's and θ's.

We conclude with some sample computations using the above algorithm.

Example 1.

Differential Equation: $y'' = (x^2 + 3)y$

Step 0: $r = x^2 + 3$, $L = \{1\}$,

$t = 1$ [no poles except at ∞], $o(\infty) = -2$.

Step 1:

$d_{fix} = -\frac{1}{2}$, $\theta_{fix} = 0$,

$d_0 = -3$, $\theta_0 = 2x$ [for the pole at ∞].

Step 2:

$s = (-\frac{1}{2})$, $d_s = -2$ [reject],

$s = (\frac{1}{2})$, $d_s = 1$, $\theta_s = x$.

Step 3: Successful with $P = x$.

$\omega = P'/P + \theta = 1/x + x$ solves the Ricatti equation.

$y = e^{\int\omega} = xe^{\frac{1}{2}x^2}$ solves the D.E.

Example 2.

Differential Equation : $y'' = (\frac{1}{x} - \frac{3}{16x^2})y$

Step 0: $r = (x - 3/16)/x^2$,

$t = x^2$, $t_2 = x$ [one pole at 0 of order 2], $o(\infty) = 1$, $L = \{2\}$.

Step 1:

$d_{fix} = -\frac{1}{4}$, $\theta_{fix} = 1/2x$,

$d_1 = \frac{1}{2}$, $\theta_1 = 1/2x$ [for the pole $c_1=0$].

Step 2:

$s = (-1)$, $d_s = 0$, $\theta_s = 1/2x$.

[the values $s = (0)$ and $s = (1)$, had they been tried first, would have been rejected since in those cases d_s is not a non-negative integer.]

Step 3: Successful with $P = 1$.

$P'/P + \theta = 1/2x$.

ω satisfies $\omega^2 - (1/2x)\omega + (1 - 16x)/16x^2 = 0$.

$\omega = 1/4x + x^{-\frac{1}{2}}$ solves the Ricatti equation.

$y - e^{\int\omega\, dx} = x^{\frac{1}{4}} e^{2x^{\frac{1}{2}}}$ solves the D.E.

Remark: No d_0, θ_0 are computed when $1 \notin L$.

Example 3.

Differential Equation: $x^2 y'' = 2y$

Step 0: $r = 2/x^2$

$t = x^2$, $t_2 = x$ [one pole at 0 of order 2]

$o(\infty) = 2$, $L = \{1, 2, 4, 6, 12\}$.

Step 1:

$d_{fix} = 0$, $\theta_{fix} = 1/2x$,

$d_0 = 3$, $\theta_0 = 0$ [for the pole at ∞],

$d_1 = 3$, $\theta_1 = 3/x$ [pole at $c_1 = 0$].

Step 2:

$s = (-\frac{1}{2}, -\frac{1}{2})$, $d_s = 3$, $\theta_s = -1/x$.

Step 3: Successful with $P = x^3$.

$\omega = P'/P + \theta = 2/x$ solves the Ricatti equation.

$y = e^{\int\omega} = x^2$ solves the D.E.

References

[1] Boyce, William E. and DiPrima, Richard C., Elementary Differential Equations and Boundary Value Problems, New York, J. Wiley, (1965).

[2] Kamke, E., Differentialgleichungen Losungsmethoden und Losungen, New York, Chelsea Publishing Co., (1948).

[3] Kovacic, Jerald J., "An Algorithm for Solving Second Order Linear Homogeneous Differential Equations", to appear.

Implementing a Polynomial Factorization and GCD Package

P. M. A. Moore and A. C. Norman
University of Cambridge Computer Laboratory
Cambridge, England

Abstract

This paper describes the construction of a rational function package where the GCD and factorization routines are well integrated and consistent with each other and both use state of the art algorithms. The work represents an exercise in producing a service rather than an experimental piece of code, where portability, reliability and clear readable code are important aims in addition to the obvious desire for speed. Measurements on the initial version of our package showed that even though it was based on the best of previously published methods its performance was uneven. The causes of the more notable bottle necks and the steps we took to avoid them are explained here and illustrate how apparently very fine details of coding can sometimes have gross effects on a system's overall behaviour.

Introduction

The development of GCD and factorization algorithms represents one of the biggest success stories for Computer Algebra. As long as ten years ago efficient algorithms for these two tasks had been developed and implemented, and it looked as if rational function manipulation was well under control. Coded versions of the more advanced algorithms did not, however, become widely available for some time - much important work remained buried within MACSYMA, which was not at the time distributed, and within SCRATCHPAD, which still is not. Much was also done in SAC-I but the FORTRAN coding discouraged LISP enthusiasts, such as ourselves, from using it. Thus when some four or five years ago, we had need of a factorizer that could be called from REDUCE, we could not find one we felt free to use and which could be easily modified to suit our purposes. Thus we started to write our own, initially only intending to cope with easy cases and using simple algorithms. Our first package did such factorization as can be performed by decomposing polynomials into their contents and primitive parts and then their square-free factors. It split quadratics, and some cubics and quartics by using closed form expressions for their roots. Then, independent of our own interest in factorization, we came under sustained pressure from a user [1] who wanted a specific set of polynomials split over the finite field GF(3). Since generating the polynomials was a job for an algebra system it seemed easiest to satisfy demand by writing, for REDUCE, a version of Berlekamp's mod-p factorization algorithm. The existance of that package led us to experiment with a univariate version of the Hensel construction, and so to the skeleton full factorization package reported in [2]. Between then and now much of our time has been taken up with consideration of how important order is in

©1981 ACM O-89791-047-8/81-0800-0109 $00.75

polynomial data structures, and as a step towards producing an experimental rational function package for the system described in [6], we have developed our factorizer into a complete, reliable, efficient and robust package integrated with a similarly polished set of GCD routines. In the process of completing this work we have had a number of surprises, and this paper will concentrate on them. Our concern is for practical issues and our main purpose is to point out that even when elaborate and theoretically good algorithms are used, small adjustments to the code can still have an order-of-magnitude effect on the speed of an algebra system.

The Multivariate Factorizer

As soon as we had a partially working multivariate factorizer, we started to analyse its cost in terms of speed. We used as test cases the fifteen problems for which Wang quoted timings in [4] and which were designed to exercise the algorithm considerably. Having established that the IBM370/165 in use at Cambridge, ran LISP at a speed roughly equivalent to that of the DEC KL-10 that Wang used, we set ourselves the target of matching his reported results. The process of factorizing polynomials in several variables requires substituting numeric values for all bar one of the variables involved, factoring the resulting univariate polynomials and then reconstructing the full factors from what is hoped are their univariate images. When our first working factorizer turned out to be much slower than Wang's we were not surprised that measurements pinned the blame on the multivariate Hensel reconstruction code.

There are a number of refinements that can be made to a simple Hensel lifting program. The ones we now incorporate are indicated by the buzz-phrases:

one variable at a time,
parallel lifting of n factors,
completed factors removed as they are noticed,
and prediction and exploitation of sparseness.

The first three of these are adequately described in [4] while the last represents our equivalent of Zippel's [5] proposals and an application of them to p-adic rather than straight-forward modular processes. In addition the Hensel construction admits many coding optimizations. For instance, if at stage j the quantities $f_i(j)$ are the best known approximations to the factors of F, then the next step will need the residue $r(j) = F - \prod_i f_i(j)$. The simplest code would compute each r(j) from the above formula, whereas efficiency can be gained by deriving r(j) from r(j-1) and knowledge of the way the factors f_i have been growing.

We found that, although the Hensel construction is basically neat and simple in theory, the fully optimised version we finally used was as nasty a piece of code to write and debug as any we have come across.

The testing of this code was helped enormously and unexpectedly by a seemingly minor decision we made early on. When selecting integer values to substitute for variables and when choosing primes to use in modular methods, there is often no very good a priori way of selecting numbers. As is commonly practised in cases of this sort, we employed a random number generator for assistance. What seems to be less usual is our setting of the seed to a thoroughly unpredictable value based on the date and time of day. The result of this has been that each of the fifteen test examples we have used has acted as several dozen different problems because the univariate image polynomial that arises is generally different from run to run. So in trying these tests we found a broader spectrum of cases than we at first anticipated, ranging from well-behaved through degenerate to rather harder than expected. Thus running just one test file through our package has exercised much more code than might be imagined. Given that the whole factorizer is riddled with special exit conditions - for instance, an unlucky prime or the fortuitous early discovery of factors - this detailed probing has been invaluable. We are uncomfortably certain that if we had used a fully deterministic "random" number generator, we would have declared our code bug-free many months ago. Indeed when we eventually did so and encouraged unbiased users to try foxing the system with their own examples, the feed-back suggested that after revealing a couple of simple bugs, they found it quite difficult to do so.

In the course of debugging the factorizer we developed a philosophy for following the passage of input data as it flows through a system. The idea is that a package implementing a well-defined algorithm should be able to generate a commentary on its progress, expressed in high-level terms rather after the style of a worked example intended for student use. If this trace is to be easily used, it must be organised into several very short sections of material that the user can select at will. These sections could be say, up to about a page in length (dense printing to attempt to fit in more information not being permitted). Then, to allow for the revelation of progressively finer details, we introduce the notion that between each pair of lines of the top-level trace document there will be further one-page elaborations, and in turn each step noted in one of these can be investigated by looking at lower-level trace files. Thus the complete trace of our program factoring some polynomial, is rather a bushy tree consisting of one-page explanations of steps in the algorithm.

With the improved version of our multivariate Hensel code written and debugged, we did further measurements on our code. Over the set of Wang's test problems we found that we were spending most time in the factorization of univariate image polynomials, and that often the selection of valid numbers to substitute into the input polynomial cost almost as much as the final reconstruction step. These measurements suggested that we need not, at least for a while, seek further refinements to the multivariate reconstruction process.

The Univariate Factorizer

Investigation of our univariate factorizer rapidly showed that there were two quite different sorts of time-consuming behaviour to study. The first involved the treatment of false modular factors, while the second was just the speed of both Berlekamp's algorithm and the univariate Hensel construction. Since major way of reducing unluckiness is to sample more primes (hoping to find a good one), we first considered the speed of our mod-p factorizer. We were using the original formulation of Berlekamp's algorithm and conventional wisedom suggested strongly that its second phase would be costly and that it could usefully be replaced by some probabalistic algorithm. It was therefore a great surprise when we found that in practice the first part of the mod-p factoriser was limiting us. At first we thought there must be a fault in our timing methods, so we separated the Berlekamp code from the rest of our factorizer in order to study it in detail. We found that for this rather straight-forward code we could gain in speed by keeping modular numbers in the range 0 to p-1 rather than -p/2 to +p/2, but that for the smallish primes we normally use, our initial timing results were actually correct. The probabilistic methods would indeed be of great value if large primes were employed but for factorization the chances of a prime being lucky do not depend strongly on its size. Our conclusion was that the use of small primes (for which setting up Berlekamp's Q matrix is fastest) are best. We also note that we found it useful to modify our LISP compiler to generate open code for modular arithmetic (with a one-word modulus).

The remainder of our problems with the univariate factorizer centred around the difficulties with false splits. There is a very uncomfortable trade-off that has to be made between the number of modular factorizations that are performed, the way in which degree information from them gets merged, and the chances of discovering in the Hensel growth stage that the selected modular factorization did not correspond to one over the integers. For irreducible polynomials we know that it will be appropriate to take about five modular images [3], but even this number, which can correspond to a reasonable cost, does not guarentee the early detection of irreducibility. Currently we create several modular images (nine in fact) and pass them through the first half of Berlekamp's Algorithm which gives the count of modular factors in each case. Of these we pick the best few to pass through the second phase of Berlekamp which gives the modular factors themselves and then their degrees can be handed to the degree analysis code. If everything behaves well, the result will be the correct number of factors to expect after the p-adic growth, together with the associated modular images. Difficulties arise when we know the modular split to be false but cannot detect before the Hensel growth, which factors to combine to form the modular image of a full univariate factor. In this case a few more modular images can be generated and factorized in the hope of improving the situation before doing the p-adic growth.

In order to increase the chances of a straight-forward Hensel construction, particularly in the more expensive multivariate part, there are a number of restrictions on the choice of values for variable substitution and primes, in creating the modular images (see [4]). We were somewhat surprised to discover that these were severe enough to dominate the cost of modular image creation and factorization. Moreover, as the restrictions continue to fail sets of numbers, these latter are increased in size until eventually a satisfactory image is obtained. However, the substitution of even moderately sized numbers into a multivariate polynomial to give a univariate (non-modular) image tends to produce very large coefficients in the polynomial fed to the univariate factorizer. When a false split occurs in the modular factoring of a polynomial, the Hensel construction will have to grow all the way up to a coefficient bound before trying to combine factors for a true split. Unfortunately this coefficient bound is dependent on the size of the coefficients in the univariate image polynomial and therefore in many cases a false split can be rather expensive. Sadly, false splits at the modular level are far more common than we had hoped. The detection and removal of a completed factor part way through the growth process can help considerably because not only do the calculations become shorter - for example the residue is smaller - but the coefficient bound can be reduced since we are effectively factorizing a smaller polynomial. This idea of a 'dynamic' coefficient bound can be taken further if a separate bound for each coefficient in each factor is considered and adjusted as factors are discovered and divided out.

With regard to the choice of numbers for variables substitution, the random element is still present in our code but with a heavy bias towards the selection of small integers. In fact, since the values zero and one prevent blow up, an initial try is made of all possible substitutions of these numbers that yield valid univariate images. Having found a univariate image, whose coefficients hopefully are not too large, we create several modular images from it since the split generally varies from prime to prime. When it comes to a false split at the univariate level, such that there are too many factors to grow to multivariate fullness, we were relieved to find this was very much a rarity. Therefore, in such an event, we restart the whole process by creating a new univariate image, factorizing it and trying again. In cases where this has happened during test runs, one repetition of this sort has always sufficed.

One other major decision which affects the cost of factorization is the choice of the variable to keep in the univariate image. Again a conflict arises. If the lowest degree variable is chosen to minimise the chances of a false split in the univariate image, then the other variables have a good potential for causing blow up in this image so that if a false split does occur the likelihood is that it will be very expensive. In practice we experimented with various choices and in the whole found that the lowest degree variable was reasonable. In the GCD code this decision is far more critical. In some examples the cost can vary by as much as a factor of ten according to which variable is kept. As a consequence we incorporated a fairly thorough analysis of input polynomials before making a final choice.

General Remarks

One of the advantages of having a random element in our code - apart from the thorough testing mentioned previously - was to show the character of a factorization/GCD problem. A single input polynomial run through the factorizer many times often produces quite a spread of timings according to the luck of images picked. The results for two of the Wang examples are shown in the appendix. One can see the clusters of runs corresponding to false splits - the main bunch showing bad modular splitting, with the small and most expensive peaks representing overshoot in the multivariate Hensel construction. In example 8, the input polynomial is of high degree so that the chances of a false split are very high and the runs shown have in fact all produced bad splits. This variation of cost means that, strictly speaking, one should produce an average time over a number of runs for each example, which hopefully falls within a tolerable area. The table of times given shows the difference in speed from the first moment the factorizer was working to the present state. These are in fact taken from one run including all the Wang examples so they represent a 'snap-shot' of the factorizer's behaviour. Finally an example of the algorithm tracer is presented including a couple of levels of trace only.

References

[1] Damerell, M. Private Communication

[2] Norman, A.C., "Towards a REDUCE solution to SIGSAM problem 7", SIGSAM Bulletin, Nov 78, p.14

[3] Musser, D.R., "Algorithms for Polynomial Factorization", Ph.D. Thesis, Computer Science Department, University of Wisconsin, Madison 1971.

[4] Wang, P.S., "An Improved Multivariable Polynomial Factorising Algorithm", Math. Comp. 32(1978) pp. 1215-1231.

[5] Zippel, R.E., "Probabilistic Algorithms for Sparse Polynomials", Proc. EUROSAM 79 [Springer Lecture Notes in Computer Science 72, Springer-Verlag, Berlin-Heidelberg-New York, 1979], pp. 216-226.

[6] Norman, A.C. and Moore, P.M.A.,"The Initial Design of a Vector Based Algebra System", Proc. EUROSAM 79 [Springer Lecture Notes in Computer Science 72, Springer-Verlag, Berlin-Heidelberg-New York, 1979], pp. 258-265.

APPENDIX

TABLE OF TIMES

Problem Number	July 1979 Times (secs)	Our Current Times (secs)	Ratio
1	0.68	0.512	1.33
2	1.27	0.513	2.48
3	3.00	0.413	7.26
4	12.7	1.904	6.67
5	44.6	2.819	15.82
6	38.35	3.088	12.42
7	0.86	0.948	.91
8	> 5 mins	55.241	> 5.43
9	*	6.316	?
10	86.17	7.631	11.29
11	> 120 secs	9.678	>12.40
12	0.27	0.249	1.08
13	2.5	0.836	2.99
14	15.4	3.411	4.51
15	4.2	2.27	1.85

Notes

1) * - problem 9 persistently uncovered a number of bugs and at this stage was still producing some.

2) Number 7 shows an apparent decline in performance but this is so small that it can be taken as an effect of the 'snap-shot' and randomness.

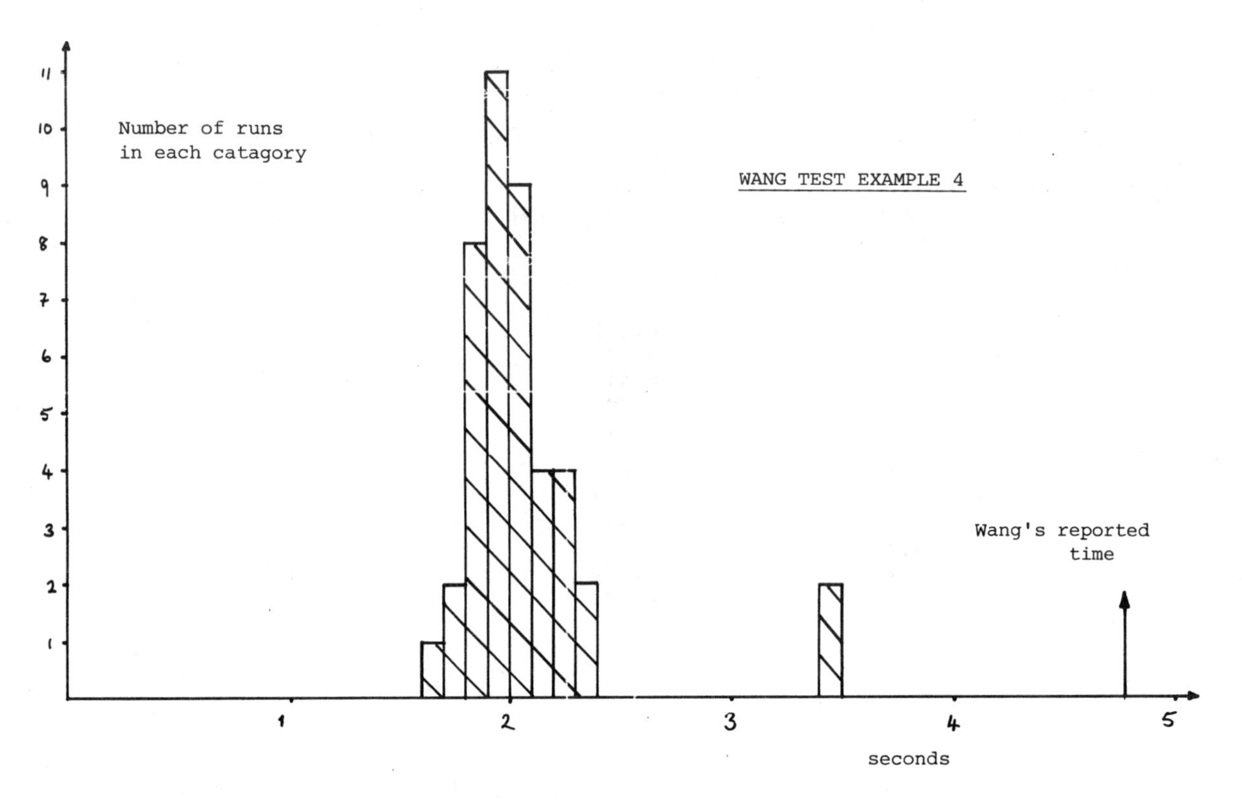
Number of runs
in each catagory
WANG TEST EXAMPLE 4
Wang's reported
time
seconds
1
2
3
4
5
6
7
8
9
10
11

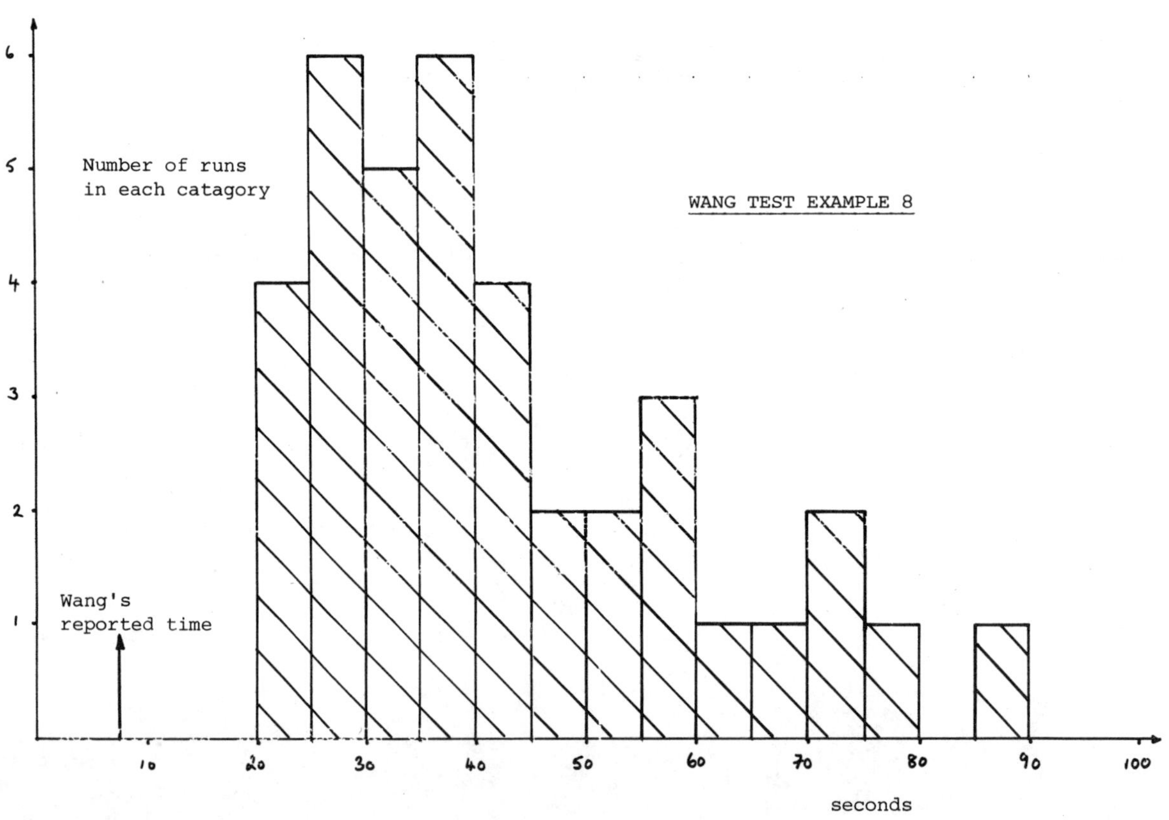
Number of runs
in each catagory
WANG TEST EXAMPLE 8
Wang's
reported time
seconds
10
20
30
40
50
60
70
80
90
100
1
2
3
4
5
6

An Example of the Algorithm Tracer

Top Level

```
                4  8       2  7          8     5  3      7           4           6      4
FACTOR : Y *Z   + X *Z  + ((Y   - Y )*X  + Y *X + 3*Y  + 5)*Z  + ((Y

     5      3  3          2    5          9  6       11       8   4          8
 - Y)*X  + Y *X  + 3*X )*Z  + ( - Y *X  + (Y   - Y )*X  + (3*Y  + 5

  4          3          7         3            4            5  8       7     4     6
*Y  - 5*Y)*X  + (3*Y  + 5*Y )*X + 15)*Z  + ( - Y *X  + (Y  - Y )*X

      4  5         3  3    3           12   7        5  6          11        7
 + 3*Y *X  + 3*Y *X )*Z  + ( - Y   *X  - 5*Y *X  + (3*Y   + 5*Y  - 5

   4    4         4  3          3        2            8  9        7  6          8  7
*Y )*X  + 15*Y *X  + 15*Y *X)*Z  + ( - Y *X  + 3*Y *X )*Z - 5*Y *X

       7  4
 + 15*Y *X
Chosen main variable is Z
Numeric content = 1
Polynomial is primitive wrt Z                    ──► Level 2.1
The polynomial is square-free.
Final result is:

  2  4       2
(Z *Y  + Z*X  + 5)
  2       3
(Z  + X*Y )
  2      3
(Z  - X *Y + 3)
  2     3  4
(Z  + X *Y )
```

Level 2.2

```
We now determine the leading coefficients of the
factors of U by using the factors of the leading
coefficient of U and their (square-free) images
referred to earlier:
 2
Z  + 5 with l.c.: 1
 2
Z  - 8 with l.c.: 1
 2
Z  + 16 with l.c.: 1
    2                       4
16*Z  + Z + 5 with l.c.: Y
By exploiting any sparsity wrt the main variable in the
factors, we can try guessing some of the multivariate
coefficients.
We have completely determined the following factor:
 4  2      2
Y *Z  + X *Z + 5
```

```
From now on we shall refer to this polynomial as U.
We now create an image of U by picking suitable values
for all but one of the variables in U.
The variable preserved in the image is Z
The leading coefficient of U is non-trivial so we must
factor it before we can decide how it is distributed
over the leading coefficients of the factors of U.
So the factors of this leading coefficient are:

 4
Y
The chosen image set is:  X=1; Y=-2;
and chosen prime is 19
Image polynomial (made primitive) =
    8     7        6        5          4         3           2
16*Z  + Z  + 213*Z  + 13*Z  - 1343*Z  - 88*Z  - 10680*Z  - 640*Z -

3200
The image polynomial mod 19, made monic, is:
 8      7      6      5       4      3      2
Z  + 6*Z  + 5*Z  + 2*Z  + 17*Z  + 4*Z  + 7*Z  + 17*Z + 9
and factors of the primitive image mod this prime are:
 2
Z  + 5
 2
Z  + 11
 2
Z  + 16
Z + 9
Z + 16
Next we use the Hensel Construction to grow these modular
factors into factors over the integers.
The full factors of the image polynomial are:
 2
Z  + 5
 2
Z  - 8
 2
Z  + 16
    2
16*Z  + Z + 5
```

Note on Probabilistic Algorithms in Integer and Polynomial Arithmetic

Michael Kaminski

Institute of Mathematics
The Hebrew University of Jerusalem
Jerusalem
Israel

Introduction

For many computational problems it is not known whether verification of a result can be done faster than its computation. For instance, it is unknown whehter the verification of the validity of the integer equality x*y=z needs fewer bit operations than a computation of the product x*y. It is sometimes much easier, however, to speed up the computation probabilistically if just the verification of the result is involved.

In this paper we present linear probabilistic algorithms for verification of the validity of the integer equalty

$$f_1(x_1,\ldots,x_N)=f_2(x_1,\ldots,x_N)$$

for rational functions f_1 and f_2, which can be of the form of a rational combination of rational functions. (Sometimes it is more convenient to write the equality in the form $f(x_1,\ldots,x_N)=0$, where f is equal to f_1-f_2.) An algorithm for verification of multivariate polynomial identities over any given integral domain (in [6] integral domains are infinite) is obtained from the algorithm for the integers by a simple modification.

Algorithms

We begin with an observation, which is a natural generalization of the ideas of [6] and [7]:

Let R be a Euclidean ring with the norm $\|\ \|$, $0\neq x\in R$, $A=\{a_1,\ldots,a_m\}$, and for some fixed $k<m$ and for any k elements of A: $b_1,\ldots,b_k$, $\|L.C.M.(b_1,\ldots,b_k)\|>\|x\|$. (We denote by $L.C.M.(b_1,\ldots,b_k)$ the least common multiple of $b_1,\ldots,b_k$.) Then for an element a randomly chosen from A the probability that $x\not\equiv 0 \pmod a$ is greater than $\frac{m-k}{m}=1-\frac{k}{m}$.

The proof is evident: there are no more than k-1 elements of A which exactly divide x. If there were k such elements $b_1,\ldots,b_k$, then $x\equiv 0 \pmod{L.C.M.(b_1,\ldots,b_k)}$, therefore x=0. This contradicts the statement above.

We shall use the following norms: $\|z\|=\lg_n|z|$ for integers in n-ary notation. (We suppose that integers are in binary notation. In the general case all proofs are similar.) If P is a polynomial of degree n, then $\|P\|=n$. We also call x the "length of x".

If f is a rational function of N variables, and $x_1,\ldots,x_N$ are fixed values of the arguments, then the "complexity of $f(x_1,\ldots,x_N)$" is defined as the minimum (over all representations of f) of the maximum length of the intermediate results in the calculation of $f(x_1,\ldots,x_N)$.

Now, to verify probabilistically the validity of the integer equality $f(x_1,\ldots,x_N)=0$, where the complexity of $f(x_1,\ldots,x_N)$ is equal to n, it suffices to find a set of integers $A=\{a_1,\ldots,a_m\}$ which will sutisfy the following conditions:

1) There exists a number $k=k(n)$ such that for any subset $\{b_1,\ldots,b_k\}$ of A, $\|L.C.M.(b_1,\ldots,b_k)\|>n$, and $1-\frac{k}{m}$ is "sufficiently large".

2) For any $a\in A$, every $x_i \pmod a$ - the remainder of x_i upon division by a, can be computed in O(n) bit operations.

3) For any $a\in A$, $f(x_1 \pmod a),\ldots,x_N \pmod a)) \pmod a$ con be computed in O(n) bit operations.

4) It is possible to choose an element a randomly from A in O(n) bit bit operations.

Before describing such a set A we want to note that sometimes a computation of the remainder can be performed faster than the division itself.

For example, let $\|a\|=n$ and $\|b\|=\lg n-\lg\lg\lg n-\lg\lg\lg\lg n$. The best known algorithm for dividing a by b uses $O(n*\lg\lg n*\lg\lg\lg n)$ bit operations. A linear algorithm for the computation of the remainder of the division of a by b can be as follows:

a) Compute $\varphi(b)$ (the Euler function of b). This can be done in O(n) bit operations, because $b<n$. Note that $\varphi(b)<b<\frac{n}{\lg\lg n*\lg\lg\lg n}$.

b) Compute $c=a(\bmod 2^{\varphi(b)}-1)$. This can also be done in O(n) bit operations, because of the form of the number $2^{\varphi(b)}-1$ (see [1]). Note that $\|c\|<2^{\varphi(b)}-1<\varphi(b)<\frac{n}{\lg\lg n*\lg\lg\lg n}$.

c) Compute $c \pmod b$ which is equal to $a \pmod b$ because $b|2^{\varphi(b)}-1$. Using the fast multiplication algorithm of [1] this can be done in

$$O(\frac{n}{\lg\lg n*\lg\lg\lg n}*\lg\lg n*\lg\lg\lg n)=O(n)$$

bit operations.

Using the idea of the example above we define A as follows: $A=\{2^p+1 \mid p$ is prime and $p<C\sqrt{n}\ln n\}$ for $4<C<\frac{\sqrt{n}}{\lg^2 n*\lg\lg n}$. We need the upper bound on C in order to perform the multiplication, division and G.C.D. algorithms in O(n) bit operations.

It is easy to prove that A satisfies the conditions above:

1) We shall prove first that if p_1 p_2 are different primes, then

$$D.C.D.(2^{p_1}+1,2^{p_2}+1)\leq 3.$$

Let p divide $G.C.D.(2^{p_1}+1,2^{p_2}+1)$, then $2^{2p_1}=2^{2p_2}\equiv 1 \pmod p$, therefore $2^{2G.C.D.(p_1,p_2)}=2^2\equiv 1 \pmod p$. Thus $p\leq 3$.

Further if $p_1,\dots,p_k$ are different primes, then

$$L.C.M.(2^{p_1}+1,\dots,2^{p_k}+1) \geq \frac{2^{\sum_{i=1}^{k} p_i}}{3^k} \geq 2^{k^2/4},$$

because p_i's are positive integers. Thus $\|L.C.M.(2^{p_1}+1,\dots,2^{p_k}+1)\| \geq \frac{k^2}{4}$. To satisfy the inequality of 1) we need $k > 2\sqrt{n}$, and $2\sqrt{n}$ primes are sufficient. This is why we use primes which are less than $4\sqrt{n}\ln n$. Now we are able to estimate $1-\frac{k}{m}$ as follows: $k=2\sqrt{n}$, $m \approx \frac{C\, n\ln n}{\ln(C\, n\ln n)} > C\sqrt{n}$, $1-\frac{k}{m} > 1-\frac{2}{C}$.

2) A computation of $x_i \pmod{2^p+1}$ can be performed in O(n) bit operations, see [1].

3) A computation of $f(x_1 \pmod{2^p+1}, \dots, x_N \pmod{2^p+1}) \pmod{2^p+1}$ can be performed in O(n) bit operations, because the length of the arguments does not exceed $\frac{n}{\lg^2 n * \lg\lg n}$.

4) A random prime of length lgn can be generated in $O(\lg^{\varepsilon} n)$ bit operations (see [4]); or, alternatively, we can choose a number i randomly between 1 and $C\sqrt{n}$ and calculate the i-th prime in $O(i^{1.125+\varepsilon})$ bit operations (see [2]).

<u>Example</u>. Verification of the validity of the integer equality $x_1/x_2+x_3/x_4=x_5*x_6$, for $\|x_i\| < m$. The maximum length of the intermediate results does not exceed n=4m.

An algorithm could be as follows:

Choose a prime p randomly between 1 and $4\sqrt{n}\lg n$ and compute $x_i' = x_i \pmod{2^p+1}$. Then compute (using the fast G.C.D. algorithm of [1]) $y_1 \equiv x_1/x_2 \pmod{2^p+1}$, $y_2 \equiv x_3/x_4 \pmod{2^p+1}$ and $y_3 = x_5 * x_6 \pmod{2^p+1}$. y_1 and y_2 exist for almost all p, except 2lgm primes, because x_2 and x_4 have no more than lgm different prime factors. The only difficulty might be when both 2^p+1 and one of the arguments x_2, x_4 have the factor 3. In this case, perform all the computations modulo $(2^p+1)/3$: to compute $x \pmod{(2^p+1)/3}$ compute $y = x \pmod{2^p+1}$ and then $y \pmod{(2^p+1)/3}$ which is equal to $x \pmod{(2^p+1)/3}$. The last computation can be performed in O(n) bit operations, because $y < 2^p+1$. Test whether $y_1+y_2 \equiv y_3 \pmod{2^p+1}$. If the answer is "no", the equality is, of course, invalid. In the othe case deside that the equality is valid. The error probability is less than 0.5.

In the case of an univariate polynomial ring we define A as $\{x^p+1 \mid p$ is prime and $1 < p < C\sqrt{n}\ln n\}$ for a suitable C. The proof can be easily carried out as in the case of integers. Schwartz in [6] uses A= $\{x-a\}$, where a ranges over a certain subset of the field, and we test the identity in some algebraic extension of the field. This is the reason why in spite of the finiteness of the field we can perform safficiently many tests of the identity.

Note that in the case of polynomials over integers, with the maximum length of coefficients m, the length of the intermediate results is O(m+lgn) in our algorithm and O(m+nlgn) in the algorithm of Schwartz ([6]). In both cases we apply the linear probabilistic algorithm for the verification of the validity of integer equalities in order to test the given polynomial identity probabilistically, and the number of bit operations involved is proportional to the sum of the lendth of the polynomial coefficients. This is true also for multivariate polynomials, see below.

<u>Example</u> of an application of the probabilistic algorithm for the verification of univariate polynomial identities:

For the two given polynomials over Z_2, $f(x)$ of degree n and $g(x)$ of degree exceeding $n-\frac{n}{\lg^2 n*\lg\lg n}$, we want to check if g exactly divides f, and if the answer is "yes" to find the quotient h. (Note that the degree of h does not exceed $\frac{n}{\lg^2 n*\lg\lg n}$.)

We can assume that x does not divide g, and then we can compute

$$h=\frac{f}{g}(\text{mod } x^{\frac{n}{\lg^2 n*\lg\lg n}})$$

in O(n) algebraic operations, using the fast G.C.D. algorithm [1] and the fast polynomial multiplication algorithm [5]. Evidently, if g exactly divides f, then the quotient must be h. Then we verify, by the probabilistic algorithm, if $h*g=f$:

As above, for a random prime p, we compute f, g and h modulo x^p+1 and check if $(g(\text{mod } x^p+1))*(h(\text{mod } x^p+1))\equiv(f(\text{mod } x^p+1))(\text{mod } x^p+1)$.

Finally we give a similar probabilistic test of the identity $P(x_1,\ldots,x_N)=0$ for an N-variate polinomial P of degree n (i.e. the degree of P in each variable does not exceed n). For this we need the following result from [6]:

Let A consist of m elements and $n<m$, then if P is not identically to zero, the number of N-tuples of elements of A which are zero of P is at most $\frac{m^{N+1}}{n}$. This means that for an $(a_1,\ldots,a_N)$ randomly chosen from A^N, the probability that $P(a_1,\ldots,a_N)=0$ is less than $\frac{n}{m}$.

We test the identity $P(x_1,\ldots,x_N)=0$ modulo $x_1^{p_1}+1,\ldots,x_N^{p_N}+1$, for random primes p_i between 1 and $C\sqrt{n}\ln n$ ($4<C<\frac{\sqrt{n}}{\lg n}$ for integral domains containing sufficiently many roots of unity, and $4<C<\frac{n}{\lg n*\lg\lg n}$ for others), in other words $A=\{w|w \text{ is a } p\text{-th root of unity}, p<C\sqrt{n}\ln n\}$. The cardinality of A is greater than $\frac{Cn}{2}$ (see above).

<u>Example</u>. Probabilistic test of the identity $P*Q=R$ for N-variate polynomials P,Q and R of the degree not exceeding n (in each variable separately) over some field F (the maximum degree of intermediate results does not exceed 2n):

According to our algorithm we choose primes $p_1,\ldots,p_N$ randomly between 1 and $4\sqrt{n}\ln n$ and compute P,Q and R successively modulo $x^{p_1}+1,\ldots,x_N^{p_N}+1$, i.e. we make the substitution $x^{p_i}=-1$ for every variable x_i. The number of algebraic operations involved is proportional to the number of terms of the polynomials.

Using the fast multiplication algorithm for $F[x_1,\ldots,x_N]$, we can multiply the "remainders" of P and Q in fewer than n^N algebraic operations. Then we compute

$$P(\bmod\ x_1^{p_1}+1,\dots,x_N^{p_N}+1)*Q(\bmod\ x_1^{p_1}+1,\dots,x_N^{p_N}+1)-R(\bmod\ x_1^{p_1}+1,\dots,x_N^{p_N}+1)(\bmod\ x_1^{p_1}+1,\dots,x_N^{p_N}+1).$$

If the last result is not equal to zero, then P*Q=R; otherwise we deside that P*Q=R. The error probability is less than 0.5.

REFERENCES

1. Aho A.V., Hopcroft J.E., Ullman J.D. "The Desing and Analysis of Computer Algorithms", Addison Wesley, 1974

2. Pollard J. "Theorems on Factorisation and Primality Testing", Proc. Cambrige Phylos. Soc. 76(1974), 521-528

3. Rabin M.O. "Probabilistic Algorithms", Algorithms and Complexity: New Directions and Recent Results", Traub J.F., ed., Academic Press, New-York, 1976, 21-39

4. Rabin M.O. "Probabilistic Algorithms for Testing Primality", J. Number Theory 12 (1980) 128-138

5. Schonhage A., "Schnelle Multiplikation von Polynomen uber Korpen der Charakteristic 2", Acta Informatica, 7 (1977) 395-398

6. Schwartz J.T., "Probabilistic Algorithms for Verification of Polynomial Identities" Symbolic and Algebraic Computation, Ng E.W., ed., Springer Verlag, Heidelberg (1979), 200-215

7. Yao A.C., "A Lower Bound to Palindrome Recognition by Probabilistic Turing Mashines", STAN-CS 77-667

8. Zippel R.E., "Probabilistic Algorithm for Sparse Polynomials", Ph.D. thesis, Massachusetts Institute of Technology (1979)

A Case Study in Interlanguage Communication: Fast LISP Polynomial Operations Written in 'C'

Richard J. Fateman[1]

Computer Science Division
and
Center for Pure and Applied Mathematics
University of California
Berkeley CA. 94720

ABSTRACT

It is shown that a simple program, written in the "C" programming language, can be interfaced to a Lisp algebraic manipulation system with substantial performance improvement as a result.

1. Introduction

Numerous papers have been written comparing various algebraic manipulation systems. Almost without exception, these papers praise the Lisp-based systems for their flexibility, and condemn them for their slowness. Even systems which have been written with special purpose "efficient" representations in Lisp, tend to be slower than other systems written in assembler, or even Fortran, *when flexibility is not essential.*

It is our opinion that Lisp is one of the best languages for problem solving when data structures and algorithms are either poorly understood, or are highly modified during computation. It also has an advantage in being the historical language of choice for sophisticated algorithms (factorization, integration) in algebraic manipulation systems.

On the other hand, certain algorithms have been written in Lisp primarily because there was no alternative (other than assembler) for implementation. In such a case, one Lisp-based algebraic manipulation system (Macsyma [6] on the DEC-10 computer) attempts to achieve efficiency through the used of a complex declaration-oriented compiler. This can produce good quality code, even in Lisp's traditionally weakest area, arithmetic expression evaluation [10].

In this paper we explore another avenue: we show it is possible to partition algebraic manipulation systems into "lisp-like" parts and "non-lisp" parts. It is especially attractive to take this approach on large virtual-address space systems supporting a variety of languages in which there are convenient interlanguage call sequences. This work was done on one such system, the VAX 11/780 computer, using the VM/UNIX operating system.

It is our opinion that such a partitioning makes it unnecessary to have a complex Lisp compiler devoted to efficient arithmetic expression evaluation and array handling, since other languages can be used in those situations where Lisp is poorly equipped for the task.

The "C" programming language is used in the UNIX operating system for the implementation of system programs including the Lisp run-time interpreter and environment. Thus it combines access to pointers and numerical routines, and can be used in cases where Lisp seemed to be the only higher-level language possible in the past.

2. The Experiment

This study examined array-based polynomial representations used from Lisp for multiplication. We do not pretend to model the complete situation encountered in Macsyma[2], which among other tasks, performs polynomial arithmetic. Instead we restrict our study to univariate polynomials with coefficients in a finite field.

Two sets of routines were written in C. One set was constructed using the traditional high-school polynomial multiplication algorithm, which for two polynomials of degree m and n takes $(m+1)(n+1)$ coefficient multiplications.

The inner-most loop of this program is a scalar cross-product, and could be performed in various vector-oriented processors by pipe-lined instructions or parallel execution.

The second was based on the use of the Fast Fourier Transform (FFT). A so-called discrete FFT rather than a complex floating-point FFT is used, as is traditional by now in algebraic manipulation.

In order to compare different languages using the same algorithms, three other routines, all written in Lisp, were compared. One set was based on the sparse list-based routines used by Macsyma. The second was written especially to deal with dense univariate polynomials in a finite field using the same array data structures as the C programs, and using the high-school arithmetic technique. The third was a discrete FFT polyno-

1. Work reported herein was supported in part by the U. S. Department of Energy, Contract DE-AT03-76SF0034, Project Agreement DE-AS03-79ER10358.

2. Macsyma's rational function representation uses lists and is particularly well suited to sparse polynomial arithmetic with coefficients which are themselves polynomials in other variables, or arbitrary precision integers. However, a number of allegedly efficient algorithms reduce the domain of interest to dense polynomials whose coefficients are "single-word" integers and where the coefficient arithmetic is performed in a finite field. Thus we are addressing a situation which does occur in practice, and could be used as the basis for replacing the standard sparse manipulation routines by appropriate mappings.

mial multiplication program taken from Bonneau's thesis.

In fact, the uniformly fastest of the three Lisp programs was the standard Macsyma-based routine.

The implementation and timings were done on Macsyma, running on the DEC VAX 11/780 computer. The host Lisp system "Franz Lisp" [3] especially when run under the UNIX operating system, provides the appropriate mechanisms to call from Lisp to other languages (C, Fortran 77, Pascal).

The algorithms being compared are described in the next two sections.

3. High School Algebra Polynomial Multiplication

In common with all the routines using arrays (that is, all but the built-in MACSYMA routines), all data is copied, by Lisp programs, from a linked list into an array, and then after the manipulation, from the array back to a list. The Lisp and C coded programs are given in Appendix 1 for this task, which is quite straightforward.

4. The FFT for Fast Polynomial Multiplication

The FFT [2] has been recognized as a component which could be used for fast polynomial operations over a finite field (e.g. [9]). When used with the Chinese Remainder Algorithm (see [5] pp. 253-256) it makes possible fast operations over the ring of integers. For example, multiplication of two dense degree n polynomials can be performed in time $O(n \log n)$ rather than $O(n^2)$, by the classical method.

Unlike many other "fast" algorithms, the FFT appears to be, in many cases, practical even for small inputs. In numeric applications it appears to be more accurate as well.

Nevertheless, the FFT has not, to the author's knowledge, been used as a default technique to multiply polynomials in an algebra system. The size of problem for which the FFT paid off was usually too far away from the typical problem of interest, or the restriction to FFT-supporting moduli too inconvenient. For an analysis based on numbers of coefficient arithmetic operations for various methods of computing polynomial powers, including the FFT, see [4]. For a detailed analysis of multiplication, see [7].

Richard Bonneau [1] showed a variety of polynomial operations which could be sped up using the FFT, including multiplication, exponentiation, division, greatest-common-divisor, resultant, and substitution. In spite of the lower computational complexity of FFT-based algorithms versus other known algorithms, the substantial overhead and the relatively poor performance for sparse polynomials, made these algorithms unattractive for the standard Macsyma system.

Here we reduced this overhead by coding the central array operations in the "C" language instead of Lisp. Specifically, we coded polynomial multiplication where the FFTs and element-by-element multiplications were both coded in C.

4.1. Algorithms for the FFT

There are several papers which describe the relationship between FFTs and polynomial multiplication ([9],[7] and its references), so that we will not dwell on this. It is appropriate to mention the techniques used for the FFT by Bonneau, since those appeared to be the best attainable within MacLisp on the PDP-10.

Bonneau used two variants of the FFT algorithm: the classic Cooley-Tukey decimation in time algorithm, and its complement, the decimation in frequency algorithm. In the process of computing the decimation in frequency algorithm, the result is first obtained in "bit reversed order". That is, each resultant term is reordered as if its index had been bitwise reversed. Thus term four of an order three (eight element) FFT would show up in position one. This result must then be reordered to yield the classic FFT result. The decimation in time algorithm operates the other way around, by first reordering the input, and then computing the FFT. Bonneau avoided the reordering by using the decimation in frequency algorithm to do the FFTs, without reordering afterward, and then using the decimation in time algorithm (without the initial reordering) to do the inverse FFT (IFFT).

Bonneau's algorithms are recursive and use Lisp lists rather than the more typical arrays: arrays are generally not well supported in Lisp. The C routines use the decimation in frequency algorithm, without recursion and with reordering, for both forward and inverse FFTs, because the elimination of recursion saves more time in the C implementation than not doing the reordering.

The decimation in time and decimation in frequency algorithms require a power-of-two number of terms, so when a transform of a non power-of-two size is desired, it must be upgraded to the next higher power of two. Other types of FFTs, including S. Winograd's recent improvements were examined briefly. It appeared unlikely that for our applications, restricting ourselves to products of degree 1023, that there would be much advantage to coding more elaborate routines.

4.2. Details

In order to retain the properties of exact calculations, the arithmetic operations of the FFT must be done using modular arithmetic. The modulus M should be chosen to be as large as possible, yet allow the use of normal computer arithmetic. For a K point FFT, M must be chosen so that there exists a number ω such that $\omega^K \equiv 1 \mod M$. Since the moduli used by Bonneau were designed for a 36-bit machine, new moduli less than 2^{30} were computed so that a signed addition of two legal numbers would not overflow on the VAX, a 32-bit word computer. The first step in finding a suitable modulus is to find a prime P such that $P-1$ is divisible by K. In order to facilitate the calculation of ω, we also require that 3 be a generator of the field $\bmod P$, thus $3^{P-1} \equiv 1 \mod P$. Under these circumstances, $\omega \equiv 3^\beta$, where $(3^\beta)^K \equiv \omega^K \equiv 1 \equiv 3^{P-1}$ (all $\bmod P$), so $\beta \equiv (P-1)/K$. A program was written which, given a starting value, computes a somewhat smaller suitable prime P along with the associated ω. It can thus be used to generate any number of suitable moduli, so that the Chinese Remainder Algorithm can be used for polynomials with large positive or negative coefficients (even though the coefficients here are restricted to positive numbers somewhat less than 2^{30}).

4.3. General comments on the FFT efficiency

The 1024 point FFT takes 0.5 seconds in the C-coded version, compared to about 100 seconds for Bonneau's (compiled) LISP routine.[3] Thus it is not worth describing in detail any tests with the FFT coded in Lisp.

A dense multiplication producing a 1023 degree result takes just over 3 seconds, which is about twice as long as the three FFTs which must be done. The extra time is taken by Lisp routines to convert between lists and arrays and the component-wise multiplication. Using the FFT for a 1024 term "sparse" multiplication (that is $x^{511} \cdot x^{512}$) is just over 2 seconds. This is not much faster than the dense multiplication, and much slower

3. A Lisp optimizing arithmetic compiler such as that available for PDP-10 Maclisp could produce more efficient code. Our VAX system is in fact designed to interface with arithmetic routines written in other languages as demonstrated here.

than a sparse algorithm, which would perform this polynomial multiplication in essentially one coefficient operation. It is a property of the FFT that it is relatively oblivious to sparsity, a disadvantage compared to algorithms which capitalize on sparseness. In general, one could chose different algorithms for different domains by a heuristic "sparseness testing" procedure which can run in time linear in the size of the inputs.

5. Results

The comparison was done between a component of the Macsyma rational function system "ptimes1", and other routines. For two polynomials of s and t terms (non-zero terms, as distinct from degrees) "ptimes1" uses $O(st)$ coefficient operations, and uses $O(st^2)$ comparisons for sorting terms in the answer.

"Fftmult" uses $O(n \log n)$ operations where n is the next higher power of 2 greater than the degree of the product.

"HsCmult" uses the high-school algorithm, but written in C. It uses $O(n^2)$ multiplications.

The variants indicated earlier which were written in Lisp, but based on the algorithms of the C routines, were uniformly slower than the system program ptimes1. They are therefore not included in the table following.

As indicated in reference [4], there are many aspects to efficiency in this problem, and the program "ptimes1" has an advantage when the number of terms in the two inputs differs drastically, or when the degree of the answer far exceeds the number of non-zero terms. The table below explores only one of many possible directions that can be tested.

Tests were performed by squaring $p_n = (x+1)^n$ in a finite field, for various n. Times are reported in seconds, and Lisp "garbage collection" is *not* included. Short times were averaged over several runs.

n	**ptimes1**	**fftmult**	**hsCmult**
1	0.010	0.015	0.007
3	0.027	0.017	0.010
4	0.042	0.036	0.017
7	0.110	0.038	0.019
8	0.158	0.068	0.020
15	0.455	0.070	0.038
16	0.466	0.161	0.045
31	2.267	0.136	0.093
32	2.483	0.247	0.097
63	9.750	0.258	0.293
64	9.900	0.520	0.290
96		0.575	0.575
127		0.615	1.042
128		1.141	1.033
256		2.683	3.766

As would be expected, **fftmult** exhibits a step-like behavior, growing nearly linearly when aggregated, while **ptimes1** and **hsCmult** are roughly n^2. Since the "C" coded programs do not use Lisp arithmetic and data structures, they almost never require garbage collections. The **hsCmult** program is extremely fast for low degrees, and except for some degrees where **fftmult** gets a close fit to a power of 2, is best for polynomials of degree less than 96. At that point, the FFT takes over.

Also note that the FFT is only about a factor of two slower than **hsCmult**, and often better than the Lisp-coded routine. The crossover point where **fftmult** becomes faster than **ptimes1** is at the multiplication of two polynomials each of degree 3, which is to say for equal-degree dense modular polynomials, it is essentially superior except for trivial cases.

6. Conclusions

It seemed reasonable to return to **ptimes1** and see if it had been coded in some fashion as to be particularly inefficient, since it was originally coded for the PDP-10, and might be deficient on the VAX, which, for example, has a more elaborate and expensive CALL instruction. In spite of various minor improvements, we could not alter the speed significantly.

Changing the data structure from lists to arrays, but still using the (VAX) Lisp system, slowed the program down even further. While another Lisp system might provide better array-handling from Lisp, but it is highly unlikely that the array-handling and arithmetic would be as convenient and fast as that used by C.

As a result of this experiment, we conclude that

(1) Use of "C" or other non-Lisp language routines in conjunction with a Lisp-based algebraic manipulation system is quite practical.

(2) The use of the FFT is practical given an appropriate language for its implementation in the context of a Lisp-based algebraic manipulation system. It could be used in other algorithms for the benefit of dense polynomial calculations.

7. Further Work

We are considering other applications in other common algebraic routines -- the basic pseudo-remainder step in the modular greatest-common-divisor algorithm, and the linear system solving in Zippel's sparse GCD [12].

There are of course substantial prospects for interaction with numerical libraries written in non-Lisp languages (usually Fortran). The use of our VAX Macsyma system in conjunction with such libraries wil be dealt with in a subsequent paper.

8. Acknowledgments

The FFT programs described here were written by Neal McBurnett as a class project for Algebraic Algorithms, CS292S, Winter Quarter, 1980 at the University of California, Berkeley. Other programs were written by the author. In each case, Keith Sklower assisted in the coding and analysis.

9. References

1. Richard J. Bonneau, *Polynomial operations using the fast Fourier transform*, Ph.D. thesis, Dept. of Mathematics, Mass. Inst. of Tech., Cambridge, Mass., 1974.

2. J. W. Cooley and J. W. Tukey, "An algorithm for machine calculation of complex Fourier series," *Math. Comp. 19 (1965) pp. 297-301.*

3. John K. Foderaro, *Franz Lisp Manual*, 3rd Berkeley Software Distribution, Computer Science Division, EECS, University of Calif.,Berkeley. (1979)

4. Richard J. Fateman, "Polynomial multiplication, powers, and asymptotic analysis, some comments," *SIAM Journal on Computing*, (3), no. 3, September 1974, pp. ~~145-155.~~ p196-213

5. Donald E. Knuth, *The Art of Computer Programming, Vol. 2 / Seminumerical Algorithms, Addison Wesley, 1969.*

6. The Mathlab Group, *Macsyma Reference Manual, Version 9.* MIT Laboratory for Computer Science, 1977.

7. Robert T. Moenck, "Practical fast polynomial multiplication," *Proc. ACM 1976 Symp. on Symbolic and Algebraic Comp.*, August, 1976, pp.136-148.

8. David A. Moon, *Maclisp Manual*, Lab for Comp. Sci. Mass. Inst. of Tech., 1974.
9. J. M. Pollard, "The fast Fourier transform in a finite field," *Math. Comp.* 25 no. 114 (1971), pp. 365-374.
10. Guy L. Steele Jr., "Fast arithmetic in Maclisp," *Proc. 1977 MACSYMA Users' Conf.* NASA CP-2012, July, 1977, pp. 215-224.
11. S. Winograd, "On computing the discrete Fourier transform," *Proc. Nat. Acad. Scien. USA*, vol 73, no. 4 April, 1976, pp. 1005-1006.
12. R. Zippel, "Probabilistic Algorithms for Sparse Polynomials," *Symbolic and Algebraic Computation*, Lecture Notes in Computer Science 72, Springer Verlag pp. 216-226.

Appendix

A C polynomial multiplication program

```
#include <stdio.h>

#define MIN(a,b) (a > b ? b : a)
#define MAX(a,b) (a > b ? a : b)

/* cheapmul does an n^2 multiplication of 2 dense array polynomials
modulo modulus, a single-word prime.  Callable from Lisp. */

cheapmul(a, dega, b, degb, c, modulus)
int a[], b[], c[]; /* c had better allocated large enough */
int *dega, *degb, *modulus;
{
    register int i, j, temp, high;
    int mod;
    mod = *modulus; /* for faster reference */
    for (i = *dega + *degb; i >= 0; i--) {
      temp = 0;
      high = MIN(*dega,i);
      for (j = MAX(i-*degb, 0); j <= high; j++)
            temp = mmuladd(a[j], b[i-j], temp, mod);
      c[i] = temp;}}

/* use two vax instructions to produce (a*b +c) mod m.
 * courtesy Keith Sklower */

mmuladd (a, b, c, m)
int a, b, c, m;
{
    asm ("emul    4(ap),8(ap),12(ap),r0");
    asm ("ediv    16(ap),r0,r2,r0");
}
```

On the Application of Array Processors to Symbol Manipulation

R. Beardsworth, Dept. of Computer Studies,
University of Leeds, Leeds LS2 9JT, ENGLAND.

Abstract

In the past general purpose programs for symbol manipulation have been written for traditional Von Neumann machine architectures. The design and implementation of a simple prototype symbol manipulation system for the ICL Distributed Array Processor (DAP), is described. The system is restricted to monovariate polynomials with single precision integer coefficients. The algorithms and data structure are discussed and the design of a more general system for multivariate polynomials using arithmetic with semi-infinite precision is considered.

1.0 INTRODUCTION

The purpose of this paper is to investigate the feasibility of using the ICL Distributed Array Processor (DAP) to perform symbol manipulation. The paper illustrates that with careful choice of the data structure, an algebra system for monovariate polynomial calculations can easily be written to run on the DAP. It is also shown that by suitable choice of the data structure it is possible to devise algorithms for more general algebra systems for the DAP.

2.0 DATA STRUCTURE

The ICL DAP is a Single Instruction Multiple Data architecture capable of performing calculations on vectors and matrices of numbers. The architecture has a plane of 64x64 processors, each with 4096 bits of store, and three special 1 bit registers, the C, Q and A registers. These registers form planes of bits used in calculations for Carry, Quotient and Activity control [3]. In addition the processor-processor interconnections allow data to be broadcast along rows and columns of store.

Given this architecture it was decided that for the purposes of the investigation a system capable of monovariate polynomial calculations with a maximum power of 63 would suffice. As a further restriction the coefficients were limited to single precision integers. This ensures that timings taken are indicative of the speed of the polynomial algorithms and not of the time consumed in multiple precision arithmetic.

The first problem was to choose a data structure around which to design the system. As the maximum power of the polynomials were to be restricted to 63, the coefficients could be stored in half of a plane of the DAP in 2900 FORTRAN mode (see Fig. 1). The other possible choice was to hold 64 polynomials in 32 planes, each row representing the 64 terms of the polynomials, this being the matrix mode, (see Fig. 2).

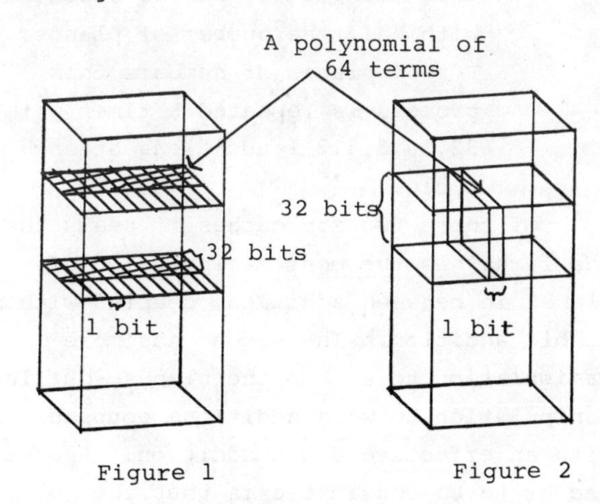

Figure 1 Figure 2

Of these two schemes the former although inefficient in storage, has the advantage that for addition no manipulation of bit patterns is needed prior to the vector addition of the C and Q planes, and as in both schemes the same amount of manipulation for multiplication is needed the former scheme was chosen for the data structure.

A collection of such contiguous planes forms the work space for polynomials, together with a section of store consisting of 128 work planes for temporary storage in multiplication, (referred to as the Wk-planes), and a plane consisting of the numbers 0-63 for use in differentiation.

3.0 ALGORITHMS

The algorithms considered in this investigation involve only the following polynomial operations; addition, negation, subtraction, multiplication, division and differentiation. Of these operations only division needs the presence of rational coefficients to ensure that the algorithm works; it will therefore be discussed in terms of integer result coefficients.

3.1 Addition, Subtraction and Negation

Because of the way the coefficients are stored addition, subtraction and negation require very little preprocessing. For addition the two polynomials to be added are fetched into the C and Q planes and a vector addition performed, taking 8 cycles to generate 32 bits, as 4 bits can be generated each cycle. This also allows a very simple check for overflow to be made as the top two columns of bits of the answer are easily accessible. Simple algorithms can also be devised for negation and subtraction.

3.2 Multiplication

The two polynomials are converted from the vector representation to the matrix representation. One of them is converted to a 32 plane deep matrix whose rows are all equal to the polynomial, and the other similarly converted to a matrix whose columns are the same as the polynomial, these are stored in Wk-planes 0-63. The generation of the 4096 sub-results is then performed in parallel resulting in a matrix of coefficients with those coefficients corresponding to equal exponents in the diagonals, these are in Wk-planes 64-127.

To add the corresponding coefficients in parallel, it was necessary to rearrange the coefficients. There were two approaches which suggested themselves :-

1. Copy the result plane into another set of planes, and by appropriate shifts with plane geometry, position corresponding coefficients at the same matrix coordinates. Then perform a matrix addition of the two planes and repeat the process 6 times (see Fig. 3).

H							
G	H						
F	G	H					
E	F	G	H				
D	E	F	G	H			
C	D	E	F	G	H		
B	C	D	E	F	G	H	
A	B	C	D	E	F	G	H

Figure 3A

H'							
G'	H'						
F'	G'	H'					
E'	F'	G'	H'				
D	E	F	G				
C	D	E	F				
B	C	D	E				
A	B	C	D				

Figure 3B

Fig. 3(a) shows the position of coeffs with equal exponents as the same letter. After once round MA2-MA5 the position of coeffs are shown in fig. 3(b) with a prime showing that corresponding coeffs have been added.

Algorithm MA

```
MA1 Initialise:- I <-- 32;
        J <-- 0;
        Clear Carry plane
MA2 Start Loop:-
        Q-plane <-- Wk-plane 127-J
MA3 Shift:-
        Shift Q-plane by I bits West
        then I bits South with plane
        geometry
MA4 Add:-
        Add C,Q and Wk-plane 127-J
        returning answer to Wk-plane
        127-J
MA5 Test end of inner loop:-
        J <-- J+1
        IF J≠32 GOTO MA2
MA6 Next cycle:- IF I=1 GOTO MA9
MA7 Loop:- I <-- I/2;
        J <-- 0;
        Clear Carry;
        GOTO MA2
MA8 Answer is in row 63
        of Wk-planes 86-127
MA9 EXIT
```

2. Convert the matrix mode numbers into 64 planes of 64 vector mode coefficients (some of which are zero) so that vector addition of consecutive pairs of planes can be carried out the C and Q planes. The sub-results are then taken from the Q plane and placed in consecutive planes so that the process can be repeated with half the number of planes. For 64 planes of numbers this process is repeated 6 times with 32,16,8,4,2,1 additions at each cycle.

Of these two approaches it seems that the first has the most manipulation of planes in between addition, coupled with a 32 bit addition. The second has more manipulation to set up the planes, but less manipulation between additions coupled with an effective 8 bit addition. Another factor to be considered is that the number of additions in the first algorithm is 6 and the number in the second is 63. Both algorithms were coded, and the first algorithm was seen to execute more quickly, so it was included in the complete multiplication routine.

To gain an impression of the speed of this routine, timings for the multiplication of two 64 term polynomials were taken for the simple system and these were then compared with a simple polynomial system, LETHAL [1] running on an AMDAHL 470 (equivalent in speed to the top IBM machines). The speed of LETHAL is comparable with CAMAL [4]. The results of this test are shown below. As can readily be seen the simple system on the DAP goes extremely quickly on this test.

Time for the AMDAHL 470	143.5 msecs.
Time for the DAP	1.9 msecs.

Table 1

Times for the 64x64 term multiplication.

3.3 Division

As division requires the ability to represent rational numbers for a general algorithm, only exact division of polynomials with integer coefficients in the result will be considered. As with the normal division algorithm, division is regarded as repeated subtraction of the divisor from the dividend. As subtraction is so easy to perform the division routine incorporates the subtraction code, the addition of the generated sub-results into the answer is only a matter of positioning the sub-results into the appropriate row of the result plane. The gains to be made here are from the parallel multiplication of the whole divisor by the ratio of the top terms which consists of a shift upwards in index and a multiplication of the coefficients by the appropriate number. A general algorithm for division follows.

Algorithm DV

```
DV1 Find highest power with non-zero
       coefficient in divisor
       P <--  Power;
       C <--  Coefficient;
       Plane 0 <--  Dividend
DV2 Start loop: Find highest power with
       non-zero coefficient in plane 0
       DP <-- Power;
       DC <-- Coefficient;
       Plane 1 <-- Divisor
DV3 Check for end:- IF DP = 0 GOTO DV10
DV4 Find ratio of coefficients and
    difference in powers:-
       R <-- DC / C
       D  -- DP - P
DV5 Shift :- Shift plane 1 D bits upward
DV6 Multiply :-
       Multiply shifted plane 1 by R
DV7 Subtract :-
       Subtract plane 1 from plane 0,
       placing result back in plane 0
DV8 Generate answer :-
       Answer plane, row D <-- R
DV9 Loop:- GOTO DV2
DV10 Finish:- EXIT
```

3.4 Differentiation

Differentiation is performed by the multiplication of the polynomial by the plane containing the numbers 0-63, followed by a shift of the whole plane by 1 bit, corresponding to a decrease in exponent. The multiplication is in fact done by repeated shift additions using the 0-63 plane as a mask of where to do the additions.

4.0 FUTURE DAP SYSTEM

A more general system for the DAP would provide for the manipulation of multivariate polynomials with multiple precision coefficients. With the introduction of multivariate polynomials it is no longer sufficient to use position as a governing factor in determining the exponent of a result. One way round this problem is to use the "packed exponents" method of CAMAL and LETHAL. Here the indices are packed into one or more consecutive words, the same method could be used on the DAP but a more efficient method can be arranged. As the DAP is completely bit orientated the simplest method of packing the indices is to use consecutive planes of store and use say 8 planes per variable (for maximum order of 255), thus the data structure would only use the exact amount of space required for the problem, and there would be no blank spaces as in the last word of the LETHAL structure for one term.

For the multiplication of two polynomials the problems become those of partitioning the polynomials and the collection of like terms at the end. The problem of polynomial partitioning requires the use of essentially the same principle as the partitioning of matrices for matrix multiplication; it should not therefore cause any difficulty. The problem of sorting out like terms in the result is more challenging. Most symbol manipulation systems calculate the position of a term as it is being produced, in the DAP however the sorting will be done at the end of the calculation. Some fast methods exist for sorting on the DAP, and the one of most use will probably be Batchers Bitonic Algorithm [5]. Wherever sets of coefficients corresponding to equal exponents occur the addition of these will be set aside until sufficient number are ready and then a matrix addition can be carried out. During addition a similar searching for the largest exponents must obviously take place.

Multiple precision arithmetic will be available for this system. Given large enough numbers great use can be made of matrix mode arithmetic. The only question remaining to be answered is whether large numbers should be held with a number base equal to a power of ten, or as numbers with a base equal to a power of two. The former makes printout out the answer very easy, while the latter makes calculations more simple!

5.0 CONCLUSIONS

It has been shown how the architecture of the DAP means that it is reasonably straightforward to implement a simple symbol manipulation system. The speed of the system is considerably greater than CAMAL for relatively large problems. However for some problems containing small polynomials, the performance of the system was decreased, as shown by the following results.

	AMDAHL	DAP
(1+x)↑32	45	59
CHEBYSHEV	10	44

TABLE 2

Times for Series Generation in msecs.

Where the Chebyshev polynomials were calculated to order 20, and the powers were done by repeated multiplication.

It is likely that a major problem is going to be to decide when it is advantageous to use the DAP, however research into an algebra system using both the host and the DAP is in progress and it is felt that this could have fruitful results.

6.0 ACKNOWLEDGEMENTS

I would like to thank my supervisors, Dr. P.M. Dew and Prof. J.P. Fitch for their useful comments on the paper.

7.0 REFERENCES

1. Beardsworth R., LETHAL:A Fortran Vector Based Algebra System, Dept. Computer Studies Report No. 134. Leeds University (1980).
2. Reddaway S.F., DAP a Distributed Array Processor, First Annual Symposium on Computer Architecture, Florida (1973).
3. International Computers Ltd., DAP:APAL Language, ICL Technical Publication 6919, (1979).
4. Fitch J.P., An Algebraic Manipulator, Ph.D. thesis, Cambridge Univ. (1971).
5. Flanders P.M., Internal Sorting Using Batchers Bitonic Algorithm, DAP Newsletter No. 5, (1980).

The Optimization of User Programs for an Algebraic Manipulation System

P.D. Pearce, R.J. Hicks

School of Electronic Engineering and Computer Science,
Kingston Polytechnic, England, KT1 2EE.

ABSTRACT

This paper attempts to list some optimising transformations for user programs for an Algebraic Manipulation System. To investigate optimisation of both computer time and space, a general purpose system REDUCE has been chosen for study. The optimising transformations may be applied manually. However, the authors hope to automate the process. Examples using various optimisations are included and clearly show the benefit of the process.

1.0 INTRODUCTION

Users of Algebraic Manipulation Systems frequently find that they are unable to obtain answers from programs (written in some user interface language, U) that are both syntactically and algorithmically correct, through lack either of space or of computer time [Pearce 79]. There are many ways in which programs in U may be optimised by transforming them to more efficient programs in U. At present no documentation exists to guide the inexperienced user in writing efficient algebraic programs. In fact, a detailed knowledge of the workings of the Algebra System is necessary to achieve efficiency. It seems unreasonable for a physicist, say, with an algebraic problem to have to grapple with more than the task of writing a correct program. Even with an understanding of a particular Algebra System, many optimisations would be very tedious to incorporate and would obscure the algorithm. Some optimisations are independent of the Algebra System, but many are system dependent. To investigate optimising transformations for programs in U, REDUCE has been chosen for study. Any changes to the Algebra System are not considered.

Optimisations for programs written in a numerical language, e.g. FORTRAN are well documented (e.g. [Busam 69, Schneck 73]) and

concentrate on time optimisation, this being the most significant problem in this area. Optimisations such as avoiding gcd calculations, are peculiar to algebraic programs and we call them algebraic type optimisations. REDUCE optimisations are transformations that optimise REDUCE programs but are not generally applicable to other Algebraic Manipulators.

The output from REDUCE statements may limit the optimisations possible if program equivalence is to be maintained (i.e. that the programs produce exactly the same output). It is desirable to keep the output to the minimum, thus allowing maximum optimisation.

In Section 3 the execution times and space usage of both unoptimised and optimised programs are compared.

"For many years compilers have contained sections which are supposed to "optimize" the code produced from computer programs. As has often been noted, this term is a misnomer because a really "optimal" solution to the "optimization" problem would involve throwing away the original program and producing in its place the best possible program to perform the desired task." [Katz 78]. This is equally applicable to algebraic programs.

The advantage that can be gained by looking at problems and rearranging the computation has been mentioned by [Sundblad 72] and is discussed in [Campbell 76] and [Campbell 79]. Campbell's approach is to pre-process symbolic computation problems in order to compress the data if the system runs into storage difficulties.

2.0 OPTIMISATION TECHNIQUES FOR REDUCE PROGRAMS

2.1 DEFINITIONS [Hearn 74, Stoutemyer 78/79]

REDUCE PROGRAMS

Throughout the paper the term "REDUCE programs" is used for algebraic mode programs for the REDUCE Algebraic Manipulator.

CONSTANTS

(1) A numeric constant may be either real or integer.
(2) An algebraic constant is a name that is never assigned a value. Its value is its own name.

VARIABLES

A variable may be either global or local. A local variable can be formed in three places :-
(1) A type statement within a compound statement. The initial value of the variable is NIL.
(2) A dummy argument for a procedure declaration. All procedure arguments are passed by value.
(3) A FOR statement loop control variable.

All other names excluding constants are global variables and have as their initial value their own name, unless they are declared in a type statement in which case their value is NIL.

We will use the term "unbound" to describe the condition of variables, in a manner reminiscent of LISP variables.

A variable is unbound while its value is itself. All other variables are bound. When an unbound variable is given a value, then at that point in the execution of the program, the variable becomes bound. An unbound variable may only be a global variable not declared in a type statement.

Bound variables fall into two categories :-
(1) A "loosely" bound variable is one which has been assigned a loosely bound expression, i.e. an expression containing at least one unbound variable.
(2) A "tightly" bound variable is one which has been assigned a tightly bound expression, i.e. an expression containing only constants and tightly bound variables.

By "expression", we mean the expression that is actually assigned to the variable at run time.

2.2 NUMERICAL TYPE OPTIMISATIONS

The following numerical type optimisations give optimisation of REDUCE programs :-
(1)Folding [Loveman 77, Rohl 75, Waite 76]
(2)Propagation of constants, scalars and expressions [Cocke 70, Loveman 77, Waite 76]
(3)Goto propagation [Loveman 77]
(4)Elimination of common expressions [Allen 72, Rohl 75, Waite 76]
(5)Retaining calculated values [Cocke 70, Rohl 75]
(6)Loop splitting
(7)Combination of loops [Allen 72, Loveman 77]
(8)Expansion of conditionals [Allen 72, Loveman 77]
(9)Elimination of dead variables [Cocke 70, Kou 77]
(10)Elimination of redundant calculations [Allen 72, Cocke 70, Loveman 77]
(11)Reduction in order of multi-dimensional arrays [Loveman 77]
(12)Test replacement / loop modification [Cocke 70, Feign 79, Loveman 77]
(13)Recognition of symmetry [Kibler 77]
(14)Elimination of recursion [Wegbreit 76]
(15)Simplification [Bagwell 70, Loveman 77, Moses 71, Standish 76]
(16)Code motion [Allen 72, Cocke 70, Loveman 77, Rohl 75, Waite 76].

Care must be taken about the binding of variables when applying these techniques to REDUCE programs to maintain program equivalence.

The optimisations given below, although giving optimisation for numerical programs, have the opposite effect on REDUCE programs :-
(1)Elimination of multi-dimensional arrays by linearising array references [Loveman 75]. The explicit arithmetic used to calculate the array reference is very much slower than the implicit arithmetic used in the multi-dimensional array access.
(2)Devolution [Loveman 77]. Evolution of structures gives an optimisation as can be seen by examining the LISP generated by REDUCE.
(3)Reduction in operator strength [Cocke 70, Feign 79, Loveman 77]. Increasing the strength of operators in REDUCE gives optimisation. For example Y:=2*I is calculated more rapidly than Y:=I+I. The effect is exaggerated when I is a large algebraic expression because addition requires the large data structure, I, to be scanned twice.
(4)Unrolling [Allen 72, Cocke 70, Loveman 77]. Loop control in REDUCE is comparatively fast so that unrolling does not usually give a significant optimisation. It may in fact slow down the program due to the loop increment being faster than the explicit add.
(5)Rearrangement of computational order as for numerical systems [Rohl 75, Waite 76]. The rearrangement that gives optimisation will depend on the particular Algebra System. For example, Horner's Rule gives optimisation in FORTRAN, but not in REDUCE.

2.3 ALGEBRAIC TYPE OPTIMISATIONS

2.3.1 TIME VERSUS SPACE

As expressions swell so the data space used by the program increases, frequently very rapidly. The space for intermediate expressions in a calculation which are no longer in use may be reclaimed by the garbage collector. Garbage collection reclaims space at the expense of time. A user running a program in minimal space may well not have sufficient time to execute the program to its conclusion, because most of the time is spent doing garbage collection. It may well be possible to improve the run time of this program very considerably at the expense of little extra store.

2.3.2 ELIMINATION OF DEAD VARIABLES

If a variable, V say, contains a value, then V may be cleared when no longer used. This optimisation is very important in algebraic manipulation. The data store used is reduced and consequently the time spent in garbage collection.

2.3.3 AVOIDING GCD CALCULATIONS

Whenever large expressions with large denominators are added or subtracted the lcm is required to avoid excessive growth of both numerator and denominator. When large expressions have been added, subtracted, multiplied, divided etcetera, division of both numerator and denominator by their gcd helps to control the size of results. These calculations are very time consuming, but if ignored blow-up may well occur [Barton 72]. The policy of the Algebra System towards these gcd and lcm calculations is part of its policy towards simplification.

For the REDUCE System the user controls

gcd calculations by setting an execution flag to "on" or "off".

In some calculations it is possible to rearrange the calculation so that a large number of gcd calculations are avoided.

e.g. (1)

$$\sum_{i=1}^{m} \frac{1}{(p+q)^{i}\,(r+s)^{i}} \text{ may be rearranged as}$$

$$\frac{\sum_{i=1}^{m} (p+q)^{m-i}\,(r+s)^{m-i}}{(p+q)^{m}\,(r+s)^{m}}$$

e.g. (2)

$$\sum_{s=0}^{m} \frac{(a+b)^{s-m-1} - (a-b)^{s-m-1}}{s-m-1}$$

may be rearranged as

$$\frac{\sum_{s=0}^{m} \frac{(m+1)!}{(m+1-s)} (a^{2}-b^{2})^{s} \left((a+b)^{m+1-s} - (a-b)^{m+1-s}\right)}{(m+1)!\ (a^{2}-b^{2})^{m+1}}$$

when s-1 <= s-m-1 <= -1
which gives m+1 >= m+1-s >= +1.

$\frac{(m+1)!}{(m+1-s)}$ always gives an integer value.

2.3.4 RECOGNITION OF SERIES

Once series are recognised as a well known type with a known sum, for example an arithmetic progression, then the code that generates the series may be replaced by the formula for the sum of the series.

2.3.5 CODE MOTION

The nature of Algebra Systems allows a complete class of optimisations to be performed by code motion over and above those available in numeric systems. For all code motion the binding of variables must be considered.

With algebraic programs, a much wider scope is available for calculation rearrangement.

e.g. In $\sum_{i=1}^{10} \int x^{i}\,dx$ we may evaluate $\int x^{i}\,dx$ as

$\frac{x^{i+1}}{i+1}$ and then execute $\sum_{i=1}^{10} \frac{x^{i+1}}{i+1}$

saving nine calculations of the integral.

Another example, in REDUCE, follows :-

```
(1)  FOR I:=1 : N SUM FN(I);
```

may be changed to give

```
(2)  NEWVAR := FN(V);
     FOR I:=1 : N SUM
         << V:=I; NEWVAR >>;
```

In (2), V is a new unbound variable and is used to replace I in FN(I). NEWVAR is a new loosely bound variable used to hold FN(V). The binding of variables common to (1) and (2) is the same. In (2) the expression FN(I) is calculated once only and then values of I are inserted.

Problems with space may arise in (2). When FN(V) is evaluated the resultant expression with V a variable may be considerably larger than with V taking a numerical value. The resultant shortage of space might cause excessive garbage collection. Even disregarding the time spent in garbage collection, there may be an increase in time if FN(V) is a much larger expression than FN(I). Whether (1) or (2) is an optimisation or vice-versa will depend on the particular function FN.

An interesting problem to consider is the best order for the calculation of

$$X := \sum_{I=1}^{N} \sum_{J=1}^{N} F(I,J)$$

Some possible orders are:-

```
(1) X := FOR I:=1 : N SUM
         FOR J:=1 : N SUM F(I,J);
(2) X := FOR J:=1 : N SUM
         FOR I:=1 : N SUM F(I,J);
(3) NEWVAR := FOR I:=1 : N SUM F(I,V);
    X := FOR J:=1 : N SUM
         << V:=J; NEWVAR >>;
(4) NEWVAR := FOR J:=1 : N SUM F(V,J);
    X := FOR I:=1 : N SUM
         << V:=I; NEWVAR >>;
(5) NEWVAR := F(V,W);
    X := FOR I:=1 : N SUM
         FOR J:=1 : N SUM
         << V:=I; W:=J; NEWVAR >>;
```

In a numerical system both (1) and (2) will give the same result in the same time using the same amount of space, apart from the influence of round-off. In a numerical system (3), (4) and (5) are not possible.

Also, in REDUCE, substitution statements should be moved to a position in the code just before they have an effect on the data structures. For example, a substitution for a variable, P say, need not appear until the variable P occurs in the program. Any substitutions applied unnecessarily will slow the execution of the program.

The scope for code motion in algebraic programs is wide reaching. Whenever a variable is given a value, the question arises as to the "best" position in the program for the assignment. e.g. In the program

```
A:=P+Q$  B:=R*S$  C:=A+B$  C;
```

any ordering of the first three statements will give the same output.

The structure of the language may restrict the positions of some assignment statements :-
e.g. In the program

```
M:=1$ X:=FOR I:=1 : M PRODUCT I;
```

there is no scope for moving the assignment M:=1 because the terminal value of a FOR loop must be a number.

It may be best to use this optimisation by executing the program, and if a problem occurs with either space or time, then to try an alternative position for an assignment.

2.3.6 OPTIMISATION OF EXPRESSIONS

DATA STRUCTURE

Knowledge of the data structure used by an Algebra System allows expressions to be put into their most efficient form.

The data structure used in REDUCE is described in [Beardsworth 79]. In REDUCE the ordering of variables affects the size of the data structure for expressions. This not only affects the data space used but also the execution time taken [Pearce (I)].

SIMPLIFICATION [Loveman 77, Standish 76]

Simplification is dependent on the particular Algebra System. Any simplification that is performed as an optimisation should conform to the politics of the system [Moses 71] and in fact give exactly the simplification provided by the system.

(1) SUBSTITUTION

In REDUCE, expressions may be simplified at execution time by the application of substitution statements.

In some cases it may be desirable to simplify the program by applying substitutions to expressions in the code prior to execution of the program. This may cause substitutions to take place where they would not normally occur at run time, and consequently not maintain program equivalence. This optimisation could only be performed with programmer interaction. The following is an example where applying substitution before evaluation gives a different result :-

```
*LET X**2 + Y**2 = 1$
*A := X**2 + Y**2 - X**2;
```

In REDUCE substitution always follows evaluation.

(2) REARRANGEMENT

The interactive nature of REDUCE allows it to be used to simplify and rearrange individual expressions prior to execution. When an expression is passed to REDUCE it is evaluated and consequently simplified.

This optimisation performs operator replacement within expressions, for example it replaces I+I by 2*I.

In some cases, for example (A+B)**1000, the resultant expanded expression is very much larger than the original. In REDUCE, this then causes a considerable slow down in execution speed. In general, if simplification causes an expression to expand, it is not an optimisation.

No matrix operations or procedure calls can be given to REDUCE for this treatment since matrix operations are not commutative, and neither are variables necessarily commutative or associative with procedure calls.

For simplification by rearrangement using REDUCE to be fully used, the values of all the execution flags must be known.

2.4 REDUCE OPTIMISATIONS

The following transformations optimise REDUCE programs for the DEC System 20.

2.4.1 CODE REDUCTION

(1) SPLITTING OF COMPOUND STATEMENT

Where a compound statement at the top level is larger than necessary, it may be split into two or more compound statements, thus decreasing the store used both to convert the statement to a parse tree and to hold the resultant parse tree.

(2) ELIMINATION OF PROCEDURES

When a procedure is "dead", i.e. no longer required, then the code for that procedure may be reduced by redefining the procedure as

```
PROCEDURE name ; ;
```

A further small reduction in space for code may be made by overlaying procedures.

(3) REMOVAL OF SUBSTITUTION STATEMENTS

A substitution that replaces a bound variable, V say, may be eliminated if V is always bound within the scope of the substitution since V will not occur in the data structure for any expression.

(4) ELIMINATION OF REDUNDANT BEGIN / END AND << / >>

When BEGIN and END or << and >> are used unnecessarily, they may be eliminated.

(5) RETURN STATEMENT

The variable, V say, in a RETURN statement may be replaced by its defining expression and the assignment to V removed, provided that V is local to the procedure.

2.4.2 REPLACING BEGIN / END BY << / >>

Since << and >> are more efficient than BEGIN and END, they should be used wherever possible. The RETURN statement must be removed and the value of the RETURN statement given to the << >> construct.

2.4.3 ECHO

The OFF ECHO statement inserted at the beginning of an input file containing a REDUCE program, inhibits the printing of the program and saves a considerable amount of time.

2.4.4 SWITCHING OF EXECUTION FLAGS

The restricted movement of an execution flag switch within a program is code motion, as described above. The addition of statements to switch flags may be desirable to obtain an answer to a problem, but the output may be different from that produced by the original program.

3.0 EXAMPLES OF OPTIMISATION OF PROGRAMS

Two problems are first programmed in the style of a user unaware about efficiency and subsequently after optimisations have been applied. The programs for these problems were run on a DEC System 2060. The programs themselves and details of the optimisations outlined below are described in [Pearce (II)].

3.1 PROBLEM 1

The problem is to evaluate the determinant of an N x N matrix whose general form is :-

$$\begin{bmatrix} \frac{1}{H+1} & \frac{1}{H+2} & \cdots & \cdots & \frac{1}{H+N} \\ \frac{1}{H+2} & \frac{1}{H+3} & \cdots & \cdots & \frac{1}{H+N+1} \\ \cdots & \cdots & \cdots & \cdots & \cdots \\ \frac{1}{H+N} & \frac{1}{H+N+1} & \cdots & \cdots & \frac{1}{H+2N-1} \end{bmatrix}$$

Three REDUCE programs for the solution of this problem are :-
(1) The unoptimised program, P1A
(2) The partially optimised program, P1B
(3) The optimised program, P1C.

The optimisations used to transform program P1A to P1B are :-
(1) Elimination of dead variables
(2) Elimination of common expressions
(3) Code motion
(4) Evolution
(5) Test replacement / loop modification
(6) Replacing BEGIN / END by << / >>.

The symmetry of the matrix is used to optimise program P1B giving program P1C.

The programs were run in 124 K words of data store with different values of N. The run times (excluding system set up time and print time) and the number of garbage collection calls follow :-

	TIME (SECONDS)			GARBAGE COLLECTION CALLS		
N	P1A	P1B	P1C	P1A	P1B	P1C
1	.23	.27	.28			
2	.31	.34	.35			
3	.48	.54	.51			
4	.89	.95	.85			
5	1.74	1.77	1.48			
6	3.67	3.32	2.64			
7	7.24	6.39	4.77	1	1	1
8	13.18	11.27	8.54	2	2	1
9	22.69	19.46	14.91	5	4	3
10	37.76	32.64	24.80	9	7	6

For N >= 6, the optimisations give a significant improvement in time and data space usage.

3.2 PROBLEM 2

The problem is to derive the expression for the differential cross-section for the ionisation of hydrogenic atoms due to collision with a single charged particle. Different values of the quantum numbers N, N1, N2 and M give different cases. A full description of the problem may be found in [Pearce 76]. Some programs for the solution of this problem are :-
(1) Unoptimised program for M=0, P2A
(2) Partially optimised program for M=0, P2B
(3) Optimised program for M=0, P2C
(4) Unoptimised program for M=1, P2U
(5) Partially optimised program for M=1, P2V
(6) Optimised program for M=1, P2W
(In all these cases N1=N2=0 and N=N1+N2+M+1).

To optimise the program it must be repeatedly scanned as one set of optimisations opens the program up to further optimisations.

To transform programs P2A to P2B and P2U to P2V, the following optimisations are applied :-
(1) Simplification by application of substitutions to expressions in the code prior to execution of the program
(2) Avoiding gcd calculations
(3) Scalar propagation
(4) Folding
(5) Code motion
(6) Elimination of redundant calculations

The programs were run in 60 K words of data store. Run times (excluding system set up time and print time) and number of garbage collection calls follow :-

TIME (SECONDS)

P2A	P2B	P2C	P2U	P2V	P2W
19.88	2.78	1.69	*	11.95	10.35

GARBAGE COLLECTION CALLS

P2A	P2B	P2C	P2U	P2V	P2W
10	1	1	*	6	5

MINIMUM DATA STORE FOR PROGRAMS TO RUN (K)

P2A	P2B	P2C	P2U	P2V	P2W
42	40	<=30	*	40	<=30

(The minimum data store allocated to REDUCE is 30 K words)

* Program P2U would not run to completion in 124 K words of data store.

The most significant saving was made in converting program P2A(P2U) to program P2B(P2V) by using the two algebraic optimisations: simplification using application of substitutions and avoiding gcd calculations. The optimisation of program P2B to program P2C gave such good results in comparison to the optimisation of program P2V to program P2W because M=0 gives a simple case of the problem.

4.0 CONCLUSIONS

The power of optimisation, and in particular algebraic type optimisation is clearly shown by the measurements in Section 3.

A number of algebraic optimisations may be

applied to REDUCE programs prior to execution. However, in certain situations with optimisations such as code motion, switching of execution flags, simplification and rearrangement of computational order, it may not be possible to decide until execution of the program the best arrangement of the code. For example, in many cases where code motion might be applied it is difficult to know the best position for a piece of code to give the maximum optimisation.

To obtain an answer from a REDUCE program, the programmer may have to try several rearrangements of the code. A supervisory program that stepwise refines and executes a program would automate the task usually performed by the programmer.

As quoted from [Katz 78] in the above Introduction, program synthesis gives an "optimal" solution to the "optimisation" problem. An understanding and use of stepwise refinement would aid program synthesis.

Optimisation of LISP programs has been attempted. Some optimisations given in Section 2.4 might be better incorporated into the LISP System than as part of the REDUCE optimiser. The results in Section 3 show that an optimiser for REDUCE programs would be a very useful tool. The optimisations described may be applied by hand. However, this process is very tedious and the algorithm used in the program may become opaque.

The authors hope to provide software to aid optimisation of programs for REDUCE.

ACKNOWLEDGEMENTS

Thanks are due to Professor J.P. Fitch and Mr D. Powell-Evans for their encouragement.

REFERENCES

Allen F E, Cocke J 1972, A catalogue of optimizing transformations. In Design and Optimisation of Compilers. Ed Rustin R. Prentice Hall. Englewood Cliffs, N.J. pp 1-30.

Bagwell J T Jr 1970, Local Optimizations. SIGPLAN Notices July 1970, Proceedings of a Symposium on Compiler Optimisation, pp 52-65.

Barton D, Fitch J P 1972, A review of algebraic manipulative programs and their application. Computer Journal Vol 15 No 4, pp 362-381.

Beardsworth R 1979, Lethal, A FORTRAN Vector Based Algebra System. Department of Computer Studies, University of Leeds, England.

Busam V A, Englund D E 1969, Optimization of Expressions in FORTRAN. Communications of the ACM Vol 12 No 12, pp 666-674.

Campbell J A 1976, Compact Storage for Computations Involving Partitions. SIGSAM Bulletin No 40 (Vol 10 No 4), pp 46-47.

Campbell J A 1979, Symbolic Computing with Compression of Data Structures: General Observations, and a Case Study. Symbolic and Algebraic Computation, Springer-Verlag, pp 503-513.

Cocke J, Schwartz J T 1970, Programming Languages and Their Compilers. Courant Institute of Mathematical Sciences, New York University.

Feign D 1979, A Note on Loop "Optimization". SIGPLAN Notices Vol 14 No 11, pp 23-25.

Hearn A C 1974, REDUCE 2 User's Manual. University of Utah.

Katz S 1978, Program Optimization Using Invariants. IEEE. Trans. of Software Eng. Vol SE-4 No 5, pp 378-389.

Kibler D F, Neighbors J M, Standish T A 1977, Program Manipulation via an Efficient Production System. SIGPLAN Notices Vol 12 No 8. pp 163-173.

Kou L T 1977, On Live-Dead Analysis for Global Data Flow Problems. JACM Vol 24 No 3, pp 473-483.

Loveman D B, Faneuf R A 1975, Program Optimizations - Theory and Practice. SIGPLAN Notices Vol 10 No 3, pp 97-102.

Loveman D B 1977, Program Improvement by Source-to Source Transformations. JACM Vol 24 No 1, pp 121-145.

Moses J 1971, Algebraic Simplification - A Guide for the Perplexed. SYMSAM/2, pp 282-300.

Pearce P D 1976, PhD Thesis. University of Southampton.

Pearce P D 1979, Optimising Users' Programs. SIGSAM Bulletin No 50 (Vol 13 No 2). pp 11-12.

Pearce P D, Hicks R J (I), Data Structures and Execution Times of Algebraic Mode Programs for REDUCE. To be published.

Pearce P D, Hicks R J (II). The Application of Optimisation Techniques to Algebraic Mode Programs for REDUCE. To be published.

Rohl J S 1975, A Introduction to Compiler Writing. MacDonald and Jane's/American Elsevier.

Schneck P B, Angle1 E 1973, A FORTRAN to FORTRAN Optimising Compiler. Computer Journal Vol 16 No 4, pp 322-330.

Standish T A, Kibler D F, Neighbors J M 1976, Improving and Refining Programs by Program Manipulation. Proceedings of the 1976 ACM. National Conference, pp 509-516.

Stoutemyer D R 1978/1979, REDUCE Interactive Lessons, REDUCE Newsletters 2, 5, 6, and 7. University of Utah.

Sundblad Y 1972, SYMBAL (Solution of Problem 2). SIGSAM Bulletin No 24, pp 18-19.

Waite W M 1976, Compiler Construction, An Advanced Course, Chapter 5.E, Optimization. Editors Bauer F L, Eickel J. Springer-Verlag, pp 549-602.

Wegbreit W 1976, Goal-Directed Program Transformation. IEEE. Trans. of Software Eng. Vol SE-2 No 2, pp 69-80.

Views on Transportability of LISP and LISP-based Systems

Richard J. Fateman[1]
Computer Science Division
and
Center for Pure and Applied Mathematics
University of California
Berkeley, CA 94720

ABSTRACT

The availability of new large-address-space computers has provided us an opportunity to examine techniques for transferring programming systems, and in particular, Lisp systems, to new computers. We contrast two approaches: designing and building a Virtual Machine implementation of Lisp, and (re)writing the system in a "portable" programming language ('C'). Our conclusion is that the latter approach may very well be better.

1. Introduction

The recent introduction of low-cost (under $150k) medium-scale computers with large address spaces (12-1000 megabytes) and appropriate memory management hardware, has made it possible to consider building popularly-priced huge Lisp-based systems.

For the past decade or more, the standard Lisp host for most laboratories has been the Digital Equipment Corporation's DEC * PDP*-10 or DEC-20. In these systems Lisp has been implemented to provide about 1 megabyte of address space or 256k Lisp cells, not counting binary program space.

Moving to a machine with one or more orders of magnitude more address space than this 1 megabyte limit is important to several programming systems which have "cut their teeth", so to speak, on the PDP-10. Among these are theorem provers, planning and natural language systems, and other "symbol manipulation" tasks. One of special interest to us is the algebraic manipulation system MACSYMA. Many MACSYMA users believe they have been thwarted by the address-space limits on the PDP-10.[2]

In some cases, providing a larger address-space would merely reveal another problem, namely that the problem being posed was computationally intractable, and the user ran out of space before he ran out of patience. Nevertheless, there are clearly going to be important problems that can be solved by increasing the address space available to a MACSYMA user.

Several groups of programming systems researchers have been struck by the possibility of constructing reasonable, cost-effective, production Lisp machines out of the next generation of large-address computers, but have taken different approaches to this construction. In this paper we describe one such effort and give our current view about maintaining a programming base by transporting a Lisp system to a new machine. We outline our rationale and approach, which, we feel, goes somewhat ***against*** the "common wisdom" of the desirability of Virtual Machines.

2. Which Lisp? Which Machine?

We will describe our efforts in relation to a Lisp system for the DEC VAX-11 systems. We are aware of other large-address-space machines and (among VAX owners) other major Lisp systems under development. MIT's Laboratory for Computer Science (under Joel Moses) and USC's Information Sciences Institute (under Robert Balzer) are developing Lisps. Our Lisp (named *Franz*) has been running since January, 1979, and is distributed to more than 60 sites. As of now (April., 1981), the MIT and ISI Lisps are unavailable. In any case, we will draw our illustrations primarily from our experience with our implementation of *Franz Lisp* on the DEC VAX 11/780, and contrast that with the approaches of MIT and ISI.

Franz Lisp is modelled after PDP-10 Maclisp, since that is the language used for MACSYMA. We believe the transportation of any programming system should be planned with an understanding of the continued support requirements for an existing programming base. We expect that users of PDP-10 Interlisp would prefer a VAX-Interlisp, and users of REDUCE and its "Standard Lisp" would differ in specification details, but not significantly. We feel it is unwise and probably unnecessary that large Lisp-based systems be ***totally*** portable between environments, since this also requires the defects of previous systems to be carried to or simulated on new implementations.[4] We are not concerned that there are differences in input/output and such details; we doubt that the differences between an Interlisp and Maclisp system would provide great difficulties to a competent programmer.

---

1. Work reported herein was supported in part by the U. S. Department of Energy, Contract DE-AT03-76SF00034, Project Agreement DE-AS03-79ER10358.

2. DEC, PDP-10, VAX, and UNIX are trademarks.

3. While it has been possible for 8 years or more to run Lisp programs and in particular, a version of MACSYMA, in the Multics environment on a Honeywell 6000 series computer, and on an MIT Lisp Machine, either of which has a very large address space. Unfortunately these machines are in other ways unsuitable because of unavailability and/or efficiency and cost.

4. It is not clear whether this simulation will be done in the case of Interlisp on the VAX.

3. The Berkeley Lisp

3.1. Goals

As has been indicated, the Berkeley VAX Lisp system had, as its primary goal, the support of a particular Lisp-based system, MACSYMA, originally written in PDP-10 Maclisp. Enhancements to Maclisp on the PDP-10 and to MACSYMA continue to be made, and much of our recent effort has been to keep tracking the changes to this system.

A second goal was to provide an interface to the "rest of the world" from Lisp, in a natural way. That is, our intention was to allow the programmer to write his/her program in the most suitable language available, Fortran, "C", Pascal, assembler, or other languages as they become available. In particular, program libraries available for the solution of engineering and applied mathematics problems had to be directly accessible to the user of a symbolic system. By contrast, it is not possible to use Fortran code in the PDP-10 Maclisp system except through the rather extraordinary route of translating the Fortran to Lisp! Some functionality is lost in the translation.

A third goal was, to the extent possible, to make as much of the system machine independent, without sacrificing efficiency. It was *not* a goal to "make the system as machine independent as possible," but to determine those parts of the Lisp system which should be made machine-dependent, and write them appropriately. It appeared that the UNIX* operating system provided many of the features needed, and the VAX enhancements would add to its convenience and efficiency. In fact, the first version of *Franz* was written for a much simpler version 6 UNIX system running on a DEC PDP-11[5].

We feel that subject to further improvements in the code generation of the compiler, and the implementation of frills which are not central to MACSYMA but convenient to have available, the Lisp system is quite satisfactory. It has been in active use at a number of sites since approximately February, 1979, just a few months after our first VAX computer was delivered. Researchers at several of these sites have supplied us with various packages adapted from other Lisp dialects.

3.2. Features

We will not characterize Franz Lisp in this relatively brief space except to indicate that most of Maclisp's linguistic features are available. We prefer to philosophize about "features" in general.

Many people think about Lisp systems in terms of esoteric features (Does your lisp support xxx? Where xxx may be "parallel processing," "closures," "spaghetti stacks," "upward funargs," "reader macros," " DWIM," particular iteration facilities, ...).

Others think in terms of the time and space efficiency of compiled code, the ability to expand data space dynamically, the ability to declare and use new data types, the ability to field asynchronous interrupts, support for inter-process communication, the ability to perform fast arbitrary-precision integer arithmetic, or fast floating-point arithmetic.

We distinguish between these two classes of features by our general belief that the first set can be simulated in many Lisps, either directly in Lisp, or with the addition of some few primitives in an "extra-lingual" extension.

5. In fact, given a reasonable set of tools including a 'C' compiler, other operating systems would have been appropriate: The extent to which our portability is independent of UNIX was dramatically illustrated when several of us, plus David Kashtan of SRI International not only brought up *Franz* on a rather different time-sharing system on the VAX, namely DEC's VMS, but brought up a full-fledged MACSYMA in about 3 weeks.

The second set is not of that nature, since these features, by and large, depend on the basic Lisp design. These features may be difficult to simulate or add to a pre-existing system; they rarely can be added by the use of Lisp alone, at least without a significant loss in efficiency. Thus, in our view a system which provides this second set of capabilities, has a basically sound data structure design, is parameterized in a fashion to allow extensions, has a reasonable run-time environment, and a simple way of adding new hooks, can be expected to provide most forms of exotica of the first kind, eventually. We feel that Franz is designed reasonably to provide this second set of features, and includes some of the first set too.

Certainly other capabilities will have to be addressed in the long run: One obvious problem will be the development of effective garbage collection strategies for huge virtual address machines. Some recent studies of Franz/Macsyma provide some especially relevant suggestions. (refs. 14, 15, 4)

4. Approaches to Portability

The concept of software portability is rather vague, running from "enough information is available to guide a required particularization" to "executable without any modification." See, for example, (ref 5). The general approach to producing software that is highly portable is to use a subset of features in a common programming language such as Cobol or Fortran. Unfortunately, the more portable a program is intended to be, the fewer the features that can be used. In talking about total systems, some features cannot be simulated, but must be present in the underlying system. Variations in identifying the level of primitives, and the type of support, are, in our opinion, important factors in the successful construction of a transportable system.

In the sections below we will distinguish between the Virtual Machine approach and a Non-Virtual Machine approach to portability.

4.1. The Virtual Machine, Defined

The Virtual Machine (or VM[6]) approach attempts to define the programming environment as an onion-like object. Viewed from outside a particular layer all interior layers are invisible. Thus the Lisp users see the layer constructed by the Lisp system program, the Lisp system program sees the layer provided in the "Lisp Virtual Machine" and the LVM sees the layer provided by the operating system. Additional interior layers can be defined by the machine architecture and micro-architecture, and carried to an extreme, bits, digital signals, etc. Other layers can be added outside this object, for example, providing a MACSYMA interface on top of Lisp.

The Lisp Virtual Machine approach to transportability, then is one where all transactions between the Lisp user and the support system must go through the Lisp VM layer, which will then be grafted from one "onion-core" to another. The VM layer can be implemented as a suite of operations or "macro-instructions," typically defined by templates or subroutines in assembler language, although it is certainly worth considering writing this layer either in a higher-level language (Fortran and Pascal have been used for such purposes), or in micro-code.

6. Note that VM is sometimes used to stand for Virtual Memory. We feel Virtual Memory is a **Very Good Idea**. We are not attacking **that** VM, but the Virtual Machine approach, as we see it being used currently.

4.2. System Implementation in a Higher-Level Language

In terms of alternatives to the VM approach, we can try to jump right in and implement a Lisp system, say by writing it in assembler, as has been done for most systems on the PDP-10, or we could use an allegedly "portable" language. We discuss this approach here.

Franz was implemented by choosing a higher-level language which is supported as a system-programming language (on several different computers) and writing a run-time system for Lisp. This run-time system is *not* a Lisp Virtual Machine because it does not attempt to define and encapsulate all modes of communication between the Lisp system and the operating system and hardware. The implementation language chosen was 'C', a language demonstrably able to support language development in UNIX, and thus a language which we felt would not require extension to do the right things[7]. A moderate amount of code was written in Lisp itself, and a few pages of assembler were used to define "bignum arithmetic" discussed below. By contrast, had we chosen Fortran, we would have had to invent numerous features, among them stacks, storage allocation, and pointers. In Pascal (used as the basis for a VAX Interlisp virtual machine by Prof. W. Havens at the University of Wisconsin, Madison) there *are* pointers but the Pascal type-checking has to be finessed. One major advantage to 'C' by comparison is that there is essentially one (portable) compiler defining the language, and thus standards committees and the like do not affect us. One could view the 'C' support system as a Virtual Machine, although this is certainly not a rigorously defined system, depending on system calls for most of its power.

The main line of the Lisp interpreter and storage allocator was written as a class project by three Berkeley undergraduates, in the Spring of 1978. This was done on a PDP-11/70 UNIX system. When the first of Berkeley's VAX 11/780's arrived, this rudimentary Lisp system was transported from the PDP 11/70, and work begun to provide the functionality required to run MACSYMA. These facilities included a name-space of functions comparable to that available in Maclisp, a comparable resolution of loose ends in the definition of functions (e.g. (CAR NIL) = NIL), and a similar input/output structure. These objectives were, in retrospect, rather easy to meet in an interpreter as long as the requirement was to provide the same answers, but not necessarily the same efficiency in the same places. A more difficult task was satisfying the requirement that the source files used to define the MACSYMA system on the PDP-10 and the VAX should be identical, character for character. This task was dependent on the operating systems and in particular, the file system conventions on two distinct computers. A very few machine dependent pieces of Lisp code were "conditionalized" for the different machines. A rather elaborate system of macro-expansions, reader-macros, and compile/load/run time directives had to be completely modelled. Complicating these issues were the efforts of people at MIT to simultaneously update the same files for use by a Multics version of MACSYMA, and the on-going transportation of facilities to the MIT Lisp Machines.

Nevertheless, we consider the transportation of MACSYMA to the VAX via Franz Lisp quite successful. We believe transportation to other systems, especially those which support virtual memory, the UNIX standard input/output library and a 'C' compiler, will be relatively simple. Transportation to a system not supporting these features can proceed by porting 'C' and other parts of the UNIX system first. This was the approach taken to move Franz to VAX/VMS.

7. We have, in fact, made a very small change to VAX load modules (call masks) to avoid saving and restoring stack pointers unnecessarily.

The principal cost to transport MACSYMA to a 16-bit micro computer supporting 'C' and UNIX would be rewriting the Franz Lisp compiler (named Liszt). This program, which has undergone several versions, most recently was re-written by John Foderaro at Berkeley. It has definite machine dependent parts, since it generates VAX assembler language. Since it is rather short, we believe it can be understood and moved with only moderate effort.

Alternatively, one could use a "machine-independent" Portable Lisp Compiler (Griss and Hearn) as a base, or use an "L-code" system developed by William Rowan at Berkeley, for Franz. We had previously examined the Griss/Hearn compiler and felt that adopting it would not provide a faster route to a Maclisp-compatible compiler.

4.3. The Virtual Machine Approach to Lisp

There are several instances of Virtual Machine Lisps, e.g. those listed in the references by Griss etal, Havens, White, Deutsch. There are no doubt others.

To review: the basic notion in each case is that a well-defined language, given a set of formal specifications, can be implemented by building a shell (or layer of an onion) around the particular architecture of any given machine. This shell defines certain primitives which are then used exclusively by the language constructs, the compiler, etc. If the primitives are simple enough and carefully chosen, the shell, or virtual machine can be moved between architectures, with only moderate loss in efficiency and with only a small amount of machine dependent code.

The principal advantages one would expect from the Virtual Machine implementation of a Lisp, are listed below.

It should be easier to debug programs by comparison with a standard. Once one system is running, any different performance can be attributed to different code in the VM, which must then be debugged to conform.

It should be easier to move properly chosen primitives than a "whole language" and its support system.

The implementor need not grapple with the total complexity of either the old or the new environment. Some particularly ugly features can be totally plastered over.

The criticisms which we level at the VM approach above do not represent attacks on these advantages; we are in full agreement with the VM proponents in these areas. The difficulties lie in properly choosing the primitives, deciding which particular features are ugly and should be plastered over, and which features should be provided a place of honor.

Furthermore, we are concerned about efficiency, and the factual ease of re-implementing a VM base; we worry about the difficulty of programming a Lisp using a predetermined VM, that can continue to satisfy all users and make full use of the environment.

We provide some details in the next few sections to support our opinions on this.

4.4. Advantages and Disadvantages of Our Approach

The advantages and disadvantages of our approach to Lisp implementation are basically those which result from a commitment to use the operating system available and implementation tools and hardware at hand, as nearly as possible. Thus we use the same loader formats, the same assembler for the compiler output, and the same input/output routines. (The binary to decimal floating-point conversion is still different from PDP-10 Maclisp, because of this commitment. We use double precision, among other distinctions.) The following sections deal with specific points of comparison which have been raised.

4.4.1. Character Handling

The UNIX operating system does not fully support the interrupt-character structure described in the Interlisp and PDP-10 Maclisp manuals without the programming of a rather elaborate "front-end" character handler. We felt that in the early implementation stages, it was appropriate to provide a primitive asynchronous interrupt handling facility, but inappropriate for us to spend time on a more elaborate one, since other users of UNIX would surely like to use this too: we were unwilling to force everyone else to use Lisp in order to use good interrupt handling. In fact, such a facility has since been implemented and has been distributed in recent ("4BSD") UNIX/VM systems distributed by Berkeley.

4.4.2. Proofs of Correctness

Another consequence of the use of the operating system and the hardware as we have chosen to do it, is that it may be relatively more difficult to formally verify the correctness of an implementation, or to prove the equivalence of two different implementations. We do not bemoan this particularly; there is no formal specification of PDP-10 Maclisp outside the several-inch thick assembler listing plus a formal specification of the PDP-10 operating system. A formal specification in some other language might be just as unwieldy. It is not clear that one could provide a check against such a specification at reasonable cost.

A formal specification of Macsyma (other than the program listing in combination with the Maclisp listing) is not likely to be developed either.

4.4.3. Cursor Positioning

As an illustration of a difficulty with a virtual machine proposal, consider a specification of a program to optimize display cursor positioning in a terminal-independent fashion. That is, the program must compute, given a current-position and a desired position, a sequence of characters to reposition the cursor with the least delay. It is possible to write such a program entirely in Franz by reference to an on-line file of "terminal capabilities." Alternatively, one could use the program already available in the system merely by calling it from Lisp. The use of such an external program would not be encouraged in a virtual machine Lisp since it hardly seems primitive, and thus on aesthetic grounds one would not have the use of such a program. On practical grounds, such a program might not be available on each implementation of the Virtual Machine, and it would thus be unwise to depend upon it. From our point of view, the Virtual Machine is modeling the operating system more than the language; the UNIX cursor-positioning program itself is portable, having been written in the same implementation language as Franz, and is thus easily specified in the supra-Lisp Virtual Machine.

Cursor positioning, if you have not thought about it, is rather tricky, It is possible to "read cursor position" from a relatively few terminals. Thus the computer must have a model of the display that is in some sense above the user-level input/output interactions with Lisp, and which keeps track of the display even when Lisp is not directly in control.

4.4.4. Fortran and Other Languages

One could implement, as has been done in PDP-10 Maclisp, and perhaps also on the MIT Lisp Machine, a program which translate Fortran into Lisp, and then, by means of a clever compiler, provide adequate Fortran execution speed. This retains control of the computer within the Maclisp environment, and is the only way to use Fortran on the Lisp Machine, since there is no real system "outside" Lisp on that machine. There are several problems with this approach on a commercial machine.

First, the Fortran-Lisp route almost inevitably produces poorer code; the highly tuned and usually reliable Fortran supplied by a manufacturer's compiler may generate instructions totally alien to the usual Lisp environment: deals naturally with double-precision complex numbers, format statements, and substantial mathematical subroutine libraries.

Second, any changes to the Fortran standard, and improvements in code generation, etc. would not be reflected automatically in the Fortran-to-Lisp translator.

By contrast, our approach allows for the compilation, linkage and execution of Fortran (or 'C' or Pascal) source code at run time; the compiler which is used is independent of Franz. Franz permits full access to the Fortran run-time libary, IMSL, Linpack, the augment pre-compiler in which interval arithmetic is implemented, etc. The Multics version of Maclisp is also capable of calling non-Lisp routines; in fact their approach resembles ours but with PL/I substituted for 'C' and more elaborate conventions for passage of data types with coercions to strings when necessary (ref. 6).

A recent example of the usefulness of the inter-language communication facility was the implementation of a Fast Fourier Transform program (in a finite field) used for a clever polynomial multiplication algorithm. By using 'C', a speed-up by factors of 100 or more could be obtained over the program written in Franz Lisp directly. It is unreasonable to expect **any** Lisp compiler to produce code nearly as good as a good 'C' compiler, on non-Lisp-like programs.

4.4.5. How to Implement Bignums

Another illustration of an advantage in our approach is the implementation of an essential data structure for MACSYMA, namely arbitrary-precision integers, or "bignums" as they are usually called. These are ordinarly implemented as linked lists, and the programs for manipulating them can be written in Lisp, 'C', or assembler. There happen to be a few very nice machine instructions on the VAX to aid in this. These instructions would have to 'stick through' the virtual machine -- and in fact, would have to 'stick through' a 'C' program too, so we wrote these parts of Franz in assembler. Particularly convenient are the extended integer arithmetic instructions, and potentially useful (although not used in our current package), is ADWC, "Add with Carry".

4.5. Is the Compiler Transportable?

The compiler for Franz is written in Lisp, as are nearly all Lisp compilers. It is a compiler which has specific information for generating VAX code, and thus is not directly portable. However, we believe the code generation portions can be re-targeted to another machine which supports 'C' with effort no more than that needed to re-implement a virtual machine on an another computer. Since there is some kind of Turing-equivalence argument here, there remains some question about how strong a claim on "ease" can be made. Certainly a compiler (one only) could be made to compile to a virtual machine, and then the effort to move the compiler is the same as the effort to move the virtual machine; the quality of code generated in such a situation is hard to judge, but we believe that the nature of the underlying machine architecture would affect the desirability of using the virtual machine in various ways, and that inserting another layer of "information hiding" at this level is detrimental. In fact, we believe moving a Virtual Machine Lisp Compiler would be just as much trouble, when one considers that the sensible approach would be to implement the code generation via expansion of machine-dependent macro instructions.

5. Other Machines, Other Times

In some recent experimental machines, most notably the MIT Lisp machine, there is a virtual machine implemented in hardware and microcode which represents a high quality support for the implementation of Lisp-based systems, graphics, real-time interaction, editing, etc. What is not supported is a Fortran compiler or any of the rather numerous numerical routines expected in the virtual machine of a symbolic/numeric support system such as we are building on the VAX. We do not feel the Lisp machine microcode interposes an undesirable layer of information hiding at the virtual machine level -- partly because it is easily altered -- but mostly because the hardware is designed to support the micro-coding of a Lisp virtual machine. Few, if any, commercially available micro-codable (or otherwise) machines come close to supporting this careful choice between generality when needed, but specificity, when appropriate for efficiency. Unfortunately, the small number of Lisp Machines, their separation from the main stream of numerical software, and unconventional floating-point arithmetic, means that production quality numerical libraries may be difficult to obtain or develop. (Very interesting and in some cases, quite robust, programs exist for the Lisp machine. Converting these research projects to a satisfactory state for a less fault-tolerant audience appears to be a major task now under way, by at least one commercial firm.)

The NIL project at MIT is attempting to provide an MIT-Lisp-Machine style Virtual Machine on the VAX but cannot rely on the possibility of utilizing microcode for efficiency. Although there is writable control store on the VAX, the 11/780 model makes it quite uneconomical to use by imposing substantial time penalties to enter such code (ref. 12). If there is a correction to this, then presumably the NIL virtual machine could be re-implemented; or the Franz compiler could be changed; or the "C" code generation could be altered. Or all of the above.

6. Portability to other machines

We have begun the examination of several machines in the hope of finding re-targetable UNIX operating system support for Franz and MACSYMA. A very low cost VAX would undoubtedly provide the most logical migration path, but the VAX is a much more complicated architecture than is needed for UNIX or Franz. There are prospects for machines which implement a far simpler instruction set with no loss in compactness. Johnson in (ref. 8) proposes an architecture more suitable for the language "C", Patterson in (ref. 10) describes a reduced instruction set computer that is being benchmarked on numerous programs, some of which have been heavily influenced by the requirements we see for Lisp. While more fully engineered computers would be required for a production system for Lisp, the mere presence of a large instruction set and features such as memory mapping are no assurance of satisfactory performance: Regardless of the other instructions, the critical operations are those of data movement and subroutine calling. (ref. 3)

7. Conclusions

Time will tell how effective various Lisp systems will be on the VAX; side-by-side comparisons of them will eventually be possible. Popularity will of course also be affected by non-technical issues such as manufacturer or users-group support, the "not-invented-here" syndrome, inertia, etc.

In the longer run, making Lisp systems available on yet lower cost machines will benefit from our experiences at this time.

While there appear to be advantages in the concept of the Virtual Machine for character-for-character transportability of programs written in Lisp, it is our opinion that a Lisp Virtual Machine approach is not particularly satisfactory for efficiency and usefulness; it appears to us that to date no systems have been entirely successful using this approach (refs. 1, 6, 9).

On the other hand, we believe that our use of the UNIX environment as a Virtual Machine (roughly speaking), has provided useful functional approaches to language and system implementation. Thus it seems primarily a question of where one should cut the "onion" -- at the operating system, or a level above. Our success has been based on the design of the UNIX environment, and the use of a higher-level language for those parts that are appropriately machine independent. All the hooks to the operating system were retained. Just as the advocate of structured programming may occasionally use a well-placed "goto" we were willing to descend to the level of assembler for those rare operations which profiling revealed were bottle-necks, and for which better alternatives existed to the code generated by the high-level language.

References

1. Robert Balzer and D.Dyer [personal communications]
2. L. Peter Deutsch, "Experience with a Microprogrammed InterLISP System," *Proc. MICRO-11*, IEEE NY 1978.
3. John K. Foderaro, "Franz Lisp Manual," in *Berkeley UNIX Reference Manual Vol. 2c*. 1979.
4. John K. Foderaro and Richard J. Fateman, "Characteristics of a Large Lisp Program," in preparation.
5. P.A. Fox, A. D. Hall, and N. L. Schryer, "The Port Mathematical Subroutine Library," *ACM Trans. on Math Soft. 4, 2* (June 1978), 104-126.
6. B. Greenberg, personal communication.
7. Martin L. Griss, Robert R. Kessler and Gerald Q. Maguire Jr., "TLISP -- a 'Transportable-Lisp' Implemented in Pcode," in *Symbolic and Algebraic Computation*, E.Ng (ed.), Lecture Notes in Computer Science 72, Springer Verlag. 1979. (490-502).
8. Stephen C. Johnson, "A 32-Bit Processor Design," Computing Science Tech Rep. #80, Bell Laboratories, Murray Hill, N. J. April 2, 1979, 11 p.
9. William Havens [personal communication]
10. David A. Patterson, "Design for a Reduced Instruction Set Computer (RISC)," working papers, U.C. Berkeley EECS department.
11. Warren Teitelman, *INTERLISP Reference Manual*, Xerox Palo Alto Res. Center, Palo Alto, CA. (1978).
12. Richard Tuck, "Micro-code tools for the VAX-11/780," Masters Thesis, EECS Dept., Univ. of Calif, Berkeley, 1979.
13. Daniel Weinreb, David Moon *Lisp Machine Manual* Second Preliminary Version, MIT AI Laboratory, January, 1979 (283 pages).
14. John L. White (and others) NIL correspondence on the MACSYMA MIT-MC computer; NIL design documents.
15. John L. White, "Address/Memory Management for Gigantic LISP Environments," in Proc. 1980 Lisp Conference, Stanford University.

Algebraic Constructions for Algorithms

by

S. Winograd*
IBM Watson Research Center
P. O. Box 218
Yorktown Heights, NY 10598

(Extended Abstract)

In the last decade or so there has been an outgrowth of research in algebraic complexity of computations which showed how to derive algorithms systematically. One of the features of these derivation is their reliance on algebraic and symbolic constructions. I would like, in this paper, to survey some of the symbolic and algebraic constructions which are used, and then draw some conclusions on the implications of this development for symbol manipulation systems. Let me start by sketching some of these derivations.

1. Polynomial Multiplication (Chinese Remainder Theorem):

We want to compute the coefficients t_0, t_1, t_2 of the polynomial

(1.1)

$$T(u) = t_0 + t_1 u + t_2 u^2$$
$$= (x_0 + x_1 u)(y_0 + y_1 u) = R(u)\ S(u).$$

It is easy to verify that

(1.2)

$$T(u) = T(u) \mod (u^2 + u) + x_1 y_1 (u^2 + u)$$
$$= T'(u) + x_1 y_1 (u^2 + u).$$

By the Chinese Remainder Theorem, we can write

$T'(u)$ as:

$$(1.3)\quad T'(u) = T_1(u)\ (1 + u) + T_2(u)\ (u),$$

where

(1.4)

$$T_1(u) = T(u) \mod u$$
$$= (R(u) \mod u)(S(u) \mod u) \mod u$$
$$= x_0 y_0$$

(1.5)

$$T_2(u) = T(u) \mod (u + 1)$$
$$= (R(u) \mod (u + 1))(S(u) \mod (u + 1)) \mod u + 1$$
$$= (x_0 - x_1)(y_0 - y_1)$$

Putting it all together we obtain: $T(u) = x_0 y_0 (1 + u) + (x_0 - x_1)\ (y_1 - y_0) u + x_1 y_1 (u^2 + u)$. Identifying coefficients of powers of u we get the following algorithm, which uses only three multiplications:

(1.6)

$$t_0 = x_0 y_0 = x_0 y_0$$
$$t_1 = x_0 y_1 + x_{y0}$$
$$= (x_0 - x_1)(y_1 - y_0) + x_0 y_0 + x_1 y_1$$
$$t_2 = x_1 y_1 = x_1 y_1.$$

---

*This research was partially supported by NSF Grant No. ECS-7921291.

This procedure for deriving the algorithm can clearly be generalized. However, the details of the derivation become more cumbersome as the degrees of the polynomial increase.

2. FIR Filters (Transportation):

We will start with the algorithm (1.6), and use it to derive a new algorithm, which we will call the transpose, or dual, or (1.6) [1].

Let us multiply the first identity of (1.6) by z_0, the second by z_1, and the third by z_2. Having done that let us sum the three identities and obtain:

$$\begin{aligned} & z_0x_0y_0 + z_1(x_0y_1 + x_1y_0) + z_2x_1y_1 \\ (2.1) \quad & = (z_0 + z_1)x_0y_0 + z_1(x_0 - x_1)(y_1 - y_0) \\ & + (z_1 + z_2)x_1y_1 \end{aligned}$$

If we now equate coefficients of the x_i's we get a new set of identities:

$$(2.2) \quad \begin{aligned} z_0y_0 + z_1y_1 &= (z_0 + z_1)y_0 + z_1(y_1 - y_0) \\ z_1y_0 + z_2y_1 &= (z_1 + z_2)y_1 - z_1(y_1 - y_0). \end{aligned}$$

The new problem, that of computing $z_0y_0 + z_1y_1$, and $z_1y_0 + z_2y_1$ is the transpose of the original problem, and the new algorithm is the transpose of the original algorithm.

An n-tap FIR filter is the computation of

$$(2.3) \quad w_j = \sum_{i=0}^{n-1} x_{i+j}\, h_{(n-1-i)} \quad j = 0,1,2,\ldots$$

So a transpose of the computation of t_0, t_1, and t_2, is computing two outputs of a 2-tap Fir filter. In general, the problem of computing m outputs of an n-tap FIR filter is a transpose of the problem of computing the coefficients of $T(u) = (\sum_{i=0}^{m-1} x_iu^i)(\sum_{j=0}^{n-1} y_ju^j)$.

3. FIR Filters (Tensor Product Construction):

We will now consider larger filtering problems than that of 2 outputs of a 2-tap FIR filter. Let us take the problem of 8 outputs of an 8-tap FIR filter. One way is to derive the algorithm using the procedure outlined in the previous section. Another method of derivation, known as the tensor product construction, is an outgrowth of the construction of Strassen [2] for matrix multiplication.

Eight outputs of an 8-tap FIR filter is the computation of

$$(3.1) \quad w_j = \sum_{i=0}^{7} x_{i+j}\, h_{7-j}, \quad j = 0,1,\ldots, 7$$

which we can write using matrical notation as $\underline{w} = X\underline{h}$, where X is a Henkel matrix. Let us partition X into 4×4 blocks, so (3.1) is now written as:

$$(3.2) \quad \begin{pmatrix} \underline{w}_0 \\ \underline{w}_1 \end{pmatrix} = \begin{pmatrix} X_0 & X_1 \\ X_1 & X_2 \end{pmatrix} \begin{pmatrix} \underline{h}_1 \\ \underline{h}_0 \end{pmatrix}$$

Using the identities (2.2) we have

(3.3)

$$\begin{aligned} \underline{w}_0 &= X_0\underline{h}_1 + X_1\underline{h}_0 \\ &= (X_0 + X_1)\underline{h}_1 + X_1(\underline{h}_0 - \underline{h}_1) \\ \underline{w}_1 &= X_1\underline{h}_1 + X_2\underline{h}_0 \\ &= (X_1 + X_2)\underline{h}_0 - X_1(\underline{h}_0 - \underline{h}_1). \end{aligned}$$

The crucial observation is that $(X_0 + X_1)\underline{h}_1$, $(X_1 + X_2)\underline{h}_0$, and $X_1(\underline{h}_0 - \underline{h}_1)$ are again the product of a Henkel matrix by a vector, i.e., are FIR-filter type calculation. We can, again, partition these matrices, or use any other algorithm for these calculations, to obtain the final algorithm. The one fly in the ointment, when we carry out this procedure by hand, is that the amount of symbolic computation which has to be performed grows rapidly.

4. Fourier Transform (Generator of a Group):

We will not describe the construction of the algorithm for computing the Fourier Transform here - it takes too much space for an extended abstract - but we will describe one aspect of the derivation of the algorithm which uses algebraic manipulation. The reader who is interested in more details is referred to [3].

The construction of the algorithms for Fourier Transform uses all the algebraic and symbol manipulation described above. But before they can be applied the inputs and outputs have to be reordered. Let me describe the reordering for computing the Fourier Transform of a prime number of points, specifically of 5 points.

Because 5 is a prime number, the set of nonzero integers, with operation multiplication modulo 5, form a group. This group is isomorphic to z_4, the group of addition modulo 4. One such isomorphism is:

(4.1) $1 \to 0,\ 2 \to 1,\ 3 \to 3,\ 4 \to 2$

where the integers on left are in M_5, and those on the right in z_4. This isomorphism is completely specified once we find that 2 is a generator of M_5. In the derivation of the algorithm we have to order the inputs a_0, a_1, a_2, a_3, a_4, and the outputs A_0, A_1, A_2, A_3, A_4 in a way which reflects this isomorphism. That is a_0, a_1, a_2, a_4, a_3, and A_0, A_1, A_2, A_4, A_3.

Similar kind of considerations occur in the derivation of multi-dimensional Fourier Transform on p x p x ... x p points, where p is a prime number [4]. There it is needed to find a generator of $GF(p^k)$ in order to specify the reordering which is needed to derive the algorithm.

5. Multidimensional FT (Field isomorphism):

In a yet unpublished result, E. Feig shows a new method for deriving an efficient algorithm for two-dimensional Fourier Transform on $2^n \times 2^n$ points. Again, we will not describe the algorithm here. What is important for our purposes is the derivation of some reordering which is a part of the new algorithm.

Let P(u) be an irreducible polynomial over the rationals, and let $r_1, r_2, \ldots, r_n$ be the roots of P(u) (in the field of complex numbers). Let Q(u) be another polynomial over the rationals, not necessarily irreducible, with roots $s_1, s_2, \ldots, s_m$. We further assume that all the s_i's are distinct, and that all of them lie in the field $F_1 = \mathbb{Q}(r_1)$ of the rationals extended by r_1. We define the n isomorphic fields $F_i = \mathbb{Q}(r_i)$, $i=1,2,\ldots, n$, and let us consider the field isomorphisms $\sigma_i: F_1 \to F_i$ given by $\sigma_i(r_q) = r_i$, $i=1,2,\ldots, n$. Each σ_i permutes the m quantities $s_1, s_2, \ldots, s_m$. These n permutations are an essential part of Feig's algorithm.

To be more specific, Feig's algorithm uses $P(u) = u^n + 1$, $Q(u) = u^n - 1$ or $Q(u) = u^n + 1$, where $n = 2^\ell$. In this case the permutations π_b are given by $a \to a'$ where

(5.1) $a' = (2b + 1)\, a \mod n,\ b = 0,1,\ldots, n-1,$

and

(5.2) $2a' + 1 = (2b+1)(2a+1) \mod 2n,$
$b = 0,1,\ldots, n-1.$

In either case, the determination of the permutation calls for symbol manipulation capabilities.

6. Conclusions:

The five algorithms described above are numerical algorithms - they take numbers as inputs and produce numbers as outputs. What should make them interesting to symbol manipulation community is that the derivation of the algorithm uses symbol manipulation and algebraic construction. Present symbol manipulation systems have the potential of aiding the algorithm designer, but more development work is needed for them to realize this potential. J. Cooley, of IBM Research, successfully used SCRATCHPAD to help him in the design of algorithms of the type described in sections 1 and 2.

To be able to fully automate the derivation of this kind of algorithm the symbol manipulation system needs to have the capabilities to perform the necessary symbolic and algebraic manipulation. It also needs the capability of taking the resulting identities and automatically transform them into a program in FORTRAN or other language. The reader may not agree with this conclusion, but I hope that this paper has enough evidence to convince the reader not to reject the conclusion without further thought.

REFERENCES

(1) J.E. Hopcroft and J. Musinski, Quality applied to matrix multiplication, SIAM J. on Comp. Vol. 2, No. 3, 1973, pp. 159-173.

(2) V. Strassen, Gaussian elimination is not optimal, Num. Math., Vol. 13, 1969, pp. 354-356.

(3) S. Winograd, Arithmetic Complexity of Computations, CBMS-NSF Regional Conference Series, No. 33, 1980, SIAM.

(4) L. Auslander, E. Feig, and S. Winograd, New Algorithms for the multi-dimensional discrete Fourier transform, submitted to IEEE Tran. on Acc. Sp. and Sign. Proc.

A Cancellation Free Algorithm, with Factoring Capabilities, for the Efficient Solution of Large Sparse Sets of Equations

J. Smit.
Twente University of Technology
Department of Electrical Engineering, EF9274
P.O.Box 217
7500AE Enschede
The Netherlands.

ABSTRACT.

Symbolic solutions of large sparse systems of linear equations, such as those encountered in several engineering diciplines (electrical engineering, biology, chemical engineering etc.) are often very lengthy, and received for this reason only occasional attention. This places the designer of a new and probably more succesfull symbolic solution method for the hard problem to find a representation which is suitable in the corresponding engineering areas, while still being neat and compact. It is believed that this problem has been solved to a great deal with the introduction of the new Factoring Recursive Minor Expansion algorithm with Memo, FDSLEM, presented in this paper.
The FDSLEM algorithm has important properties which make the implementation of an algorithm which can generate the approximate solution of a perturbed system of equations relatively straightforward.
The algorithms given can operate on arbitrary sparse matrices, but one obtains optimal profit of the properties of the algorithm if the matrices have a certain fundamental form, as is illustrated in the paper.

1. INTRODUCTION.

Large sparse systems of linear equations have to be solved in an efficient way with symbolic solution techniques in several diciplines. One of the problems with the required solutions is that a fully expanded representation of the results leads more than often to unacceptably long expressions. It is however not possible to use a fully unprocessed result from any of the popular algorithms, like the Gaussian elimination algorithm, for the solution of a linear system of equations, because such algorithms produce results with numerous internal cancellations. These cancellations have generally two natures, they are multiplicative and additive. A mere expansion of the steps of a Gauss Elimination algorithm in a straight line code could, only with severe limitations, be used to prepare the solution of a system of linear equations for numerical processing. Such an expansion of a code requires that intermediate results are not substituted directly, rather one subsitutes the numerical value, held by the expression in the current context of the algorithm. This technique could, again with severe limitations, be used to solve a system of linear equations with the aim to obtain a symbolic result.

©1981 ACM O-89791-047-8/81-0800-0146 $00.75

It should be noted however that there is relatively little profit to obtain from a symbolic solution which will be used for numerical processing as the algorithm to perform the required steps during the numerical processing is relatively short and effective. The possibility to perform pivotal interchanges when the code is generated dynamically is not totally unimportant in this respect.
The availability of a symbolic solution instead of a classical numerical solution can be desired for several reasons like:

- Further symbolic processing.
- Human interpretation of the result.
- Derivation of special properties like (symbolic) sensitivities etc.
- Derivation of a trucated result.
- Further numerical processing.
- Increased numerical stability of results processed with floating point instructions.
- Numerical processing without error conditions due to pivoting conditions.
- A matching of parameters in the system of linear equations to a numerical constraint using Newton iteration.

to name only a few.
It is however of utmost importance that the obtained results are compact. The result should be free of internal cancellations, in order to obtain the desired compact result, as intermediate expression swell will too soon lead to useless long and cumbersome results. The reduction to a canonical form is required if the algorithm itself is not cancellation free. An expanded form is in most cases the most obvious and an effective, but very expensive, way to remove the intermediate cancellations. In addition to the high cost of such algorithms, one sees that the used expansion of intermediate results leads to (considerably) longer formulas than strictly necessary, as will be shown in this paper. Other techniques have been exploited succesfully to keep symbolic results short, like the partially factored representation used by Hearn [1] and by Brown [2].
The algorithms presented in this paper perform a certain ordered evaluation, combined with a renaming of intermediate results, based on the actual sparsity structure of the sparse matrices involved in the solution. Most of the techniques introduced in this paper date back to the period 1693-1800, like the use of determinants, first introduced by Leibnitz in 1693 [3], Cramers rule in 1750 [4], the (near) discovery of nested minors by Bezout in 1764 [5], the introduction of keys representing minors by Vandermonde in 1772 [6]. Several additions and extensions of these old ideas were however needed in order to come to the current algorithms.

2. A discussion of some known algorithms.

Cramers rule can be used to solve a system of linear equations, using determinant algorithms to obtain the definite solution. Determinants can be calculated with several well known algorithms, however most of these algorithms are, in contrast with traditional determinant algorithms not free of internal additive and multplicative cancellations. The best known algorithm in this respect is the elimination algorithm.

2.1. Elimination algorithms.

Successive transformations can be used to eliminate a single variable from a system of equations Ax=b. The first transformation can be arranged to make all matrix entries in the first column under the main diagonal of the coefficient matrix zero, the second transformation to make all matrix entries in the second column under the main diagonal of the coefficient matrix zero, and so on. This process, which is given in matrix notation in (2.1,1) is known as the forward elimination step.

$$T_{n-1} T_{n-2} \ldots T_2 T_1 Ax = T_{n-1} T_{n-2} \ldots T_2 T_1 b \qquad (2.1,1)$$

The transformation of the coefficient matrix into upper triangular form, as used in the elimination algorithm, can also be used to calculate the value of the determinant of the coefficient matrix, as the determinant of a triangular matrix is known to be equal to the product of all matrix entries on the diagonal (the trace).
Following this algorithm one can calculate the determinant of the matrix in (2.1,2) with the elimination steps:

$$\begin{vmatrix} a1 & b1 & c1 \\ a2 & b2 & c2 \\ a3 & b3 & c3 \end{vmatrix} = \frac{1}{a1^2} \begin{vmatrix} a1 & b1 & c1 \\ 0 & \begin{vmatrix} a1 & b1 \\ a2 & b2 \end{vmatrix} & \begin{vmatrix} a1 & c1 \\ a2 & c2 \end{vmatrix} \\ 0 & \begin{vmatrix} a1 & b1 \\ a3 & b3 \end{vmatrix} & \begin{vmatrix} a1 & c1 \\ a3 & c3 \end{vmatrix} \end{vmatrix} = \qquad (2.1,2)$$

$$= \frac{1}{a1^2 \begin{vmatrix} a1 & b1 \\ a2 & b2 \end{vmatrix}} \begin{vmatrix} a1 & b1 & c1 \\ & \begin{vmatrix} a1 & b1 \\ a2 & b2 \end{vmatrix} & \begin{vmatrix} a1 & c1 \\ a2 & c2 \end{vmatrix} \\ & & \begin{vmatrix} \begin{vmatrix} a1 & b1 \\ a2 & b2 \end{vmatrix} & \begin{vmatrix} a1 & c1 \\ a2 & c2 \end{vmatrix} \\ \begin{vmatrix} a1 & b1 \\ a3 & b3 \end{vmatrix} & \begin{vmatrix} a1 & c1 \\ a3 & c3 \end{vmatrix} \end{vmatrix} \end{vmatrix} \qquad (2.1,3)$$

Two multiplicative factors: a1 and $\begin{vmatrix} a1 & b1 \\ a2 & b2 \end{vmatrix}$ are easily cancelled in (2.1,3), so it follows that the result after elimination can be written as: (2.1,3)

$$\frac{1}{a1} \begin{vmatrix} \begin{vmatrix} a1 & b1 \\ a2 & b2 \end{vmatrix} & \begin{vmatrix} a1 & c1 \\ a2 & c2 \end{vmatrix} \\ \begin{vmatrix} a1 & b1 \\ a3 & b3 \end{vmatrix} & \begin{vmatrix} a1 & c1 \\ a3 & c3 \end{vmatrix} \end{vmatrix} = \frac{1}{a1} \begin{vmatrix} (a1b2-b1a2) & (a1c2-c1a2) \\ (a1b3-b1a3) & (a1c3-c1a3) \end{vmatrix}$$

$$\begin{aligned} &= (a1b2a1c3-a1b2c1a3-b1a2a1c3+b1a2c1a3)/a1 \\ &\quad -(a1c2a1b3-a1c2b1a3-c1a2a1b3+c1a2b1a3)/a1 \end{aligned} \qquad (2.1,4)$$

The additive cancellation: $\frac{b1a2c1a3}{a1} - \frac{c1a2b1a3}{a1} = 0$ gives now:

$$\begin{aligned} &= (a1b2a1c3-a1b2c1a3-b1a2a1c3)/a1 \\ &\quad -(a1c2a1b3-a1c2b1a3-c1a2a1b3)/a1 \end{aligned} \qquad (2.1,5)$$

This result allows a multiplicative cancellation resulting in:

$$\begin{aligned} &+ (a1b2c3-b2c1a3-b1a2c3) \\ &- (a1c2b3-c2b1a3-c1a2b3) \end{aligned} \qquad (2.1,6)$$

The result presented is (of course) the same as the result obtained with a traditional determinant algorithm, however to obtain the cancellation free result given in (2.1,6), one needs to perform exact cancellations in the course of the algorithm.

The Gaussian Elimination algorithm as applied to numerical calculations, proceeds only up to formula (2.1,3).

It can be seen now clearly that the elimination fails if one of the factors in the multiplicative cancellation is zero. This is the well known result that the elimination can only be performed if all principal minors on the main diagonal are nonzero. [7]
The cancelling factor A1 in (2.1,5) is present in all terms except for the additively cancelling terms. The accuracy of the calculation decreases considerably if the magnitude of the multiplicative (cancelling) factor, which could not be cancelled in earlier stages of the elimination, has a value which is small compared to the magnitude of the additively cancelling terms. This implies that the elimination algorithm is inaccurate if the cancelling factors are relatively small or if the cancelling terms are relatively large. Elimination algorithms use therefore certain pivoting strategies in order to keep the errorgeneration relatively low. It will however be clear that such algorithms are, by necessity, far from complete if one wants to avoid the calculation of determinants during the elimination.
Multistep elimination algorithms can be used successfully to overcome the growth of intermediate results if determinants of matrices with integer matrix entries have to be calculated [8]. These algorithms can also be used successfully to generate cancellation free determinants from matrices with dense symbolic polynomials as entries [9], [10], [11]. However the need for cancellation of common terms and common factors implies that intermediate results have to be processed heavily, in order to obtain the desired exact determinant. This need for internal processing of results is yet an indication that the classical bound for the complexity of Gaussian elimination, which states that the elimination requires O(n**3) elementary operations, is no longer true if there is a need to process anything but real numbers, like: integers, Gaussian integers, rational numbers or symbolic expressions such as polynomials.
The internal processing of an elimination algorithm makes the expansion [12] of intermediate results necessary in order to remove the inherent internal cancellations. This removes good chances to generate a result which is effectively shorter than the expanded form of the result.

2.2 Polynomial interpolation algorithms.

Univariate interpolation algorithms are based on the idea that the univariate polynomial P(x) of degree n can be reconstructed if the value of P(x) is known at n+1 distinct values for x. The interpolation can be extended to the calculation of determinants which generate multivariate polynomials P(x,y,z,...).
Relatively many publications consider the construction of symbolic transferfunctions with interpolation techniques [13], [14], in all these publications one sees that the interpolating values are calculated with an elimination algorithm which operates on floating point data with floating point arithmetic. Such techniques are very sensitive to error propagation, resulting in possibly erroneous results, especially when one considers an increasing number of variables.

2.3 Classical determinant algorithms.

Classical determinant algorithms can be easily shown to be free of additive and multiplicative cancellations. That is the reason why these algorithms get some special attention here. It has been shown elsewere [15] that algorithms based on the Recursive Minor Expansion algorithm, given in (2.3,1) can be succesfully adapted for the efficient evaluation of sparse matrices.

$$\text{Det} = \begin{cases} \sum_{j=1}^{n} a(i,j)\text{x}sign(i,j)\text{x}minor(i,j) & i \text{ arbitrary} \\ & n>0 \\ 1 & n=0 \end{cases} \qquad (2.3,1)$$

Where minor(i,j) is the determinant of the given matrix with row i and column j deleted, and sign(i,j) gives -1 if the number of permutations to bring i,j to the 1,1 position is odd and sign(i,j) is +1 if this number of permutations is even.

Reference [15] gives more details about an algorithmic description of the DSLEM algorithm, the datastructures used in it and comparative data about its performance when applied to large sparse matrices.

Let us therefore only recall the most important properties P1..P4 of the DSLEM algorithm:

- P1. The (F)DSLEM algorithm develops the determinant with respect to the row or column with the lowest number of nonzero matrix entries, using a datastructure called HISTOGRAM [15] for efficient selection of rows and columns. This property is of fundamental importance for the adaptive nature and the corresponding efficiency of the algorithm.
- P2. Numerical minors like those encountered in the Kirchhoff equations [31] are kept nested when they have more than one nonzero matrix entry in each row and column. This property ensures that the resulting determinant is kept free of additive cancellations.
- P2t.The property P2 implies that the additive truncation error (6.1,3) can be effectively avoided.
- P3. The DSLEM determinant algorithm uses a Memo table [32] to store already calculated minors, such that recomputation of lower minors becomes unnecessary. This important feature is exploited in the FDSLEM algorithm to obtain a result in a structured and nested form [15].
- P4. There are no divisions in the FDSLEM algorithm.
- P4t.The property P4 implies that the multiplicative truncation error (6.1,4) cannot occur.

and concentrate first only on the new aspects of the nested representation of the result which is so characteristic for the FDSLEM algorithm.

3. The Factoring Double Sparse Laplace Expansion algorithm.

The FDSLEM algorithm is based on the DSLEM algorithm which utilizes the freedom in (2.3,1) to select an arbitrary row or column from the matrix to take the row or column with the lowest number of nonzero matrix entries (P1 see also [15]). The important implication of this algorithm is that the DSLEM algorithm will never develop any minor which could have a zero complementary minor due to the fact that any possible row or column in the complementary minor is empty (has no nonzero matrix entries). Furthermore the algorithm uses keys which give a unique identification of minors, with some similarity to the keys introduced in [6] and a representation for these keys as well as a hashing technique, introduced by Griss [16]. Minors which have not been computed before are stored in the hashed memo table (P3). Minors which still have to be computed are taken from the hashed memo table if possible. This provides an effective mechanism which prevents recomputation of lower minors.

The FDSLEM algorithm adds to these features a mechanism to manipulate the keys in the hash table, mapping implicitly found minors to a lower minor if possible, and the occurences of variables bound with these keys to see if simple factorisations of these variables are possible. The FDSLEM algorithm does not perform a full decomposition of the matrix into irreducible components in each elementary step of the development of the determinant, as this would make the expansion of the determinant prohibitively expensive. Instead the information present in the keys and the hash table is used to obtain similar results from previous steps in the development of the determinant, thereby obtaining an algorithm which can do some factorisations which cannot always be done with a matrix decomposition algorithm, because the FDSLEM algorithm uses automatic simplification techniques on intermediate resuls, but it might as well be that a factorisation could be obtained in some cases from the far more expensive combination of the DSLEM algorithm and a full matrix decomposition algorithm, where the cheaper FDSLEM algorithm is unable to find this factorisation.

4. A discussion of some computed results.

A detailed discussion of the FDSLEM algorithm will be delayed, in order to obtain first a better understanding of the ways in which the computed results can be presented. The analysis of the electrical network of figure 4,1 will be used to show some of the visible properties of the algorithm.

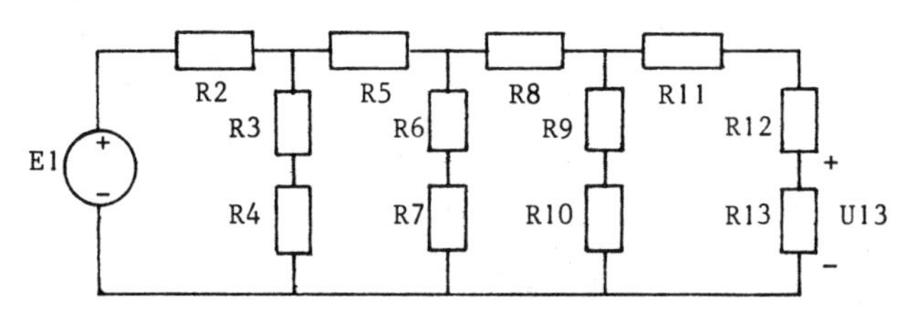

Figure 4,1

The FDSLEM algorithm writes the transferfunction:

$$\frac{U(13)}{E(1)} = \frac{\text{numerator}}{\text{DENM}} \quad \text{as:} \qquad (4,1)$$

```
M(1) = R(11) + R(12) + R(13)
M(2) = M(1) + R(9) + R(10)
M(3) = M(1)*(R(9) + R(10)) + M(2)*R(8)
M(4) = M(2)*(R(6) + R(7)) + M(3)
M(5) = M(3)*(R(6) + R(7)) + M(4)*R(5)                    (4,2)
M(6) = M(4)*(R(3) + R(4)) + M(5)
DENM = M(5)*(R(3) + R(4)) + M(6)*R(2)
M(7) = R(9) + R(10)
M(8) = R(6) + R(7)
numerator =  M(7)*M(8)*R(13)*(R(3) + R(4))
```

And with the application of the Grow factor algorithm due to Breuer [18], [19], as:

```
M(1) = R(11) + R(12) + R(13)
X(1) = R(9) + R(10)
M(2) = M(1) + X(1)
M(3) = M(1)*X(1) + M(2)*R(8)
X(2) = R(6) + R(7)
M(4) = M(2)*X(2) + M(3)                                  (4,3)
M(5) = M(3)*X(2) + M(4)*R(5)
X(3) = R(3) + R(4)
M(6) = M(4)*X(3) + M(5)
DENM = M(5)*X(3) + M(6)*R(2)
numerator =  X(1)*X(2)*X(3)*R(13)
```

The following remarks should be made concerning this result:

- The FDSLEM algorithm has used a nested structure of the output, thereby reducing both the time needed to calculate the result as well as the length of the output and hence the time for a numerical (re)evalution.

This aspect was recently used to solve certain network synthesis methods based on the matching between the generated symbolic expressions and a given numerical template expression using Newton iteration. We could utilize an exceptionally thight nesting in this case [17] as the required differentiation for the Newton method could preserve the original nesting. On top we could add a new nesting for the calculation of partial derivatives.

- In spite of the new nested representation, we see that some expressions do occur repeatedly. This situation can be improved with the aid of an algorithm like the Breuer grow factor algorithm [18]. A recent implementation of this algorithm is given by van Hulzen [19].
- It is unlikely that the Grow factor algorithm could obtain the result given in (4.3) from the expanded representation. If the same representation could be reached, then it should be noted that this could only be obtained at a considerably higher cost than the cost involved in the transformation from (4,2) to (4,3).
- The algorithm is tested in this example to see if it could recognize that resistor combinations like R(3) + R(4), R(6) + R(7) and R(9) + R(10) are placed in series. This test was passed succesfully when we generated the Kirchoff equations in the most sparse form [20], using algorithms similar to the Breuer grow factor algorithm [18] and algorithms given by Chua and Chen [21].
- The algorithm is tested for its factoring properties.
  Not only do we recognize nice patterns in the nested minors indicated with M(1)...M(6), it is also readily recognized that the numerator is a fully factored expression, consisting of 3 factors.

It should be recognized that it is not the purpose of the algorithm to find nicely factored expressions in the top level expression only. It can be of even greater importance that the top level expression is recognized to be a sum, such that both summands can be factored. The total amount of computational savings for a reevaluation of the denominator of the considered transferfunction has a considerably greater importance than the savings due to the factorisation of the numerator in the case given here. Comparing the number of additions and multiplications in the nested, expanded and the nested version with application of the grow factor algorithm for numerator and denominator, one sees remarkable differences,

Repesentation:	Nested		Expanded		Nested+Gfa.	
Operator:	+	x	+	x	+	x
Denominator:	14	8	137	414	11	8
Numerator:	3	3	7	16	0	3

showing that a good nesting of the expressions may be as important as the successfull factorisation. Nestings as given in the ladder network, are yet well known [22], but the the FDSLEM algorithm was the first implementation of an algorithm which could generate such nested results [15], [23], [24] automatically. Reference [25] gives an examle of a recent implementation of an algorithm which tries to minimize the number of multiplications involved in the solution of systems described with signal flowgraphs. This algorithm is (not yet) suitable for the solution of systems in which multiple integer coefficients like those in the coefficient matrices of Kirchhoff currentlaw and Kirchhoff voltagelaw equations occur.

5. A closer look at the FDSLEM algorithm.

It was stated that lower minors will not be recomputed, due to the use of the hashed memo table. It will now be explained that certain (higher) minors can still be recomputed, and that these minors need a special treatment when they are discovered. This phenomenon has to do with the factorisation of intermediate results. To understand what may happen it is usefull to explain first in more detail how factors are recognized with the FDSLEM algorithm. To this end a very simple example will be used.

$$\begin{vmatrix} a1 & a2 & . & . \\ . & b2 & b3 & b4 \\ c1 & c2 & . & . \\ . & . & d3 & d4 \end{vmatrix} = a1 \begin{vmatrix} b2 & b3 & b4 \\ c2 & . & . \\ . & d3 & d4 \end{vmatrix} - a2 \begin{vmatrix} . & b3 & b4 \\ c1 & . & . \\ . & d3 & d4 \end{vmatrix} \quad (5,1)$$

Develoment of any row/column with two nonzero matrix entries [a,c,d,1,3,4].

$$= - a1c2 \begin{vmatrix} b3 & b4 \\ d3 & d4 \end{vmatrix} + a2c1 \begin{vmatrix} b3 & b4 \\ d3 & d4 \end{vmatrix} \quad (5,2)$$

Development of rows/columns with one nonzero matrix entry.

$$= (a2c1Minor1-a1c2Minor1) \text{ with: } Minor1=\begin{vmatrix} b3 & b4 \\ d3 & d4 \end{vmatrix} \quad (5,3)$$

$$= (a2c1-a1c2)Minor1 = (a2c1-a1c2)(b3d4-b4d3) \quad (5,4)$$

The actual factorisation is done in the last two steps, where the special variable Minor1 is brought out of brackets.

It is of course a well known fact that this matrix can be brought into block diagonal form, trough the interchange of the rows indicated with b and c. Note however that the FDSLEM algorithm obtains the same effect on the basis of sparsity information from the matrix. A determinant algorithm which would develop the determinant with respect to the top (or bottom) row of the matrix would not have reached this result.

Now consider a certain band of a matrix according to figure 5,1.

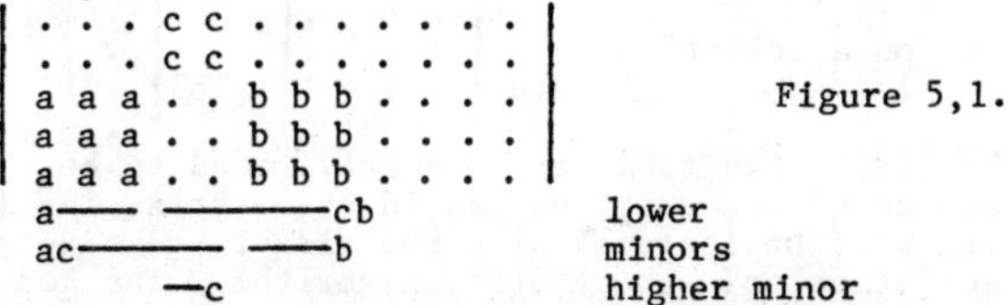

Figure 5,1.

The largest minor which can be taken from this band of the matrix given, can be regarded as having 4 lower minors, indicated with a, cb ac and b. In addition we recognize the (higher) minor indicated as c. This minor will be recomputed during the recursive expansion of the determinant. There is little to worry about recumputation, if one is interested in the expansion of a full determinant, as it has been shown in [26] that there is no preference for a certain way in which the rows of the determinant have to be processed in order to obtain a minimal number of multiplications and additions. It is for this reason that it can be considered optimal to construct the determinant from all 2x2 determinants from the bottom two rows, and next form all 3x3 determinants from the third row and all 2x2 determinants found so far, if the matrix is known to be full. This technique which is known as the Nested Minors construction, will however expand the matrix given in (5,1), thereby recomputing minors which would not have been recomputed if the development folowed by the FDSLEM algorithm, which would give a factored result, had been followed.

The uniform processing of all ,lower- and higher-minors, after the actual occurence of a factorisation extends the factoring capabilities of the FDSLEM algorithm considerably.

It remains to make some notes about the occurence of numerical matrix entries such as those encountered in the coefficient matrices of the Kirchoff Currentlaw equations and the Kirchhoff voltagelaw equations (see P2). To this extend we consider the determinant (5,5) as a possible minor from a much larger determinant.

```
|a1a2 . . . .|               |c3 . 1 .|
|b1b2b3 . . .|               |d3 . . 1|                (5,5)
| . .c3 . 1 .|=(a1b2-a2b1)   | . 1 1 .|
| . .d3 . . 1|               | . 1 . 1|
| 1 1 . 1 1 .|                        |1 1|      |1 1|
| 1 1 . 1 . 1|                  -a1b3 |1 1|+a2b3 |1 1|
                  first term       2nd term   3rd term
```

It can be easily seen that the determinant can be factored according to the first term in the expansion given, as the second and the third term are zero. This factorisation may be found by the FDSLEM algorithm, because it takes numerical values into account. An algorithm based on block diagonal decomposition would never be able to find this factorisation because the matrix entry with the zero cofactor, b3, is off diagonal.
The given example could not occur in the form given with the present NETFORM system, as it will not generate the Kirchoff equations in the format given in (5,5). Instead we will have a larger matrix (5.6) with the same determinant which can be developed directly to a column in which only one nonzero matrix is present.
The remaining matrix can now easily be seen to be block diagonal. A typical development as could have been selected by the FDSLEM algorithm is:

```
|a1a2 . . . . .|
|b1b2b3 . . . .|
| . .c3 . 1 . .|  |a1a2| |c3 1 . .|
| . .d3 . . 1 .|= |b1b2| |d3 . 1 .|-                (5,6)
| . . . . 1 . 1|         | . 1 . 1|
| . . . . . 1 1|         | . . 1 1|
| 1 1 . 1 . .-1|
          !              | . 1 . .|       | . 1 . .|
"one on a column"        | . . 1 .|       | . . 1 .|
                   -a1b3 | . 1 . 1|+a2b3  | . 1 . 1|
                         | . . 1 1|       | . . 1 1|
```

The last two terms are directly found to be zero, because of the empty column in the matrix. The first term will be factored with the aid of the factoring facilities of the FDSLEM algorithm. The example shows how the different features present in the NETFORM implementation interact to obtain an optimal performance of the algorithms.
The discussion about the aspects of the FDSLEM algorithm will be concluded with a remark about the occurrence of different minors with the same symbolic value. This situation can occur due to the fact that there is no distinction between "different" ones as matrix entries. To show what is meant I will present some minors from the development of the determinant of an actual network, which was first analysed by Mielke [27]. This network has been analysed with the aid of the NETFORM system [29] too. The Kirchhoff voltage law equations and the Kirchhoff currentlaw equations were optimized with the earlier mentioned techniques [20], [21], [28], resulting in an extremely sparse system of equations. In spite of this highly sparse representation of the corresponding equations one can still find some spurious occurences of multiple minors with equal symbolic values. Two such pairs were still present after the optimization process, one pair being:

```
   9695938986        969589878683
101a1-1 . . .     101a1-1 . . . .                  (5,7)
106 . . .a2+1     106 . .a2 .+1 .
108-1 . . .-1     108-1 . . .-1 .    Both determinants
113 .+1+1 . .     112 . . .+1 .+1    have the value:
115 . .+1+1 .     113 .+1 .-1 . .    -(Z(2)*Y(7)+1)
                  115 . .+1 . .+1
a1=Z(2), a2=Y(7)
```

Distinct symbols, with associated keys for the distinct minors, are maintained as long as there is no factorisation possible. However a characteristic symbol is reseved as an alternative name for the distinct symbols. This name will be substituted in the case that a factorisation with this symbol can be made.
Until now it was found that this mechanism is primarily active in the enhancement of the representation of the result. There were no contributions to factorisations found in the few cases which were analysed in detail with this mechanism.

6. Solutions to Perturbed systems of equations.
Another extension to the FDSLEM algorithm.

The correct symbolic solution of a perturbed system of linear equations is one of the most cumbersome problems to be solved by hand. This is especially true for the symbolic solution of systems of equations which describe the behaviour of electronic systems. Students experience the truncation rules involved in the solution process as magic, because of the experienced difference between textbook exersices and examinations. Even experienced engineers know that they have to proceed carefully, even for small problems, in order to come to the "most probable" solution in a limited number of iterations. Training and restriction to a certain class of problems can give the (dangerous) illusion that the problem becomes understood.
The remedy for the problems related to a proper truncation is, when the problem involved is recognized, relatively simple, one should use an algorithm which is free of additive and multiplicative cancellations, like the FDSLEM algorithm, in order to obtain the truncated symbolic solution of the system of equations. The determinant algorithms used to implement such a strategy are however much less friendly for human processing than the usual Gaussian elimination, especially when large sparse matrices are involved. A situation in which a computer can be very usefull.

6.1. Truncated solutions.

It is not sufficient that one utilizes a cancellation free algorithm to obtain a correct truncated result. This will be illustrated with a tableau description of an electrical network [30] on one hand and a description based on the well known nodal approach on the other hand.
Let us therefore first concentrate on the tableau approach. The system of equations which describes the elementary properties of the network elements and the involved Kirchhoff equations can be solved with the aid of Cramers rule. The sparse coefficient matrices which are involved in the solution process have most often the form given in (6.1,1).

```
-1 x1. . . . . . . . . . . . . . . . . . . .
 . .-1 x2. . . . . . . . . . . . . . . . . .
 . . . .-1 x3. . . . . . . . . . . . . . . .
 . . . . . .-1 x4. . . . . . . . . . . . . .
 .
 .                                            (6.1,1)

 .
 . . . . . coefficient matrix with entries: . .
 . . . . . +1,0,-1 . . . . . . . . . . . . . .
 . . . . . . . . . . . . . . . . . . . . . . .
```

I.e. there are n rows in the coefficient matrix which originate from the element equations. The symbolic entries in these rows come from the symbolic elementvalues, (x1,x2,x3,x4,...xn). Furthermore there are n Kirchhoff equations with matrix entries +1,0,-1, this gives a coefficient matrix, sometimes called tableau matrix [30], of

dimension 2nx2n [15], [31].
A mere recursive minor expansion [15] of the determinant involved shows that the value of the determinant is a polynomial of the first degree in x_1. The coefficients of this polynomial are first degree polynomials in x_2. The general situation, which can be derived by induction is hence that the determinant takes the form of a polynomial in x_k, the coefficiets of this polynomial being polynomials in x_{k+1}, for k=1(1)n-1. The coefficients of the polynomial in x_n are integers.
The problem of a proper truncation of this polynomial will now be described in more detail. Suppose that m of the variables involved in the coefficient matrix are known to be considerably smaller than the variables $x_{m+1}, x_{m+2}, \ldots, x_n$ and let this fact be expressed by the order relation which states that $x_1, x_2, \ldots, x_m$ also denoted as $\varepsilon_1, \varepsilon_2, \ldots, \varepsilon_m$ are $o(\varepsilon)$ with respect to $x_{m+1}, \ldots, x_n$. The Lowest Order algorithm described below gives the truncated nonzero value of the determinant within the lowest possible order.

[LO1] Initial ordernumber n=0; (6.1,2)

[LO2] calculate DET=determinant within $o(\varepsilon^n)$.

[LO3] If DET is nonzero within current order, then ready.

[LO4] n:=n+1; if n<=m then goto [LO2]

[LO5] DET=0, ready.

This algorithm seems simple and applicable to any algorithm for the symbolic calculation of the determinant. The crux is however that the algorithm can only be efficient if truncations are made during the evaluation of the determinant. This however imposes the condition on the determinant algorithm that it can handle a proper truncation. This leads naturally to cancellation free algorithms for the calculation of the determinant, as they can provide the required stablity concerning additive and multiplicative cancellations and hence avoid the associated truncation errors:

additive truncation errors: (6.1,3)
$1-(1-\varepsilon_1)=\varepsilon_1 \; o(\varepsilon)$ instead of $1-1=0 \; o(\varepsilon)$

multiplicative truncation errors: (6.1,4)
$(\varepsilon_1\varepsilon_2+\varepsilon_1\varepsilon_3)/\varepsilon_1=\varepsilon_2+\varepsilon_3 \; o(\varepsilon)$ instead of $(0+0)/\varepsilon_1=0 \; o(\varepsilon)$

The exact internal cancellations in the Gaussian elimination algorithm make truncations during the operation of the algorithm unsafe and expansion of the result necessary. It will be clear that both properties are highly unwanted.
Let me refrase here the advantages of cancellation freeness:

a. Internal cancellations can be performed in a stable way.
b. A sensible nested structure of the output can be generated.
c. Some factors in the result can be found automatically.
d. The algorithm is comparatively fast due to b and c
e. The algorithm will be faster as more of the coefficients in the sparse matrix (i.e. the variables in the polynomial) are declared to be of order $o(\varepsilon)$.

7. Truncations in the context of the FDSLEM algorithm.

The truncation mechanism build in the FDSLEM algorithm is based on the following properties:

P1t. The property P1 of the FDSLEM algorithm can be extended such that rows which might contribute to a truncation are selected from classes in the Histogram which are preferred, such that a truncation will be performed as soon as possible.
P2t. The property P2 implies that the additive truncation error (6.1,3) can be effectively avoided.
P4t. The property P4 implies that the multiplicative truncation error (6.1,4) cannot occur.

The property P1t improves the performance of the algorithm, as it gives an early truncation, which implies that nonzero cofactors involved need not be inspected. The other properties are of direct importance for the proper operation of the algorithm.

8. Generalizations and restrictions.

First we will generalize the form which can be taken by the coefficientmatrix of the system of equations as it was given in (6.1,1).

- To avoid the additive truncation error it is sufficient that lower minors be simplified. So they need not be restricted to be numerical. Example:

$$\begin{vmatrix} 1 & \varepsilon_1 & . & . \\ . & . & 1 & \varepsilon_2 \\ a & b & c & b \\ d & e & f & e \end{vmatrix} \text{ is } -1 \begin{vmatrix} b & b \\ e & e \end{vmatrix} = 0 \; o(1) \qquad (8,1)$$

$$\text{and } \varepsilon_1 \begin{vmatrix} a & b \\ d & e \end{vmatrix} + \varepsilon_2 \begin{vmatrix} b & c \\ e & f \end{vmatrix} \; o(\varepsilon) \qquad (8,2)$$

  The two terms of order $o(\varepsilon^2)$:
  $-\varepsilon_1\varepsilon_2 \begin{vmatrix} a & c \\ d & f \end{vmatrix}$ are not inspected by the FDSLEM algorithm.
- The nested form of the result, one of the highlights of the FDSLEM algorithm, puts some restrictions on the presence of matrix entries which will be truncated in the algorithm. The point is here that the nested form is only partially simplified. Example:

$$\begin{vmatrix} 1 & \varepsilon_1 & . & . \\ . & . & 1 & \varepsilon_1 \\ a & b & a & b \\ c & d & c & d \end{vmatrix} \text{ is } 0 \; o(1) \text{ and } \varepsilon_1 \begin{vmatrix} a & b \\ c & d \end{vmatrix} + \varepsilon_1 \begin{vmatrix} b & a \\ d & c \end{vmatrix} \; o(\varepsilon) \qquad (8,3)$$

  Note however that the last expression will simplify to zero if the nesting in the result is removed. So the order which was calculated to be o(e) was incorrect. The terms of second order are:

$$\varepsilon_1^2 \begin{vmatrix} a & a \\ c & c \end{vmatrix} \qquad (8,4)$$

  This simplifies once again to zero. Note that the multiple presence of equal matrix entries like a,b,c,d in the lower half of the given matrix is unpleasant for the nested structure of the result built by the FDSLEM algorithm. Thus generally speaking one should prefer a formulation of a system of equations which does not have such a multiple occurrence of matrix entries.

This invites to a short discussion about the desired form in which the equations should be presented to the FDSLEM algorithm.
Reduced systems of equations which are derived from more "fundamental" formulations of the equations of the same system are almost without value for a proper operation of the FDSLEM algorithm, as the process to obtain the reduced system of equations is (or can be formulated as) an elimination process. Elimination algorithms are however not cancellation free, and the combination of an elimination and a cancellation free algorithm will be generally inferior with respect to the application of a cancellation free algorithm on its own.

Example:
The network given in fig 8,1 can be analysed with the well known nodal analysis method:

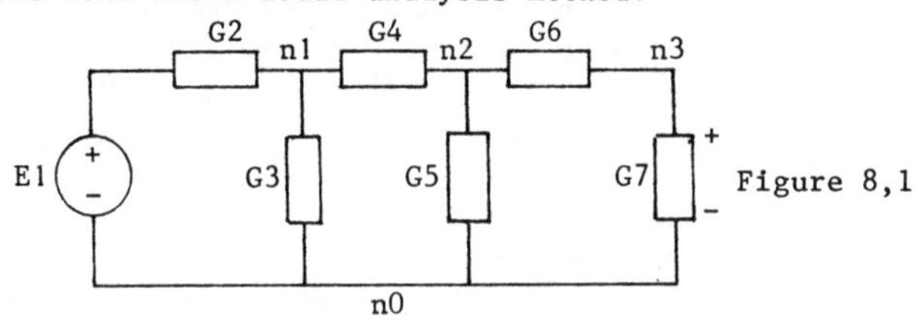

Figure 8,1

This method uses the compact matrix (8,5) which can be formulated almost entirely from inspection of the network.

$$\begin{vmatrix} y11 & y12 & 0 \\ y21 & y22 & y23 \\ 0 & y32 & y33 \end{vmatrix} \begin{vmatrix} Vn1 \\ Vn2 \\ Vn3 \end{vmatrix} = \begin{vmatrix} E1*G2 \\ 0 \\ 0 \end{vmatrix} \qquad (8,5)$$

With:
y11=G2+G3+G5
y12=y21=G4
y22=G4+G5+G6
y23=y32=G6
y33=G6+G7

Using Cramers rule one finds for Vn3:

$$Vn3=\frac{E1*G2*y21*y32}{y11*(y22*y33-y23*y32)-y12*y21*y33} \qquad (8,6)$$

$$=\frac{E1*G2*G4*G6}{(G2+G3+G4)*[(G4+G5+G6)*(G6+G7)-G6^{2}]-G4^{2}*(G6+G7)} \qquad (8,7)$$

Formula (8,7) cannot be truncated with respect to its intermediate expressions due to the exactly cancelling terms (G2+G3+G4)*G6**2 and G4**2*(G6+G7). Such truncations are however without any problem possible for the tableau description of the network equations (which involves 14 equations in 14 unknowns).
This observation is a strong recommendation for more elementary descriptions of systems of equations than was usual up to now.

9. Solving truncated systems of equations.

The memo table used to avoid the recalculation of lower determinants can also be used to store and retrieve minors common to the calculation of numerators and denominators in the application of Cramers rule. This implies that one should not restrict to Lowest Order algorithm to the calculation of one determinant. Instead one should try to calculate as much determinants as possible at a given order in the trucation process, as this may improve the efficiency and improve the nested form of the result. The direct application of Cramers rule, in which the right hand vector is inserted in the matrix to find the appropriate numerator, is very often not desired. There are two reasons for this:

1. The introduction of the vector b from the right hand side of the system of equations Ax=b will lower the probability that minors are shared by the denominator and the numerators.
2. Transferfunctions are defined as the quotient of a solved variable with respect to the corresponding excitation, an nonzero entry from the vector b. There is however no sense to associate a certain order to a given excitation.

Hence one can better restrict the application of Cramers rule to the calculation of the cofactors which are defined by the variables to be solved and the nonzero entries of b. The calculation of a full inverse, it need hardly be stated here, is of course highly unwanted if only a few variables have to be solved.

10. An illustrative example.

The network given in figure 10,1 is the input stage of a current conveyor [33], [34], or "super transistor", designed by W. de Jager [35]. Design goals were:
1. A very low input impedance Zi=-U(1)/J(1).
2. An output or "collector" current I(3) which comes as close to the input current J(1) as possible.

So we are interested in the perturbations from zero as it concerns the input impedance and the perturbations from 1 as it concerns the output current. In order to ovoid long results it was decided to solve the "base" current I(2) of the super transistor as the output current is known to be equal to the input current minus the "base" current.
Figure 10,1 gives the annotated input to the NETFORM system for the analysis of this problem in lowercase characters. The corresponding output is given in Figure 10,2 in capital characters.

```
REDUCE 2 (NETFORM: MAR-12-79) ...
operator d,q;                                   Figure (10,1)
n:=0;

procedure transistor(ne,nb,nc);                 A transistor is
begin scalar b1,b2,b3,nbi;                      merely a proce-
  n:=n+1;                                       dure.
  nbi:=node();
  nb1:=port(nb,nbi);                            Transistor structure
  b2:=port(nc,nbi);
  b3:=port(nbi,ne);
  equ u(b1);                                    Transistor equations
  equ i(b2)*d(n)-i(b1);
  equ i(b3)*r(n)-u(b3);
  parasite(d(n),1);                             d(n) is of o(e)
  factor d(n);
  return q(n);
end;

procedure transdiode(ne,nc);                    A transdiode.
begin scalar b1;
  n:=n+1;
  b1:=port(ne,nc);
  equ i(b1)*r(n)-u(b1);
  return q(n);
end;
                                                Definition of input
iim(0,1);amm(3,0);amm(4,0);                     current and output
                                                current.
transistor(1,5,4);                              Network description.
transistor(1,6,4);
transdiode(2,5);
transdiode(2,6);
transistor(7,1,5);
transistor(8,1,6);
transistor(7,3,4);
transistor(8,3,4);
calc;
                                                Figure (10,2)
M(1) := R(5) + R(6) + R(7) + R(8)               results at
M(2) := R(1) + R(2)                             o(1)
DNM := - M(1)*M(2)

M(3) := R(6) + R(8)                             results at
M(4) := R(5) + R(7)                             o(ε)
M(5) := D(1)*R(2) + R(1)*D(2)

M(6) := R(5) + R(7)                             results at
M(7) := R(6) + R(8)                             o(ε**2)
M(8) := M(6)*D(8) + M(7)*D(7)
M(9) := D(1)*R(2) + R(1)*D(2)

U(1) := (M(3)*M(4)*M(5)*J(1))/DENM              Zi=-U(1)/J(1)
I(2) := ( - M(8)*M(9)*J(1))/DENM                "base" current
I(3) := ( - M(1)*M(2)*J(1))/DENM                I(3)=J(1)
```

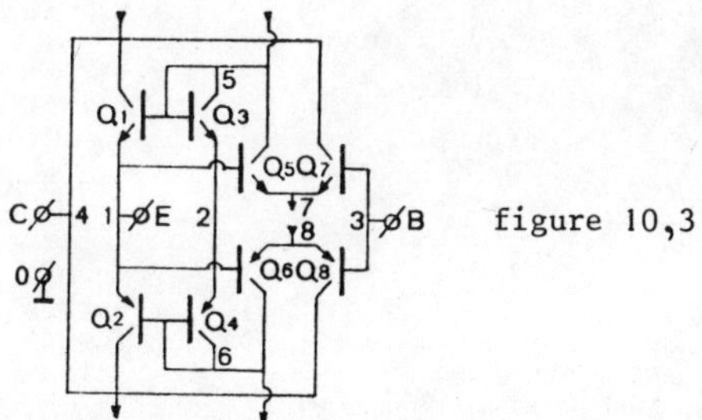

figure 10,3

Schematic diagram of the current conveyor of Figures 10,1 and 10,2.

11. Further remarks and conclusions.

Some care has to be taken with the generation of truncated symbolic solutions from systems of equations, as the number of terms which might have been truncated can be large, such that the original order assuption can still be violated. It can therefore be important to calculate a result up to a higher order. The decision to do so should however be left to the user of the system. A provision for calculations up to a higher order than the lowest order has been added to the Lowest Order algorithm (6.1,2).

In the treatment of the generation of truncated results we almost implicitly assumed that numerical coefficients are not small, such that there is no need to consider them in the truncation process. The NETFORM system does not have facilities to propagate order relations for numerical coefficients.

Truncation is also possible for systems described by SI-A, i.e descriptions in the Laplace domain or eigenvalue problems. Care has however to be taken, as can be seen from the following example:

$$\begin{vmatrix} S & 1 & . & . & . \\ . & S & 1 & . & . \\ . & . & S & 1 & . \\ . & . & . & S & 1 \\ -a & . & . & . & S \end{vmatrix} = S^5 - a$$

The characteristic equation has with: a=ε**5 five roots on a circle with radius e, located at the origin.

With a=1/100000 one finds the roots on a circle with diameter 1/10. Compared with other nonzero matrix entries in the matrix A one might think that the matrix entry with value a=1/100000 might be truncated without problems. However one should take into account all the variables involved into the truncation, so also the variable S. If this is done it becomes clear that the truncation of a with respect to S**5 will only be valid for values of S sufficiently large with respect to 1/10. The charactersitic equation has a root at zero with multiplicity 5 when one truncates the value of a with respect to the value of S**5. The truncated characteristic equation is not particularly inaccurate if the truncation constraint S >> 1/10 holds.

It is important to note that the truncation of a single matrix entry is not made on a static basis. This is much more sophisticated. First there are certain levels of "accuracy" in which the solution algorithm operates, second there is the effect that the local truncation level depends on the truncations already in effect in the recursion. It can happen that the truncation of a certain matrix entry is allowed in a certain context during the development of a single determinant, while it is not allowed to truncate the same matrix entry during the same development of the same determinant in another context. The differences in context come from the accumulated order during the evaluation of the term to which the matrix entry belongs at the respective moments.

It are precisely the above arguments which make the determination of a truncated solution of a system of linear equations so badly suited for human processing. Truncated (intermediate) results are to some extend similar to floating point numbers, both give a better result if the length of the objects grows, both suffer from the fact that they cannot give sufficiently exact results if the algorithms used are not cancellation free. Concerning the required consequent bookkeeping required for the truncated symbolic solution of large sparse systems of equations we should come to the conclusion that the task can be much better given to a computer than that it can be given to a human as the latter has considerable problems with deep recursions. The work concerning truncations of systems of equations, reported in this paper, has however also given some results for the human being as the understanding of truncated solutions has gained from the research undertaken.

REFERENCES:

[1] Hearn, A. C., An improved factored polynomial representation. Proc. Hawaii International Conf. on System Sciences, Honolulu (1977) 155.

[2] Brown, W. S, On computing with factored rational expressions. SIGSAM Bull. 8, #3, ACM, (1974) 26.

[3] Leibnitz, correspondence with de l'Hospital dated 28th of April 1693, published in: Leibnitzens Mathematische Schriften, herausgegeben von C.I.Gerhardt. 1. Abth. ii pp 229, 238-240, 245. Berlin 1850.

[4] Cramer. Introdution a l'Analyse des Lignes Courbes Algebriques. (pp. 59,60, 656-659) Geneve, 1750.

[5] Bezout, Recherches sur le degree des equations resultantes de l'evanoussement des inconnues, et sur les moyens qu'il convient d'employer pour trouver ces equations. Hist. de l'Acad. Roy. des Sciences. Ann. 1764 (pp 288-338), pp 291-295

[6] Vandermonde, Memoire sur l'elimination. Hist. de l'Acad. Roy. des Sciences (Paris), Ann. 1772, 2e partie (pp. 516-532).

[7] Wilkinson, J. H., The algebraic eigenvalue problem, Monographs on numerical Analysis, Oxford university press, 1965.

[8] Bareiss E. H., Sylvesters identity and multistep integerpreserving Gaussian elimination. Math. Comput. 22, 103 (july 1968), pp 565-578.

[9] Lipson J., Symbolic methods for the computer solution of linear equations with applications to flowgraphs, Proceedings of the 1968 summer institute on symbolic mathematical computation, R. G. Tobey, editor.

[10] Downs T., On the reduction and inversion of the nodal admittance matrix in rational form., IEEE Trans. CAS vol. CAS-21 Sept 1974, pp 592-598.

[11] Horowitz E. and Sani S.,On computing the exact determinant of matrices with polynomial entries. JACM, vol 22, No 1, Jan 1975, pp 38-50

[12] Tobey, R.G., Bobrow R.J., Zilles S., Automatic Simplification in FORMAC, Proc. of fall joint computer conference, Spartan books vol 27, nov. 1965.

[13] Lin P. M. and Alderson G. E., "Computer generation of symbolic networkfunctions - A new theory and implementation", IEEE Trans. Circuit Theory, Vol. C.T.-20, pp 48-56, Jan 1973

[14] Singal K. and Vlach J., Symbolic analysis of analog and digital circuits. IEEE Trans. CAS vol. CAS-24 Nov. 1977, pp 598-610.

[15] Smit J., New recursive minor expansion algorithms, A presentation in a comparative context., Symbolic and Algebraic Computation, Lecture notes in computer science no 72. , Goos G. and Hartmanis J. eds., Springer.

[16] Griss M. L., Efficient recursive minor expansion., Utah symbolic computation group, Report UCP-51, 1977.

[17] Hulsegge E., The application of frequency matching for the synthesis of certain electrical networks. (in Dutch) THT report 1231-KV-0281/EL. Jan. 1981.

[18] Breuer M. A., Generation of optimal code for expressions via factorization. Comm. ACM Vol 12, no 6 June 1969, pp 333-340.

[19] Hulzen J. A. van, Breuer's grow factor algorithm in computer algebra. These proceedings.

[20] Schilstra Y. G., Algorithms to minimize the representation of current and voltage law equations based on the introduction of new equations. (in Dutch) THT report 1231-KV-0181/EL. Jan. 1981.

[21] Chua L. O. and Chen L. , On optimally sparse cycle and coboundary basis for a linear graph, IEEE trans. on circ. Th., Vol CT-20, No. 5, Sept 1973, pp 495-503.

[22] Herrero J.L. and Willoner G., Synthesis of Filters, Prentice-Hall, Englewood Clifs, New Yersey.

[23] Smit J., An efficient factoring symbolic determinant expansion algorithm. THT report 1231-AM-0478/EL. Dec. 1978.

[24] Smit J., A recursive tearing technique for systems in which small as well as large elementvalues are significant. THT Report 1231-AM-0378/EL. Dec. 1978.

[25] Endy G. E. and Lin P. M., A minimization problem in systems characterized by loopless signal flowgraphs. 1980 IEEE intern. Symp. on Circ. and Syst. Proceedings, pp 365-367.

[26] Gentleman W. M. and Johnson S. C., The evaluation of determinants by expansion by minors and the general problem of substitution, Math. of comp., vol. 28, no. 126, april 1974, pp 543-548.

[27] Mielke R. R., A new signal flowgraph formulation of symbolic networkfunctions. IEEE trans. circ. and syst., vol. CAS-25, pp. 334-340, june 1978.

[28] Renkema G. H., Algorithms to minimize the representation of current and voltage law equations. (in Dutch) THT report 1231-BV-0879/EL.

[29] Smit J., NETFORM users manual, THT report 1231-AM-0381/EL.

[30] Hachtel G., Brayton R. K. and Gustavson F. G., The sparse tableau approach to network analysis and design, IEEE trans on circ. theory, Vol. CT-18, pp. 101-113, Jan 1971.

[31] Smit J., Sparse Kirchhoff equations, an effective support tool for the numeric and symbolic solution of large sparse systems of network equations. THT report 1231-AM-0681/EL Jan 1981.

[32] Griss, M. L., An Efficient Sparse Minor Expansion Algorithm, Proc. ACM 76, (1976), pp. 429-434.

[33] Sedra A. and Smith K. C., A second generation current conveyor and its applications. IEEE Trans. on Circ. Theory, Vol CT-17, 1970, pp 132-134.

[34] Bel N., A high presision monolitic current follower. Electronic circuits and systems, Jan. 1977, Vol 1, No. 2, pp 79-84.

[35] Jager W. de, and Smit J., Application, design and symbolic analysis of a current follower, Electronic circuits and systems, Jan. 1977, Vol. 1, no. 2, pp. 79-84.

Efficient Gaussian Elimination Method for Symbolic Determinants and Linear Systems

Tateaki Sasaki* and Hirokazu Murao**

*) The Institute of Physical and Chemical Research
Wako-shi, Saitama 351, Japan

**) Department of Information Science, The University of Tokyo
Bunkyo-ku, Tokyo 113, Japan

Extended Abstract

Let M be an $n \times n$ symbolic matrix:

$$M = (a_{ij}), \quad 1 \le i,j \le n. \qquad (1)$$

The conventional Gaussian elimination method calculates the determinant of M by the following iteration formula:

$$M_{ij}^{(k+1)} = (M_{kk}^{(k)} M_{ij}^{(k)} - M_{ik}^{(k)} M_{kj}^{(k)}) / M_{k-1\,k-1}^{(k-1)} \qquad (2)$$

where k, $1 \le k \le n-1$, denotes the elimination step, $k \le i \le n$, $k \le j \le n$, and $M^{(1)} = M$. $M_{00}^{(0)}$ is defined to be 1. After the (n-1)th elimination, we obtain the determinant:

$$|M| = M_{nn}^{(n)}.$$

Although the formula (2) is quite efficient for integer matrices or matrices with univariate polynomial entries, it is inefficient for multivariate polynomial entries. This is because, in calculating $M_{ij}^{(k+1)}$, we must calculate the products of multivariate polynomials: multiplication of two multivariate polynomials containing t_1 and t_2 terms, respectively, gives $t_1 t_2$ terms at most, and such a large polynomial is then divided by $M_{k-1\,k-1}^{(k-1)}$. This paper presents a method for calculating $M_{ij}^{(k+1)}$ efficiently without causing such an intermediate expression swell.

The idea is quite simple. In formula (2), $M_{ij}^{(k+1)}$ is given as the quotient of the two polynomials. Suppose the denominator $M_{k-1\,k-1}^{(k-1)}$ has a unique term u. Then the terms proportional to u in the numerator will involve most of the terms of the quotient. Hence, we try to calculate $M_{ij}^{(k+1)}$ by calculating *only* the terms proportional to u in the numerator $M_{kk}^{(k)} M_{ij}^{(k)} - M_{ik}^{(k)} M_{kj}^{(k)}$.

The above idea is actually realized in the following way. First, let us replace all the diagonal elements of M by new independent variables:

$$M \to M = \begin{pmatrix} X_1 & a_{12} & \cdots & \\ a_{21} & X_2 & \cdots & \\ \vdots & & \ddots & \\ & & & X_n \end{pmatrix} \qquad (3)$$

According to Jordan, $M_{ij}^{(k)}$ is given as (see Bareiss [1]):

$$M_{ij}^{(k)} = \begin{vmatrix} X_1 & a_{12} & & a_{1,k-1} & a_{1j} \\ a_{21} & X_2 & & a_{2,k-1} & a_{2j} \\ \vdots & \vdots & & \vdots & \vdots \\ a_{k-1,1} & a_{k-1,2} & & X_{k-1} & a_{k-1,j} \\ a_{i1} & a_{i2} & & a_{k-1,j} & a_{ij} \end{vmatrix}$$

$$k < i, j \le n. \qquad (4)$$

The $M^{(k-1)}_{k-1\,k-1}$ is the determinant of the top-leftmost $(k-1)x(k-1)$ minor (a principal minor) of M, and it is a polynomial of degree $k-1$ containing the highest degree term $X_1X_2\cdots X_{k-1}$ when viewed w.r.t. variables $X_1, \ldots, X_n$. Hence, we can divide $(X_1X_2\cdots X_{k-1})^2$ by $M^{(k-1)}_{k-1\,k-1}$:

$$(X_1X_2\cdots X_{k-1})^2 = Q^{(k-1)}M^{(k-1)}_{k-1\,k-1} + R^{(k-1)} \quad (5)$$

where $Q^{(k-1)}$ and $R^{(k-1)}$ are the quotient and the remainder, respectively, and the division is made by treating $X_1, \ldots, X_{k-1}$ as main variables. Note that $R^{(k-1)}$ does not contain any term which is proportional to $X_1X_2\cdots X_{k-1}$, although it may contain terms which are of degree greater than $k-2$ in $X_1, \ldots, X_{k-1}$.

By multiplying both sides of Eqs. (2) and (5), we get

$$M^{(k+1)}_{ij}(X_1X_2\cdots X_{k-1})^2 = (M^{(k)}_{kk}M^{(k)}_{ij} - M^{(k)}_{ik}M^{(k)}_{kj})Q^{(k-1)} + M^{(k+1)}_{ij}R^{(k-1)} . \quad (6)$$

Since $M^{(k+1)}_{ij}$ is linear in each of $X_1, \ldots, X_{k-1}$, and since $R^{(k-1)}$ does not contain any term proportional to $X_1X_2\cdots X_{k-1}$, the rightmost term in Eq.(6) does not give any term proportional to $(X_1X_2\cdots X_{k-1})^2$. Therefore, Eq.(6) means that $M^{(k+1)}_{ij}$ can be calculated by taking out all terms that are proportional to $(X_1X_2\cdots X_{k-1})^2$ from $(M^{(k)}_{kk}M^{(k)}_{ij} - M^{(k)}_{ik}M^{(k)}_{kj})\times Q^{(k-1)}$. Since $Q^{(k-1)}$ is a polynomial which is linear in each of $X_1, \ldots, X_{k-1}$, only the terms proportional to $X_1X_2\cdots X_{k-1}$ in $(M^{(k)}_{kk}M^{(k)}_{ij} - M^{(k)}_{ik}M^{(k)}_{kj})$ contribute to $M^{(k+1)}_{ij}$.

Next, let us describe an efficient method of calculating the terms proportional to $X_1X_2\cdots X_{k-1}$ selectively in $(M^{(k)}_{kk}M^{(k)}_{ij} - M^{(k)}_{ik}M^{(k)}_{kj})$. This is done by using a particular polynomial representation and by introducing a special multiplication $\boxtimes$. The representation treats $X_1, \ldots, X_n$ as main variables and is recursively defined as follows:

$$P(X_i, X_{i+1}, \ldots, X_n) = P_m(X_{i+1}, \ldots, X_n)X_i^m + P_{m-1}(X_{i+1}, \ldots, X_n)X_i^{m-1} + \cdots + P_0(X_{i+1}, \ldots, X_n),$$

$$i = 1, 2, \ldots, n. \quad (7)$$

Note that we are interested only in polynomials which are linear in each of the main variables because so is $M^{(k)}_{ij}$. Being applied to polynomials P and $\tilde{P}$ which are linear in the main variables $X_1, \ldots, X_n$ and represented as (7), our multiplication operator works as follows:

$$P \overset{k}{\underset{i}{\boxtimes}} \tilde{P} = P_1 \overset{k}{\underset{i+1}{\boxtimes}} \tilde{P}_1 X_i + (P_1 \overset{k}{\underset{i+1}{\boxtimes}} \tilde{P}_0 + P_0 \overset{k}{\underset{i+1}{\boxtimes}} \tilde{P}_1), \quad \text{for } i \le k. \quad (8)$$

This relation defines our special multiplication recursively. The $\overset{k}{\boxtimes}$ becomes a usual multiplication operator for the variables $X_{k+1}, X_{k+2}, \ldots, X_n$ and other variables.

The above multiplication opertor $\boxtimes$ allows us to calculate $Q^{(k-1)}$ in Eq.(5) efficiently. Let us decompose $M^{(k-1)}_{k-1\,k-1}$ as

$$M^{(k-1)}_{k-1\,k-1} = X_1X_2\cdots X_{k-1} + R, \quad (9)$$

where R does not contain any term proportional to $X_1X_2\cdots X_{k-1}$. Then,

$$(X_1X_2\cdots X_{k-1})^2/M^{(k-1)}_{k-1\,k-1} = X_1X_2\cdots X_{k-1} + R + R^2/(X_1X_2\cdots X_{k-1} + R). \quad (10)$$

Therefore, we have

$$Q^{(k-1)} = X_1X_2\cdots X_{k-1} - R + R \boxtimes R - R \boxtimes R \boxtimes R + \cdots + (-1)^N R \boxtimes \cdots \boxtimes R,$$
$$N = [(k-1)/2]. \quad (11)$$

In order to show a performance of our algorithm, we calculate the determinant of the following *4x4* matrix as an example.

$$M^{(1)} = \begin{pmatrix} a_1 & a_2 & a_3 & a_4 \\ b_1 & b_2 & b_3 & b_4 \\ c_1 & c_2 & c_3 & c_4 \\ d_1 & d_2 & d_3 & d_4 \end{pmatrix} \longrightarrow \begin{pmatrix} X_1 & a_2 & a_3 & a_4 \\ b_1 & X_2 & b_3 & b_4 \\ c_1 & c_2 & c_3 & c_4 \\ d_1 & d_2 & d_3 & d_4 \end{pmatrix}$$

After the first elimination, we have

$$M^{(2)}_{i,j\ge 2} = \begin{pmatrix} X_2X_1-a_2b_1 & b_3X_1-a_3b_1 & b_4X_1-a_4b_1 \\ c_2X_1-a_2c_1 & c_3X_1-a_3c_1 & c_4X_1-a_4c_1 \\ d_2X_1-a_2d_1 & d_3X_1-a_3d_1 & d_4X_1-a_4d_1 \end{pmatrix}$$

At the next elimination, since $Q^{(1)}= X_1$, we have only to calculate terms proportional to X_1

$$M^{(3)}_{i,j\ge 3} = \begin{pmatrix} (c_3X_2-b_3c_2)X_1-a_3c_1X_2 & (c_4X_2-b_4c_2)X_1-a_4c_1X_2 \\ -c_3a_2b_1+a_3b_1c_2+a_2b_3c_1 & -a_2b_1c_4+a_2b_4c_1+a_4b_1c_2 \\ (d_3X_2-b_3d_2)X_1-a_3d_1X_2 & (d_4X_2-b_4d_2)X_1-a_4d_1X_2 \\ -a_2b_1d_3+a_3b_1d_2+a_2b_3d_1 & -a_2b_1d_4+a_2b_4d_1+a_4b_1d_2 \end{pmatrix}$$

Only terms proportional to X_1X_2 are calculated at the last elimination.

$$\begin{aligned}
M^{(4)}_{4\,4} &= M^{(3)}_{3\,3} \overset{2}{\boxtimes} M^{(3)}_{4\,4} - M^{(3)}_{4\,4} \boxtimes M^{(3)}_{3\,4} \\
&= \{(c_3X_2-b_3c_2) \overset{2}{\underset{2}{\boxtimes}} (d_4X_2-b_4d_2)\}X_1 \\
&\quad + (c_3X_2-b_3c_2) \overset{2}{\underset{2}{\boxtimes}} (-a_4d_1X_2-a_2b_1d_4+a_2b_4d_1+a_4b_1d_2) \\
&\quad + (d_4X_2-b_4d_2) \overset{2}{\underset{2}{\boxtimes}} (-a_3c_1X_2-a_2b_1c_3+a_3b_1c_2+a_2b_3c_1) \\
&\quad -\{(d_3X_2-b_3d_2) \overset{2}{\underset{2}{\boxtimes}} (c_4X_2-b_4c_2)\}X_1 \\
&\quad - (d_3X_2-b_3d_2) \overset{2}{\underset{2}{\boxtimes}} (-a_4c_1X_2-a_2b_1c_4+a_2b_4c_1+a_4b_1c_2) \\
&\quad - (c_4X_2-b_4c_2) \overset{2}{\underset{2}{\boxtimes}} (-a_3d_1X_2-a_2b_1d_3+a_3b_1d_2+a_2b_3d_1) \\
&= (c_3d_4X_2-b_3c_2d_4-b_4c_3d_2)X_1 \\
&\quad + (-a_4c_3d_1X_2+a_4b_3c_2d_1-a_2b_1c_3d_4+a_2b_4c_3d_1+a_4b_1c_3d_2) \\
&\quad + (-a_3c_1d_4X_2+a_3b_4c_1d_2-\underline{a_2b_1c_3d_4}+a_3b_1c_2d_4+a_2b_3c_1d_4) \\
&\quad -(c_4d_3X_2-b_3c_4d_2-b_4c_2d_3)X_1 \\
&\quad - (-a_4c_1d_3X_2+a_4b_3c_1d_2-a_2b_1c_4d_3+a_2b_4c_1d_3+a_4b_1c_2d_3) \\
&\quad - (-a_3c_4d_1X_2+a_3b_4c_2d_1-\underline{a_2b_1c_4d_3}+a_3b_1c_4d_2+a_2b_3c_4d_1)
\end{aligned}$$

This expression is different from the determinant of $M^{(1)}$ only in the terms underlined.

Since $M_{22}^{(2)}=X_2X_1-a_2b_1$, we have $Q^{(2)}=X_1X_2+a_2b_1$ and $(M_{33}^{(3)}\boxtimes M_{44}^{(3)}-M_{34}^{(3)}\boxtimes M_{43}^{(3)})\boxtimes Q^{(2)}=M_{44}^{(4)}+(c_3d_4-c_4d_3)a_2b_1$. The extra two terms in $M_{44}^{(4)}$ are exactly canceled by the right-most term in this expression. Finally, replacing X_1 and X_2 by a_1 and b_2, respectively, we have the answer.

Our method is applicable also to linear equation. Let A and B be $n\times n$ and $n\times m$ matrices:

$$A = (a_{ij}), \quad B = (b_{ij}).$$

The conventional Gaussian elimination method for solving the equation

$$AZ = B, \quad Z = (z_{ij}), \quad 1\le i\le n, \ 1\le j\le m$$

is as follows. Let M be the following augmented matrix:

$$M = \begin{pmatrix} a_{11} & \cdot\cdot & a_{1n} & b_{11} & \cdot\cdot & b_{1m} \\ \vdots & & \vdots & \vdots & & \vdots \\ a_{n1} & \cdot\cdot & a_{nn} & b_{n1} & \cdot\cdot & b_{nm} \end{pmatrix}$$

In the first step, the left n columns of M are upper-triangularized by the same elimination method as that for determinant:

$$M \to M' = \begin{pmatrix} a'_{11} & \cdot\cdot & a'_{1n} & b'_{11} & \cdot\cdot & b'_{1m} \\ & \cdot & \vdots & \vdots & & \vdots \\ 0 & & a'_{nn} & b'_{n1} & \cdot\cdot & b'_{nm} \end{pmatrix} \tag{12}$$

This step gives the determinant of A:

$$D = a'_{nn} = |A|.$$

In the second step, the left n columns of M' are diagonalized

$$M' \to M'' = \begin{pmatrix} a''_{11} & & 0 & b''_{11} & \cdot\cdot & b''_{1m} \\ & \cdot & \vdots & \vdots & & \vdots \\ 0 & & a''_{nn} & b''_{n1} & \cdot\cdot & b''_{nm} \end{pmatrix} \tag{13}$$

This diagonalization is performed by the formula

$$b''_{ij} = (Db'_{ij} - \sum_{k=i+1}^{n} a'_{ik}b''_{kj})/a'_{ii}, \quad 1\le j\le m,$$

where i is changed successively as $n-1$, $n-2$, ..., 1. Then the solution matrix Z is given by $z_{ij} = b''_{ij}/D$. Here, we encounter the exact division again. Hence, our efficient method for elimination is applicable not only in the upper-triangularization step (12) but also in the diagonalization step (13).

The following table shows empirical test of our algorithm. The test problems are

(1) determinant of Vandermonde matrix $M_{ij} = a_j^{i-1}$,
(2) determinant of symmetric Toeplitz matrix $M_{ij}= a_{|i-j|}$,
(3) inversion of M,
$M_{ii}=A$, $M_{i+1,i}=M_{i,i+1}=B$, $M_{i,i+2}=M_{i+2,i}=C$, $M_{i,i+3}=M_{i+3,i}=D$,
(4) inversion of M,
$M_{ii}=1+x^2$, $M_{i,i+1}=M_{i+1,i}=x$.

	n	minor expansion	our method	Bareiss' method
(1)	3	7	7	8
	4	14	22	54
	5	52	116	384
	6	313	735	21434
	7	2138	4882	
(2)	3	10	10	11
	4	34	32	56
	5	177	165	331
	6	933	788	3416
	7	6321	4724	
	8	49053	31166	

Timings (in milli-seconds) of determinant calculation by three algorithms.

	n	our method	Bareiss' method
	3	42(12)	48(14)
	4	153(35)	371(58)
(3)	5	584(125)	2101(252)
	6	2275(484)	10905(1196)
	7	8524(1814)	36860(4505)
	3	31(11)	44(11)
	4	80(21)	149(28)
(4)	5	221(49)	378(77)
	6	572(131)	844(162)
	7	1382(338)	1694(331)
	8	3140(794)	3014(621)

Timings (in milli-seconds) of matrix inversion (determinant) by two Gaussian elimination algorithms.

References

[1] Bareiss, E. H., "Sylvester's identity and multistep integer-preserving Gaussian elimination," Math. Comp. 22 (1968), pp.565-578.

[2] Sasaki, T. and Murao, H., "Efficient Gaussian elimination method for symbolic determinants and linear systems," TR-81-09, Dept. of Info. Sci., Univ. of Tokyo, 1981.

Parallelism in Algebraic Computation and Parallel Algorithms for Symbolic Linear Systems

Tateaki Sasaki* and Yasumasa Kanada†

*) The Institute of Physical and Chemical Research
Wako-shi, Saitama 353, Japan

†) Institute of Plasma Physics, Nagoya University††
Chikusa-ku, Nagoya 464, Japan

Abstract

Parallel execution of algebraic computation is discussed in the first half of this paper. It is argued that, although a high efficiency is obtained by parallel execution of divide-and-conquer algorithms, the ratio of the throughput to the number of processors is still small. Parallel processing will be most successful for the modular algorithms and many algorithms in linear algebra. In the second half of this paper, parallel algorithms for symbolic determinants and linear equations are proposed. The algorithms manifest a very high efficiency in a simple parallel processing scheme. These algorithms are well usable in also the serial processing scheme.

Key Words and Phrases: algebraic computation, divide-and-conquer algorithm, modular algorithm, parallel processing, symbolic·determinant, symbolic linear system, minor expansion, Cramer's method.

§1. Introduction

A remarkable advancement has been attained in the study of parallel processing recently,[1,2,3] and many parallel algorithms were developed in the area of numeric computations. This paper is concerned with parallel processing of algebraic computations, in particular, with parallel algorithms for symbolic linear systems.

Tere are two reasons for studying parallel algorithms for symbolic linear systems. First, the algebraic calculations which are performed most frequently in the applications are calculations of symbolic determinants and solving symbolic linear equations, and the users are requesting efficient routines for these calculations. Second, calculations for symbolic linear systems are well suited for parallel processing, and a very high efficiency will be obtained by parallel processing as we shall show in this paper.

In the below, the term "*task*" is used as a full calculation to be executed by a computer, and the term "*task unit*" as a part of the calculation which is executable by one of the parallel processors of the computer.

There are basically two problems in any scheme of parallel processing. One is the cost of communications among different processors working parallelly, and the other is how to divide a *task* into *task units* which are executable highly paral-

††) Address since May 16, 1981, Computer Centre, University of Tokyo, 2-11-26 Yayoi, Bunkyo-ku, Tokyo 113, Japan

lelly. Regarding to the first problem, the ratio of the communication cost to the total computation cost decreases as the size of each *task unit* increases. On the other hand, if the size of each *task unit* is increased, parallelism among the *task units* decreases in general. Due to this reason, many researchers are concentrating their attention mainly upon low level parallelism, such as parallelism among basic arithmetic operations.

In this paper, we consider only high level parallelism, that is, the parallelism among *task units* of large sizes. We cannot always find such high level parallelism in a general algebraic algorithm. We can, however, find high level parallelisms in many algorithms for symbolic linear systems. For example, we can reduce the main part of the determinant calculation to a set of large *task units* of almost the same sizes which are executable parallelly. The communication cost is negligible in such algorighms, and we can obtain a very high efficiency.

§2. Various parallelisms in symbolic/ algebraic algorithms

Parallel processing of LISP has been discussed by many authors.[4,5,6] Among various parallelisms in processing lists, a parallelism which is quite obvious and seems to be quite effective for speeding up the processing is parallel execution of the function EVLIS. Thst is, we evaluate a set of arguments in each procedure parallelly. This scheme will be effective if the arguments themselves are large *task units* of almost the same size. However, the most LISP programs are composed of many procedures of quite different sizes, and they refer to each other many times going into deep recursion levels. Furthermore, the number of arguments in a procedure is not many: the number is only 2 or 3 on an average. Therefore, the efficiency of computation will not increase so much even if we use many parallel processors.

The point mentioned above was clarified by Yasui *et al*. of Osaka University recently.[7] These authors simulated the parallel execution of EVLIS on five common LISP programs, and found that the efficiency increased by a factor of 4 ∿ 5 on only one program and no remarkable increase of the efficiency was observed on other four programs. Furthermore, the efficiency increase was almost saturated at eight processors.

A high efficiency will be obtained if we execute divide-and-conquer algorithms parallelly. In many divide-and-conquer algorithms, a *task* is usually divided into two almost equal *subtasks* which are executable parallelly. Each *subtask* itself is divided into two smaller *subtasks*, and so on. Therfore, we have 2^k *subtasks* at the k-th recursion level, and we can execute these *subtasks* parallelly.

Suppose we have a *task* of size n which can be divided into two *subtasks* of size $n/2$, and let $f(n)$ be the number of basic operations necessary to divide the *task* into the *subtasks* and unify the results of *subtasks* to get the answer. Let $T_s(n)$ be the number of basic operations necessary to execute this *task* in the conventional serial processing scheme. Then, we have the following recursive relation:

$$(1) \quad T_s(n) = 2 \cdot T_s(n/2) + f(n).$$

If $f(n) \propto n \log_2^m n$, which is the case for many fast algorithms, the above relation gives

$$(2) \quad T_s(n) \propto \frac{1}{m+1} n \log_2^{m+1} n.$$

On the other hand, in the parallel processing scheme mentioned above, we have the following recursive relation for the time complexity function $T_p(n)$:

$$(3) \quad T_p(n) = T_p(n/2) + f(n).$$

If $f(n)=cn$, where c is a constant, we have

$$(4) \quad T_p(n) = 2cn.$$

Here, we assumed we had an enough number of processors. Similarly, if $f(n)=cn \log_2 n$, we get

$$(5) \quad T_p(n) = 2cn \log_2(n/2),$$

and if $f(n)=cn \log_2^2 n$, we obtain

$$(6) \quad T_p(n) = 2cn(\log_2^2 n - 2\cdot\log_2 n + 3).$$

For example, $f(n)=(3/2)n$ for the fast Fourier transform. Hence, $(3/2)n \log_2 n$ steps are necessary to calculate the Fourier transform in the serial computation scheme, while the computation will finish within $3n$ steps in the parallel processing scheme.

Probably only one defect of the above parallel execution of divide-and-conquer algorithms is that high parallelism is attained only at deep recursion levels. At the k-the recursion level, 2^k processors are working while only one processor is working at the top level where the size of the *task unit* is largest. Therefore, the processor working ratio is about $(\log_2 N)/N$ if we have N processors $(N >> 1)$.

The third type of the parallelism, which is the main theme of this paper, is on an algorithm which can be divided into parallelly executable large *task units* of almost the same sizes. A typical example is the modular algorithm. For example, let us consider the modular algorithm for coupled linear equations with integer coefficients. The main part of the algorithm is to solve the equations by the Gaussian elimination over Galois fields $GF(p_i)$, $i=1,\cdots,k$, where $p_1, \cdots, p_k$ are mutually distinct prime numbers. Therefore, the original *task* is divided into k almost the same *task units* which are executable parallelly and a *task unit* which is to interpolate the answers over $GF(p_i)$, $i=1,\cdots,k$.

The architecture of a machine which executes the modular algorithms parallelly is quite simple: a set of processors having their own work memories and being coupled loosely to a large common memory under a supervising processor. In fact, Yoshimura *et al.* of Toshiba Corporation realized such a machine and demonstrated that the machine was quite useful for solving coupled linear equations with integer coefficients by a modular algorithm.[8)]

§3. Data-driven computation and a minor expansion algorithm

One of the most attractive and prospective schemes of parallel computation is the data-driven computation.[2)] In this scheme, the computation is done in the following way. When a *task unit* has been executed, the result is transmitted to all the *task units* that use the result as their inputs. When a *task unit* is given all necessary inputs, it is activated to be executable irrespectively of the state of other *task units*. That is, the flow of data controls the computation and the processor working ratio is made as large as possible. Therefore, we will obtain a large throughput in principle.

Many algorithms for symbolic linear systems can be well executed in the data-driven computation scheme. A typical example is the minor expansion algorithm by Griss[9)] and Wang.[10)] In this algorithm, all different minors that are necessary to expand the determinant recursively are firstly replaced by temporal variables generated by the system. When the determinant is expanded completely, the minors are evaluated successively and substituted for the temporal variables. The computation time is mostly spent at the second step in this algorithm when executed serially. This step can be executed highly parallelly in the data-driven scheme.

In order to get a better insight into the computation, let us consider the following 4×4 matrix:

$$(7)\quad M_4 = \begin{bmatrix} a_1 & b_1 & c_1 & d_1 \\ a_2 & b_2 & c_2 & d_2 \\ a_3 & b_3 & c_3 & d_3 \\ a_4 & b_4 & c_4 & d_4 \end{bmatrix}.$$

In the first step of the algorithm, the determinant of this matrix is expanded in temporal variables as follows:

$$(8)\quad \begin{aligned} G01 &:= c_3d_4 - c_4d_3; \\ G02 &:= b_3d_4 - b_4d_3; \\ G03 &:= b_3c_4 - b_4c_3; \\ G04 &:= b_2G01 - c_2G02 + d_2G03; \\ G05 &:= a_3d_4 - a_4d_3; \\ G06 &:= a_3c_4 - a_4c_3; \\ G07 &:= a_2G01 - c_2G05 + d_2G06; \\ G08 &:= a_3b_4 - a_4b_3; \\ G09 &:= a_2G02 - b_2G05 + d_2G08; \\ G10 &:= a_2G03 - b_2G06 + c_2G08; \\ |M_4| &:= a_1G04 - b_1G07 + c_1G09 - d_1G10; \end{aligned}$$

The minors being assigned to temporal variables *G01*, *G02*, *G03*, *G05*, *G06* and *G08* can be evaluated parallelly, and after these evaluations, *G04*, *G07*, *G09* and *G10* can be evaluated parallelly.

It is interesting to point out that the procedures (8) are quite similar to a program written in the single assignment language.[11)] This language was designed to manifest the flow of data to maximize, and it is considered to be the simplest language for data-driven computation.

We can observe from the above example that the size of a *task unit* increases and the degree of parallelism decreases as the processing proceeds in the above algorithm. This point is undesirable for parallel processing, although the processor working ratio is much larger in the above algorithm than in parallel processing of a divide-and-conquer algorithm.

§4. New determinant algorithm with high parallelism

Let M be the following n×n matrix:

$$(9)\quad M = \begin{bmatrix} a_{11} & \cdots & a_{1n} \\ \vdots & & \vdots \\ a_{n1} & \cdots & a_{nn} \end{bmatrix}.$$

Following Laplace's expansion formula, we can calculate the determinant of M as

$$(10)\quad |M| = \sum_{(i_1,\ldots,i_{n1})} |M^{(i_1,\ldots,i_{n1})}|\cdot|\tilde{M}^{(j_1,\ldots,j_{n2})}|,$$

$$\begin{aligned} &n_1 + n_2 = n, \\ &1 \leqq i_1 < i_2 \ldots < i_{n1} \leqq n, \\ &1 \leqq j_1 < j_2 \ldots < j_{n2} \leqq n, \\ &\{i_1,\ldots,i_{n1},j_1,\ldots,j_{n2}\} = \{1,2,\ldots,n\}, \end{aligned}$$

where $M^{(i_1,\ldots,i_{n1})}$ is the $n_1\times n_1$ submatrix constructed from left n_1 columns and i_1, ..., i_{n1} rows of M without changing their order, $\tilde{M}^{(j_1,\ldots,j_{n2})}$ is the $n_2\times n_2$ submatrix constructed from right n_2 columns and j_1, ..., j_{n2} rows of M multiplied by the sign factor, and the summation is made over all possible sets $(i_1,\ldots,i_{n1})$. The number of summands in (10) is ${}_nC_{n1} = {}_nC_{n2}$. In the below, we write $M^{(i_1,\ldots,i_{n2})}$ and $\tilde{M}^{(j_1,\ldots,j_{n2})}$ as, respectively, $M^{(i)}$ and $\tilde{M}^{(i')}$ in short. For example, the determinant of the 4×4 matrix (7) is calculated as

$$|M_4| = \begin{vmatrix} a_1 & b_1 \\ a_2 & b_2 \end{vmatrix} \times \begin{vmatrix} c_3 & d_3 \\ c_4 & d_4 \end{vmatrix} - \begin{vmatrix} a_1 & b_1 \\ a_3 & b_3 \end{vmatrix} \times \begin{vmatrix} c_2 & d_2 \\ c_4 & d_4 \end{vmatrix} + \begin{vmatrix} a_1 & b_1 \\ a_4 & b_4 \end{vmatrix} \times \begin{vmatrix} c_2 & d_2 \\ c_3 & d_3 \end{vmatrix} - \begin{vmatrix} a_2 & b_2 \\ a_3 & b_3 \end{vmatrix} \times \begin{vmatrix} c_1 & d_1 \\ c_4 & d_4 \end{vmatrix} + \begin{vmatrix} a_2 & b_2 \\ a_4 & b_4 \end{vmatrix} \times \begin{vmatrix} c_1 & d_1 \\ c_3 & d_3 \end{vmatrix} - \begin{vmatrix} a_3 & b_3 \\ a_4 & b_4 \end{vmatrix} \times \begin{vmatrix} c_1 & d_1 \\ c_2 & d_2 \end{vmatrix},$$

if we choose $n_1 = n_2 = 2$.

The conventional minor expansion algorithms for determinant calculation use the formula (10) recursively by setting $n_1 = 1$ and $n_2 = n-1$, i.e., by expanding the determinant w.r.t. a row or a column (cf., the expansion (8)). The algorithm we propose in this paper uses the formula (10) recursively by setting $n_1 \simeq n/2$, i.e., by splitting the matrix M into two matrices of nearly the same orders. Then, as the above example shows, the major part of the determinant calculation is divided into ${}_nC_{n1}$

subcalculations, i.e., evaluation of minors $|M^{(i)}|$ and $|\tilde{M}^{(i')}|$ and their product. These subcalculations are of nearly the same sizes and executable parallelly. Furthermore, the summation in (10) can be made highly parallelly by the binary summation method:

$$\sum_{i=1}^{m} A_i = (\cdots(((A_1+A_2) + (A_3+A_4)) + \cdots + ((A_{m-3}+A_{m-2}) + (A_{m-1}+A_m)))\cdots). \tag{11}$$

When n is small, the ratio ${}_nC_{n_1}/n!$, which is the ratio of the number of summands in (10) to the number of terms of the determinant expanded completely, is rather large. Hence, it is better to use a conventional minor expansion algorithm if n is small. (In the following algorithms, we employ the conventional minor expansion algorithm if n is less than six.) Taking this notice into account, we obtain the following algorithm:

```
Parallel algorithm DET.SPLIT(M)
  % Evaluate n×n determinant |M| by splitting it        ;
  % into minors of orders n1 ≃ n/2 and n2 = n-n1.       ;
  % Input : an n×n matrix M with polynomial entries.    ;
  % Output: a polynomial D = |M|.                        ;
  if n ≦ 5 then DET.MINOR(M) else begin
    n1 ← [n/2];  n2 ← n - n1;
  % [parallel execution w.r.t. index i, i=1,...,nCn1]   ;
    begin
      construct M(i) and M~(i') ;
      D(i) ← DET.SPLIT(M(i)) × DET.SPLIT(M~(i')) ;
    end;
  % [parallel binary sum w.r.t. index i, i=1,...,nCn1];
    D ← sum D(i);
    return D  end;
```

Here, DET.MINOR in the above procedure is a determinant-evaluating procedure using the conventional minor expansion method.

The algorithm DET.SPLIT manifests a very high parallelism. In fact, if we use ${}_nC_{n_1}$ processors, we can evaluate $D^{(i)}$, $i=1,\cdots,{}_nC_{n_1}$, all at once. A high parallelism comparable to that in the above algorithm can be found only in modular algorithms, so far as algebraic algorithms are concerned. Furthermore, it is worthwhile to note that the algorithm DET.SPLIT is executable by such a simple machine as was referred to at the end of §2.

It is possible to improve the above algorithm considerably. The DET.SPLIT evaluates the same minor many times when $n \geqq 4$. The first improvement is, hence, to avoid duplicate evaluation of minors. This can be almost achieved by dividing $|M^{(i)}|$ and $|\tilde{M}^{(i')}|$, $i=1,\ldots,{}_nC_{n_1}$, into groups so that all determinants in each group contain as many same minors as possible and by evaluating all these groups parallelly. Then, we can avoid duplicate evaluation of the same minor in each group easily by such a method as mentioned in §3. This improvement will be quite effective if n is quite large. However, for many practical applications where $n \simeq 10$, the redundancy due to duplicate evaluation of minors in DET.SPLIT will not be so serious. For example, for the determinant of order 10, the duplicate evaluation is done only for minors of order less than or equal to 4, while minors of order less than or equal to 8 are multiply evaluated in the conventional recursive minor expansion algorithm.

The second improvement is for evaluating sparse determinants. The improvement is quite simple in our algorithm. We make reordering of rows and columns of M so that the nonzero elements are gathered around the diagonal line of M. Then, the number of summands in (10) decreases much. For example, suppose $n = 10$ and all elements M_{ij}, $i=9$ and 10, $j=1,2,\cdots,5$, become zero after a reordering of rows and columns. This decreases the number of summands in (10) from ${}_{10}C_5 = 252$ to ${}_8C_5 = 56$.

§5. Parallel algorithm for symbolic linear equations

Using the same idea as was used in algorithm DET.SPLIT, we can construct efficient parallel algorithms for solving symbolic linear equations. Consider the following coupled linear equations of n unknowns:

$$(12)\quad \begin{array}{l} a_{11}x_1 + \cdots + a_{1n}x_n = b_1, \\ \vdots \qquad\qquad \vdots \quad \vdots \\ a_{n1}x_1 + \cdots + a_{nn}x_n = b_n, \end{array}$$

where, we assume the coefficient matrix $M = (a_{ij})$ is not singular, i.e., $|M| \neq 0$. Let M_k be the following $n\times n$ matrix:

$$(13)\quad M_k = \begin{pmatrix} a_{11} \cdots a_{1,k-1} & b_1 & a_{1,k+1} \cdots a_{1n} \\ \vdots & \vdots & \vdots \\ a_{n1} \cdots a_{n,k-1} & b_n & a_{n,k+1} \cdots a_{nn} \end{pmatrix},$$

$k = 1, \cdots, n.$

Then, Cramer's formula gives the solution of (12) as

$$(14)\quad x_k = D_k/D, \quad k = 1, \cdots, n,$$

with $D_k = |M_k|$ and $D = |M|$.

From the viewpoint of parallel processing, D and D_k, $k=1,\cdots,n$, can be computed parallelly. The degree of parallelism in this scheme is only $n+1$. We can, however, calculate D and D_k highly parallelly in the following way. Let $n_1 = [n/2]$ and $n_2 = n-n_1$, as before. According to formula (10), we can calculate D as

$$(15)\quad D = \sum_{i=1}^{{}_nC_{n_1}} |M^{(i)}|\cdot|\tilde{M}^{(i')}|,$$

where the submatrix $M^{(i)}$ is of order n_1, $\tilde{M}^{(i')}$ is of order n_2, and the summation is made over ${}_nC_{n_1}$ different submatrices. The D_k, $k \leqq n_1$, is given by

$$(16)\quad D_k = \sum_{i=1}^{{}_nC_{n_1}} |M_k^{(i)}|\cdot|\tilde{M}^{(i')}|, \quad 1 \leqq k \leqq n_1,$$

where $M_k^{(i)}$ is the submatrix of order n_1 constructed from M_k by the same way as the construction of $M^{(i)}$ from M. Similarly, for $n_1 < k \leqq n$, we have

$$(17)\quad D_k = \sum_{i=1}^{{}_nC_{n_1}} |M^{(i)}|\cdot|\tilde{M}_k^{(i')}|, \quad n_1 < k \leqq n.$$

Note that the same minor $|\tilde{M}^{(i')}|$ appears in (15) and (16), and the same minor $|M^{(i)}|$ in (15) and (17). Therefore, if we save the expressions $|M^{(i)}|$ and $|\tilde{M}^{(i')}|$, $i=1,\cdots,{}_nC_{n_1}$, which are obtained in the evaluation of D, we can use them for the evaluation of $D_1, \cdots, D_n$. Thus, we obtain the following algorithm:

```
Parallel algorithm CRAMER.SPLIT(M,B)
% Solve the coupled linear equations Σ_{j=1}^n a_ij x_j =;
% b_i, i=1,...,n, by Cramer's method, where a_ij      ;
% and b_i are polynomials.                             ;
% Input : an n×n nonsingular matrix M = (a_ij) and;
%         an n-dimensional vector B=(b_1,...,b_n).    ;
% Output: polynomials D=|M| and D_k=|M_k|,            ;
%         k = 1, ..., n.                               ;
begin
  D ← DET.SPLIT(M), and save d^(i) ← |M^(i)| and
  d~^(i) ← |M~^(i')|, i = 1, ..., nCn1;
  for k ← 1 until n1 step 1 do
    begin
      % [parallel execution w.r.t. index i,           ;
      %  i = 1, ..., nCn1]                             ;
        D^(i) ← d~^(i)×DET.SPLIT(M_k^(i));
      % [parallel binary sum w.r.t. index i,          ;
      %  i = 1, ..., nCn1]                             ;
        D_k ← sum D^(i);
    end;
  for k ← n1+1 until n step 1 do
    begin
      % [parallel execution w.r.t. index i,           ;
      %  i = 1, ..., nCn1]                             ;
        D^(i) ← d^(i)×DET.SPLIT(M~_k^(i'));
      % [parallel binary sum w.r.t. index i,          ;
      %  i = 1, ..., nCn1]                             ;
        D_k ← sum D^(i);
    end;
  return D, D_1, ..., D_n
end;
```

If we have many processors, parallel evaluation of $D_1, \cdots, D_n$ is also possible.

The above algorithm manifests a very high parallelism as DET.SPLIT does, and it is executable by such a simple machine as was referred to at the end of §2.

The CRAMER.SPLIT will be quite efficient if $n < 10$ or so. If n becomes quite large, however, the overhead due to duplicate minor evaluation becomes serious. This overhead can be decreased much by the same method described for algorithm DET.SPLIT. In addition to this improvement, we have the following attractive scheme for decreasing the overhead.

Let $n'_1 = [n/4]$, $n'_2 = [n/2] - [n/4]$, $n'_3 = [3n/4] - [n/2]$, and $n'_4 = n - [3n/4]$. Let us divide the n columns of M into four groups and define four matrices M_I, M_{II}, M_{III}, and M_{IV} of forms $n \times n'_1$, $n \times n'_2$, $n \times n'_3$ and $n \times n'_4$, respectively:

$$
\begin{aligned}
M_I &= (M_{ij}),\ i=1,\cdots,n,\ j=1,\cdots,[n/4],\\
M_{II} &= (M_{ij}),\ i=1,\cdots,n,\ j=[n/4]+1,\cdots,[n/2],\\
M_{III} &= (M_{ij}),\ i=1,\cdots,n,\ j=[n/2]+1,\cdots,[3n/4],\\
M_{IV} &= (M_{ij}),\ i=1,\cdots,n,\ j=[3n/4]+1,\cdots,n.
\end{aligned}
$$

We first evaluate ${}_nC_{n'_1}$ minors of order n'_1 which are constructed from M_I without changing the order of the elements. Let these minors be in class I. Similarly, we evaluate ${}_nC_{n'_2}$, ${}_nC_{n'_3}$, and ${}_nC_{n'_4}$ minors constructed from M_{II}, M_{III}, and M_{IV}, respectively. Let these minors be in classes II, III, and IV, respectively. Second, we evaluate $|M^{(i)}|$, $i=1,\cdots,{}_nC_{n_1}$, by using minors in classes III and IV. Let these minors be in class III+IV. Then, we can evaluate D_k, $k=1,\cdots,[n/4]$, by using minors in classes II and III+IV. Similarly, we can evaluate D_k, $k=[n/4]+1,\cdots,[n/2]$, by using minors in classes I and III+IV, and so on.

It is clear that the number of groupes into which the n columns of M are divided may be another number than 4. In addition to the above improvement, preordering of rows and columns of M and $\vec{B}$ will reduce the total amount of computations much when the matrix M is sparse, as was explained for DET.SPLIT.

§6. Empirical study

We have not tested our parallel algorithms yet but tested only a serial computation version of the determinant algorithm. We compared three algorithms, DET.MINOR, DET.EMINOR, and DET.SPLIT'. The DET.MINOR is the well-known recursive minor expansion algorithm,[12] and DET.EMINOR is the efficient minor expansion algorithm which was implemented in REDUCE by Griss. This algorithm uses backet hashing to avoid duplicate minor evaluation. The DET.SPLIT' is the serial computation version of DET.SPLIT presented in §4.

The test problem is the following matrix of order $n-1$:

$$
M = \begin{pmatrix} a_{n-1} & 2a_{n-2} & \cdots & (n-1)a_1 \\ & a_{n-1} & \cdots & (n-2)a_2 \\ & & \cdots\cdots & \\ & & & a_{n-1} \end{pmatrix} \times \begin{pmatrix} a_1 & & & \\ 2a_2 & a_1 & & \\ \cdots & \cdots & \cdots & \\ (n-1)a_{n-1} & (n-2)a_{n-2} & \cdots & a_1 \end{pmatrix}
- \begin{pmatrix} na_n & (n-1)a_{n-1} & \cdots & 2a_2 \\ & na_n & \cdots & 3a_3 \\ & & \cdots\cdots & \\ & & & na_n \end{pmatrix} \times \begin{pmatrix} na_0 & & & \\ (n-1)a_1 & na_0 & & \\ \cdots & \cdots & \cdots & \\ 2a_{n-2} & 3a_{n-3} & \cdots & na_0 \end{pmatrix}.
$$

The determinant of M gives the discriminant of equation of degree n except for a numeric factor.[13]

The results are given in Table I. Table I showes that our algorithms are, when executed serially, not so efficient as the efficient minor expansion algorithm but as efficient as the recursive minor expansion algorithm. (note that DET.MINOR and DET.SPLIT' are the same for determinants of order less than or equal to five.) Therefore, our algorithms will be quite efficient when executed parallelly.

It is rather dangerous to say much about the serial computation version of our algorithms from only the data in Table I.

order of M	DET.MINOR	DET.EMINOR	DET.SPLIT'
4	0.169	0.165	0.177
5	1.096	0.848	1.230
6	9.52	5.65	9.60
7	95.5	42.5	94.6

Table I: Computing times for dense determinant by three algorithms: times in seconds, and GBC times excluded. Memory = 11.2 MB.

However, we may well expect that, when improvements on our algorithms mentioned in §4 and §5 will be done, the algorithms will become quite efficient even in the serial computation scheme. In particular, our algorithm for symbolic linear systems will be useful because we can systematically avoid duplicate evaluation of $|M^{(i)}|$ and $|\tilde{M}^{(i')}|$ in the calculation of D_k, $k=1,\cdots,n$. It is worthwhile to note that our algorithms are space-efficient when executed serially. This is an important feature in applications, because the determinant calculation is often necked not by time but by space.

References

[1] Enslow, P.H., "Multiprocessor organization - A Survey," *Computing Surveys*, Vol. 9, No. 1 (1977).

[2] Misunas, D.P., "Workshop on Data Flow Computer and Program Organization," *Comp. Arch. News (ACM SIGARCH)*, Vol. 6, pp. 6-22 (1977).

[3] Backus, J., "Can Programming be Liberated from the von Neumann Style? A Functional Style and its Algebra of Programs," *CACM*, Vol. 21, pp. 613-641 (1978).

[4] Friedman, D.P. and Wise, D.S., "Aspects of Applicative Programming for Parallel Processing," *IEEE Trans. Comp.*, Vol. c-27, pp. 289-296 (1978).

[5] Prini, G., "Explicit Parallelism in LISP-like Languages," *Proc. LISP Conf.*, pp. 13-18 (1980).

[6] Marti, J.B., "Compilation Techniques for a Control-Flow Concurrent LISP System," *Proc. LISP Conf.*, pp. 203-207 (1980).

[7] Yasui, H., Saito, T., Mitsuishi, A. and Miyazaki, Y., "Architecture of EVLIS Machine and Dynamic Measurements of Parallel Processing in LISP" (in Japanese), *working paper of Kigoshori-Kenkyukai, IPS of Japan*, (Dec. 1979); abstract in JIP. Vol. 2, p. 232 (1980).

[8] Yoshimura, S., Mizutani, H. and Shibayama, S., "A Parallel Processing Machine" (in Japanese), *Collected Papers on Pattern Processing System - Natl. R&D Program, Agency of industrial Sci. and Tech. of Japan*, pp. 211-222 (1980).

[9] Griss, M.L., "Efficient Expression Evaluation in Sparse Minor Expansion, Using Hashing and Deferred Evaluation," *Proc. Hawaii Intl. Conf. on System Sciences*, pp. 169-172 (1977).

[10] Wang, P.S., "On the Expansion of Symbolic Determinants," *Proc. Hawaii Intl. Conf. on System Sciences*, pp. 173-175 (1977).

[11] Tesler, L.G. and Enea, H.J., "A Language Design for Concurrent Processors," *Proc. AFIPS Conf.*, pp. 291-293 (1968).

[12] Griss, M.L., "The Algebraic Solution of Sparse Linear Systems via Minor Expansion," *ACM TOMS*, Vol. 2, pp. 31-49 (1976).

[]3] Sasaki, T., Kanada, Y. and Watanabe, S., "Calculation of Discriminants of High Degree Equations," *Tokyo J. Math.*, (to appear).

Algebraic Computation for the Masses

Joel Moses
Massachusetts Institute of Technology
Cambridge, Massachusetts 02139

ABSTRACT

Powerful algebraic manipulation systems will be available at very low costs in this decade. We are thus close to achieving the goal many of us have been after for nearly two decades, namely, availability of our systems to the masses. In this talk we shall discuss the problems we still face in making our systems useful to the hordes of potential users.

Construction of Nilpotent Lie Algebras over Arbitrary Fields

Robert E. Beck
Villanova University, Villanova, Pennsylvania

Bernard Kolman
Drexel University, Philadelphia, Pennsylvania

1. Introduction

In this paper we present a general description of a computationally efficient algorithm for constructing every n-dimensional nilpotent Lie algebra as a central extension of a nilpotent Lie algebra of dimension less than n.

As an application of the algorithm, we present a complete list of all real nilpotent six-dimensional Lie algebras. Since 1958, four such lists have been developed: namely, those of Morozov [2], Shedler [3], Vergne [5] and Skjelbred and Sund [4]. No two of these lists agree exactly. Our list resolves all the discrepancies in the other four lists. Moreover, it contains each earlier list as a subset.

A more extensive discussion of computer aspects of the algorithm is given in a paper that follows [1].

2. Central Extensions

Let $\mathfrak{g}$ be a Lie algebra over an arbitrary field F. A skew-symmetric bilinear form $B: \mathfrak{g} \times \mathfrak{g} \to F^k$ is said to be closed if it satisfies the Jacobi identity

$$B([x,y],z) + B([z,x],y) + B([y,z],x) = 0 \qquad (1)$$

for all $x,y,z \in \mathfrak{g}$. Let $C_k(\mathfrak{g})$ denote the space of all closed skew-symmetric bilinear forms. The radical rad(B) of the closed from B, is defined by

$$\mathrm{rad}(B) = \{x \in \mathfrak{g} \mid B(x,y) = 0 \text{ for all } y \in \mathfrak{g}\}.$$

Consider the short exact sequence of Lie algebras over F:

$$0 \to \mathfrak{g}_1 \to \mathfrak{g} \overset{\mu}{\to} \mathfrak{g}_2 \to 0.$$

The algebra $\mathfrak{g}$ is called an extension of $\mathfrak{g}_2$ by $\mathfrak{g}_1$. Furthermore $\mathfrak{g}$ is called a central extension if $\ker(\mu)$ is contained in the center $Z(\mathfrak{g})$ of $\mathfrak{g}$. It follows that any Lie algebra with nonzero center is the central extension of some other Lie algebra.

If B is a closed form on $\mathfrak{g}$ we define a Lie algebra structure on $\mathfrak{g} \oplus F^k$ by letting

$$[(x,u),(y,v)] = ([x,y], B(x,y)),$$

where $x,y \in \mathfrak{g}$, and $u,v \in F^k$. We denote this Lie algebra by $\mathfrak{g}(B)$. An easy calculation shows that $\mathfrak{g}(B)$ is a central extension of $\mathfrak{g}$ and

$$Z(\mathfrak{g}(B)) = (\mathrm{rad}\, B \cap Z(\mathfrak{g})) \oplus F^k .$$

We now want to consider closed forms that define central extensions whose center is exactly k-dimensional. Thus, we let $\hat{C}_k$ denote the set of all closed forms $B: \mathfrak{g} \times \mathfrak{g} \to F^k$ such that $\mathrm{rad}(B) \cap Z(\mathfrak{g}) = 0$. The following result can be established.

Proposition. (a) If $B_1 \in C_k(\mathfrak{g}_1)$ and $B_2 \in \hat{C}_k(\mathfrak{g}_2)$, $\dim \mathfrak{g}_1 = \dim \mathfrak{g}_2$, and $\alpha: \mathfrak{g}_1(B_1) \to \mathfrak{g}_2(B_2)$ is a Lie algebra isomorphism, then

$$\alpha = \begin{bmatrix} \alpha_1 & 0 \\ \alpha_3 & \alpha_4 \end{bmatrix} \qquad (1)$$

where $\alpha_1: \mathfrak{g}_1 \to \mathfrak{g}_2$ is a Lie algebra isomorphism, $\alpha_3 \in \mathrm{Hom}(\mathfrak{g}, F^k)$, $\alpha_4 \in GL(k,F)$ and

$$\alpha_3 \circ [\cdot,\cdot] + \alpha_4 \circ B_1 = B_2 \circ (\alpha_1 \times \alpha_1) \qquad (2)$$

(b) Let $B_1, B_2 \in \hat{C}_k(\mathfrak{g})$, and suppose that $\alpha: \mathfrak{g}(B_1) \to \mathfrak{g}(B_2)$ as defined in (1) satisfies (2), where $\alpha_1 \in \mathrm{Aut}(\mathfrak{g})$, and α_3 and α_4 are as in (a). Then α is a Lie algebra isomorphism.

3. Equivalence Classes in C_k

In this section we define two equivalence relations in the space C_k of closed forms with the goal of characterizing the nonisomorphic central extensions of a given Lie algebra $\mathfrak{g}$. This characterization will be used in the next section to obtain a classification of low-dimensional nilpotent Lie algebras.

Two bilinear forms B_1 and B_2 in C_k are said

to be equivalent ($B_1 \sim B_2$) if $\mathfrak{g}(B_1) \approx \mathfrak{g}(B_2)$. Since $\hat{C}_k$ is a subset of C_k, the equivalence relation ~ on C_k induces an equivalence relation on $\hat{C}_k$. It can be shown that if $B \in \hat{C}_k$, then the equivalence class determined by B in $\hat{C}_k$ coincides with the equivalence class determined by B in C_k.

A form $B \in C_k$ is said to be exact if $B = \phi \circ [\cdot,\cdot]$ for some $\phi \in \mathrm{Hom}(\mathfrak{g}, F^k)$. Let E_k denote the subspace of exact forms in C_k.

Two forms $B_1, B_2 \in C_1$ are said to be E-equivalent ($B_1 \underset{E}{\sim} B_2$) if $B_1 - B_2 \in E_1$. The E-equivalence class determined by $B \in C_1$ is denoted by $\bar{B}$. Let S_k be the collection of all k-dimensional subspaces of C_1/E_1. Let $W \in S_k$ and let $\{\bar{B}_1, \bar{B}_2, \cdots, \bar{B}_k\}$ be a basis for W. For $\alpha \in \mathrm{Aut}(\mathfrak{g})$ define

$$T_\alpha(W) = \langle \overline{B_1 \circ (\alpha\times\alpha)}, \overline{B_2 \circ (\alpha\times\alpha)}, \cdots, \overline{B_k \circ (\alpha\times\alpha)} \rangle ,$$

where $\langle x_1, x_2, \cdots, x_n \rangle$ denotes the subspace spanned by $\{x_1, x_2, \cdots, x_n\}$. It can be shown that if $W \in S_k$, then $T_\alpha(W) \in S_k$, and that the definition of T_α is independent of the choice of basis and representatives.

The mapping $(\alpha, W) \mapsto T_\alpha(W)$ defines an action of the group $\mathrm{Aut}(\mathfrak{g})$ on the set S_k since we can show that if e is the identity of $\mathrm{Aut}(\mathfrak{g})$, then $(e, W) \mapsto W$ and $(\alpha\beta, W) \mapsto T_{\alpha\beta}(W) = T_\beta T_\alpha(W)$. An orbit of this action is called a Z-orbit if every subspace in the orbit contains only E-equivalent classes $\bar{B}$ of bilinear forms such that $\mathrm{rad}(B) \cap Z(\mathfrak{g}) = 0$.

An equivalence class in $\hat{C}_k(\mathfrak{g})/\sim$ is called an I-equivalence class if it contains a bilinear form B such that $\mathfrak{g}(B)$ is an indecomposable Lie algebra. We can establish the following result which provides the basis for the algorithm described below.

Theorem There exists a one-to-one correspondence between the set of I-equivalence classes of $\hat{C}_k/\sim$ and the set of Z-orbits of the action of $\mathrm{Aut}(\mathfrak{g})$ on S_k.

4. Algorithm

For the purpose of computation we establish the following notation and conventions.

Let $\{e_1, e_2, \cdots, e_n\}$ be an (ordered) basis for the Lie algebra $\mathfrak{g}$, over an arbitrary field F, where $\{e_{n-p+1}, \cdots, e_n\}$ is a basis for the center $Z(\mathfrak{g})$ of $\mathfrak{g}$. Define the array $[c_{ij}^k]$ by

$$[e_i, e_j] = \sum_{k=1}^{n} c_{ij}^k e_k$$

Let $\{B_{12}, B_{13}, B_{23}, B_{14}, \cdots, B_{n-1,n}\}$ be an (ordered) basis for the space $\Lambda^2\mathfrak{g}$ of skew-symmetric forms on $\mathfrak{g}$, where

$$B_{rs}(e_i, e_j) = \delta_{ri}\delta_{sj} - \delta_{rj}\delta_{si} .$$

Using this indexing scheme, we obtain a canonical description of nilpotent n-dimensional Lie algebras as central extensions of smaller dimensional nilpotent Lie algebras. Specifically, if n-p is the dimension of $Z(\mathfrak{g})$ then the last n-p elements of the above basis for $\mathfrak{g}$ form a basis for $Z(\mathfrak{g})$. Moreover, as we construct $\mathfrak{g}(B)$ from $\mathfrak{g}$, the multiplication table for $\mathfrak{g}$ carries over to form part of the multiplication table of $\mathfrak{g}(B)$. The rest of the multiplication table of $\mathfrak{g}(B)$ comes from the description of B in terms of the standard basis for $\Lambda^2\mathfrak{g}$.

The following algorithm will generate all nilpotent Lie algebras of dimension n given those of dimension less than n. Every n-dimensional nilpotent Lie algebra is either decomposable or indecomposable. The decomposable n-dimensional algebras are obtained by combining the partitions of n as a sum of positive integers with the classification of the indecomposable algebras of dimension less than n. The indecomposable algebras of dimension n are determined by computing the central extensions of nilpotent algebras of dimension less than n.

Decomposable n-dimensional nilpotent Lie algebras

Since there are no decomposable nilpotent Lie algebras of dimension 2, we need only consider partitions of n which do not contain 2. For n=2 through 6 the appropriate partitions of n are given in Table 1 at the end of the paper. For each partition $(n_1, n_2, \cdots, n_k)$ of n we form all direct products $\mathfrak{g}_1 \times \mathfrak{g}_2 \times \cdots \times \mathfrak{g}_k$, where $\mathfrak{g}_i$ is an indecomposable nilpotent Lie algebra of dimension n_i. The decomposable real nilpotent Lie algebras of dimensions 2 through 6 are also given in Table 1.

Indecomposable n-dimensional nilpotent Lie algebras

To obtain all of the indecomposable nilpotent Lie algebras of dimension n, we must consider central extensions of dimension k of all nilpotent Lie algebras (both decomposable and indecomposable) of dimension n-k for $k = 1, 2, \cdots, n-1$. However the following lemma provides a bound on the dimension k of the center of an indecomposable central extension.

Lemma 2. If an n-dimensional Lie algebra $\tilde{\mathfrak{g}}$ can be obtained as a central extension of an (n-k)-dimensional Lie algebra $\mathfrak{g}$, then

$$k \le \frac{2n+1 - \sqrt{8n+1}}{2}$$

Suppose all nilpotent Lie algebras of dimension less than n are known. That is, their structure constants and automorphism groups are available. To determine the indecomposable Lie algebras of dimension n as central extensions of smaller dimensional algebras we perform the algorithm in Figure 1.

Table 2, at the end of the paper, gives the results of the algorithm for real nilpotent Lie algebras of dimension 1 through 6. In the table, the second column gives, for each algebra, the corresponding designation which has been used by the various classifiers. The last column gives the canonical description of each algebra as a central extension. In this table, the Lie algebras are listed in the following canonical order.

First, algebras are listed in order of increasing dimension. Equal dimensional Lie algebras are listed in order of increasing dimension of their centers. Finally, if $\mathfrak{g}(B)$ and $\mathfrak{g}'(B')$ have equal dimensions and their centers have equal dimensions, then $\mathfrak{g}(B)$ is listed before $\mathfrak{g}'(B')$ if $\mathfrak{g}$ precedes $\mathfrak{g}'$ in Table 1 or Table 2, or if $\mathfrak{g} = \mathfrak{g}'$ and B precedes B' in the lexicographic ordering of $\Lambda^2\mathfrak{g}$ induced by the ordering of its standard basis given above.

REFERENCES

1. Beck, R.E. and Bernard Kolman, "Algorithms for Central Extensions of Lie Algebras", Proceedings SYMSAC '81, Paul S. Wang (editor), ACM, New York, 1981

2. Morozov, V.V., "Classification of Nilpotent Lie Algebras of Sixth Order", Isvestija Vyssih Uabnyh Zaredenii, Matematika (Kazan), 4(5), 161-171, 1958.

3. Shedler, G.S., "On the Classification of Nilpotent Lie Algebras of Dimension Six", IBM Corp. Research Report, RC-1689, Sept. 29, 1966.

4. Skjelbred, T., and T. Sund, "On the Classification of Nilpotent Lie Algebras", Preprint Series, Matematisk Instituutt, Universitetet i Oslo, 1977.

5. Vergne, M.F., Variété des algebres de Lie nilpotentes, these, 3e cycle, Université Paris, 7 June 1966.

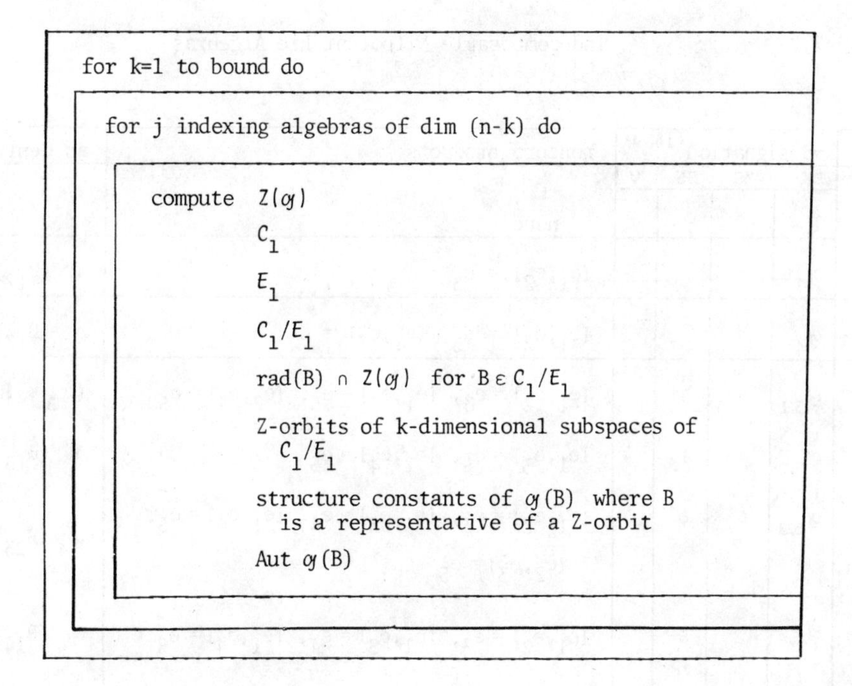

Figure 1.

Decomposable Nilpotent Lie Algebras

The notation $\mathfrak{g}_n$ or $\mathfrak{g}_{n,i}$ refers to the classification of real indecomposable nilpotent Lie algebras given in Table 2.

dimension	partition	algebra
2	(1,1)	$\mathbb{R}^2$
3	(1,1,1)	$\mathbb{R}^3$
4	(3,1)	$\mathfrak{g}_3 \times \mathbb{R}$
	(1,1,1,1)	$\mathbb{R}^4$
5	(4,1)	$\mathfrak{g}_4 \times \mathbb{R}$
	(3,1,1)	$\mathfrak{g}_3 \times \mathbb{R}^2$
	(1,1,1,1,1)	$\mathbb{R}^5$
6	(5,1)	$\mathfrak{g}_{5,i} \times \mathbb{R}$, $i=1,2,\cdots,6$
	(4,1,1)	$\mathfrak{g}_4 \times \mathbb{R}^2$
	(3,3)	$\mathfrak{g}_3 \times \mathfrak{g}_3$
	(3,1,1,1)	$\mathfrak{g}_3 \times \mathbb{R}^3$
	(1,1,1,1,1,1)	$\mathbb{R}^6$

Table 1.

Indecomposable Nilpotent Lie Algebras

dimension	designation(1)				nonzero products	as central extension
	BK	MS	SS	V		
1	$\mathbb{R}$				none	
3	$\mathfrak{g}_3$				$[e_1,e_2] = e_3$	$\mathbb{R}^2\ (B_{12})$
4	$\mathfrak{g}_4$				$[e_1,e_2] = e_3$, $[e_1,e_3] = e_4$	$\mathfrak{g}_3\ (B_{13})$
5	$\mathfrak{g}_{5,1}$	5	3		$[e_1,e_2] = e_3$, $[e_1,e_3] = e_5$, $[e_2,e_4] = e_5$	$(\mathfrak{g}_3 \times \mathbb{R})(B_{13} + B_{24})$
	$\mathfrak{g}_{5,2}$	4	1		$[e_1,e_3] = e_5$, $[e_2,e_4] = e_5$	$\mathbb{R}^4\ (B_{13} + B_{24})$
	$\mathfrak{g}_{5,3}$	6	6		$[e_1,e_2] = e_3$, $[e_1,e_3] = e_4$, $[e_1,e_4] = e_5$, $[e_2,e_3] = e_5$	$\mathfrak{g}_4\ (B_{23} + B_{14})$
	$\mathfrak{g}_{5,4}$	2	5		$[e_1,e_2] = e_3$, $[e_1,e_3] = e_4$, $[e_1,e_4] = e_5$	$\mathfrak{g}_4\ (B_{14})$
	$\mathfrak{g}_{5,5}$	1	2		$[e_1,e_2] = e_4$, $[e_1,e_3] = e_5$	$\mathbb{R}^3\ (B_{12},B_{13})$
	$\mathfrak{g}_{5,6}$	3	4		$[e_1,e_2] = e_3$, $[e_1,e_3] = e_4$, $[e_2,e_3] = e_5$	$\mathfrak{g}_3\ (B_{13},B_{23})$
6	$\mathfrak{g}_{6,1}$	15	3	12	$[e_1,e_2] = e_3$, $[e_1,e_3] = e_4$, $[e_2,e_3] = e_6$, $[e_1,e_4] = e_6$, $[e_2,e_5] = e_6$	$(\mathfrak{g}_4 \times \mathbb{R})(B_{23} + B_{14} + B_{25})$

dimension	designation[1]				nonzero products	as central extension
	BK	MS	SS	V		
6	$\mathfrak{g}_{6,2}$	17	2	11	$[e_1,e_2]=e_3, [e_1,e_3]=e_4, [e_1,e_4]=e_6, [e_2,e_5]=e_6$	$(\mathfrak{g}_4 \times \mathbb{R})(B_{14}+B_{25})$
	$\mathfrak{g}_{6,3}$	12	1	2	$[e_1,e_2]=e_3, [e_1,e_3]=e_6, [e_4,e_5]=e_6$	$(\mathfrak{g}_3 \times \mathbb{R}^2)(B_{13}+B_{45})$
	$\mathfrak{g}_{6,4}$	16	7	13	$[e_1,e_2]=e_3, [e_1,e_3]=e_5, [e_2,e_4]=e_5, [e_3,e_4]=e_6, [e_1,e_5]=e_6$	$\mathfrak{g}_{5,1}(B_{34}+B_{15})$
	$\mathfrak{g}_{6,5}$	20	13	19	$[e_1,e_2]=e_3, [e_1,e_3]=e_4, [e_1,e_4]=e_5, [e_2,e_3]=e_5, [e_2,e_4]=e_6, [e_1,e_5]=e_6$	$\mathfrak{g}_{5,3}(B_{24}+B_{15})$
	$\mathfrak{g}_{6,6}$	22	14	20	$[e_1,e_2]=e_3, [e_1,e_3]=e_4, [e_1,e_4]=e_5, [e_2,e_3]=e_5, [e_3,e_4]=e_6, [e_2,e_5]=-e_6$	$\mathfrak{g}_{5,3}(B_{34}-B_{25})$
	$\mathfrak{g}_{6,7}$	19	11	17	$[e_1,e_2]=e_3, [e_1,e_3]=e_4, [e_1,e_4]=e_5, [e_2,e_3]=e_6, [e_1,e_5]=e_6$	$\mathfrak{g}_{5,4}(B_{23}+B_{15})$
	$\mathfrak{g}_{6,8}$	21	12	18	$[e_1,e_2]=e_3, [e_1,e_3]=e_4, [e_1,e_4]=e_5, [e_3,e_4]=e_6, [e_2,e_5]=e_6$	$\mathfrak{g}_{5,4}(B_{34}-B_{25})$
	$\mathfrak{g}_{6,9}$	2	10	16	$[e_1,e_2]=e_3, [e_1,e_3]=e_4, [e_1,e_4]=e_5, [e_1,e_5]=e_6$	$\mathfrak{g}_{5,4}(B_{15})$
	$\mathfrak{g}_{6,10}$	13	5	10	$[e_1,e_2]=e_4, [e_1,e_3]=e_5, [e_1,e_4]=e_6, [e_3,e_5]=e_6$	$\mathfrak{g}_{5,5}(B_{14}+B_{35})$
	$\mathfrak{g}_{6,11}$	14	6		$[e_1,e_2]=e_4, [e_1,e_3]=e_5, [e_2,e_4]=e_6, [e_3,e_5]=e_6$	$\mathfrak{g}_{5,5}(B_{24}+B_{35})$
	$\mathfrak{g}_{6,12}$		4	9	$[e_1,e_2]=e_4, [e_1,e_3]=e_5, [e_3,e_4]=e_6, [e_2,e_5]=e_6$	$\mathfrak{g}_{5,5}(B_{34}+B_{25})$
	$\mathfrak{g}_{6,13}$	18	9		$[e_1,e_2]=e_3, [e_1,e_3]=e_4, [e_2,e_3]=e_5, [e_1,e_4]=e_6, [e_2,e_5]=e_6$	$\mathfrak{g}_{5,6}(B_{14}+B_{25})$
	$\mathfrak{g}_{6,14}$		8	15	$[e_1,e_2]=e_3, [e_1,e_3]=e_4, [e_2,e_3]=e_5, [e_2,e_4]=e_6, [e_1,e_5]=e_6$	$\mathfrak{g}_{5,6}(B_{24}+B_{15})$
	$\mathfrak{g}_{6,15}$	9	20	8	$[e_1,e_2]=e_3, [e_1,e_3]=e_5, [e_2,e_3]=e_6, [e_1,e_4]=e_6$	$(\mathfrak{g}_3 \times \mathbb{R})(B_{13},B_{23}+B_{14})$
	$\mathfrak{g}_{6,16}$	8	21	5	$[e_1,e_2]=e_3, [e_1,e_3]=e_5, [e_2,e_3]=e_6, [e_2,e_4]=e_6$	$(\mathfrak{g}_3 \times \mathbb{R})(B_{13},B_{23}+B_{24})$

dimension	designation(1)				nonzero products	as central extension
	BK	MS	SS	V		
6	$\mathfrak{g}_{6,17}$	1	18	6	$[e_1,e_2] = e_3, [e_1,e_3] = e_5, [e_1,e_4] = e_6$	$(\mathfrak{g}_3 \times \mathbb{R})(B_{13},B_{14})$
	$\mathfrak{g}_{6,18}$	7	19	4	$[e_1,e_2] = e_3, [e_1,e_3] = e_5, [e_2,e_4] = e_6$	$(\mathfrak{g}_3 \times \mathbb{R})(B_{13},B_{24})$
	$\mathfrak{g}_{6,19}$	10	22		$[e_1,e_2] = e_3, [e_1,e_3] = e_5, [e_2,e_4] = e_5,$ $[e_2,e_3] = e_6, [e_1,e_4] = -e_6$	$(\mathfrak{g}_3 \times \mathbb{R})(B_{13} + B_{24}, B_{23} - B_{14})$
	$\mathfrak{g}_{6,20}$	6		7	$[e_1,e_2] = e_3, [e_1,e_3] = e_5, [e_2,e_4] = e_5,$ $[e_1,e_4] = e_6$	$(\mathfrak{g}_3 \times \mathbb{R})(B_{13} + B_{24}, B_{14})$
	$\mathfrak{g}_{6,21}$	4	16	1	$[e_1,e_2] = e_5, [e_1,e_3] = e_6, [e_2,e_4] = e_6$	$\mathbb{R}^4\ (B_{12},B_{13} + B_{24})$
	$\mathfrak{g}_{6,22}$	5	17		$[e_1,e_3] = e_5, [e_2,e_4] = e_5, [e_2,e_3] = e_6,$ $[e_1,e_4] = -e_6$	$\mathbb{R}^4\ (B_{13} + B_{24}, B_{23} - B_{14})$
	$\mathfrak{g}_{6,23}$	11	23	14	$[e_1,e_2] = e_3, [e_1,e_3] = e_4, [e_2,e_3] = e_5,$ $[e_1,e_4] = e_6$	$\mathfrak{g}_4\ (B_{23}, B_{14})$
	$\mathfrak{g}_{6,24}$	3	15	3	$[e_1,e_2] = e_4, [e_1,e_3] = e_5, [e_2,e_3] = e_6$	$\mathbb{R}^3\ (B_{12}, B_{13}, B_{23})$

Table 2

(1) BK denotes Beck-Kolman
MS denotes Morozov-Shedler [2,3]
SS denotes Skjelbred-Sund [4]
V denotes Vergne [5]

Algorithms for Central Extensions of Lie Algebras

Robert E. Beck
Villanova University, Villanova, Pennsylvania

Bernard Kolman
Drexel University, Philadelphia, Pennsylvania

1. Introduction

It follows from [1] that every n-dimensional nilpotent Lie algebra is a central extension of a lower dimensional nilpotent Lie algebra. This paper develops algorithms to handle two problems: (1) the decomposition of a given nilpotent Lie algebra $\mathfrak{g}$ as a finite sequence of central extensions of lower dimensional nilpotent Lie algebras and (2) the construction of all n-dimensional nilpotent Lie algebras as central extensions of lower dimensional nilpotent Lie algebras.

Let $\{e_1, e_2, \cdots, e_n\}$ be an ordered basis for the Lie algebra $\mathfrak{g}$ over an arbitrary field F. Thus, any element of $\mathfrak{g}$ can be represented as an ordered n-tuple. In the case of the second problem, we also arrange the above basis for $\mathfrak{g}$ so that $\{e_{n-p+1}, e_{n-p+2}, \cdots, e_n\}$ is a basis for the center $Z(\mathfrak{g})$ of $\mathfrak{g}$.

The nonzero products in $\mathfrak{g}$ are given in the input array MULTAB of size NUMPRODS × 4, where NUMPRODS is the number of nonzero products in $\mathfrak{g}$.

If $[e_i, e_j] = \sum_{r=1}^{q} a_{I_r} e_{I_r}$, where $a_{I_r} \neq 0$ for $r = 1, 2, \cdots, q$, then the corresponding entries in MULTAB which indicate this product are

[I, J, I1, AI1]
[I, J, I2, AI2]
•
•
•
[I, J, IQ, AIQ]

Thus, if [I, J, K, A] is the only entry in MULTAB with initial components having values I and J, then $[e_i, e_j] = a\, e_k$. The first three components of each entry in MULTAB must be integers, the fourth is algebraic. We also assume that MULTAB is ordered by the lexicographic ordering of the pairs (I,J). Finally, if NUMPRODS = 0, then $\mathfrak{g}$ is abelian.

The entire multiplication of $\mathfrak{g}$, is given by its structure constants which form the $n \times n \times n$ array $C = [c_{ij}^k]$, where $n = \dim \mathfrak{g}$, and

$$[e_i, e_j] = \sum_{k=1}^{n} c_{ij}^k e_k, \qquad 1 \le i,j \le n$$

The entries in C are algebraic. Simply-developed procedures convert MULTAB to C and vice versa.

Let $\{B_{12}, B_{13}, B_{23}, B_{14}, \cdots, B_{n-1,n}\}$ be an ordered basis for the space $\Lambda^2 \mathfrak{g}$ of skew-symmetric forms on $\mathfrak{g}$, where

$$B_{rs}(e_i, e_j) = \delta_{ri}\delta_{sj} - \delta_{rj}\delta_{si} \quad .$$

Thus, any closed skew-symmetric bilinear form $B: \mathfrak{g} \times \mathfrak{g} \to F$ can be represented as an ordered $\frac{1}{2}n(n-1)$ tuple. Such forms also have a natural representation as skew-symmetric matrices. Simply developed procedures convert a bilinear form from its matrix representation to its vector representation and vice versa.

By fixing the bases for $\mathfrak{g}$ and $\Lambda^2\mathfrak{g}$, as above, we may describe the structure of $\mathfrak{g}$, its extensions, and the algebraic objects associated with them in terms of vectors and matrices with entries from the field F.

2. The Decomposition Problem

(a) Center of $\mathfrak{g}$. We first compute the center of $\mathfrak{g}$ from the structure constants of $\mathfrak{g}$. Now

$$Z(\mathfrak{g}) = \{x \in \mathfrak{g} \mid [x,y] = 0 \text{ for all } y \in \mathfrak{g}\}.$$

Note that the property which defines the center is linear in y. Thus, it suffices to only work with a basis for $\mathfrak{g}$. Let

$$x = \sum_{i=1}^{n} a_i e_i \in Z(\mathfrak{g}). \text{ Then}$$

$$[x, e_j] = 0 = \left[\sum_{j=1}^{n} a_i e_i, e_j\right] = \sum_{i=1}^{n}\sum_{k=1}^{n} a_i c_{ij}^k e_k \qquad j = 1, 2, \cdots, n$$

©1981 ACM O-89791-047-8/81-0800-0175 $00.75

Thus, $x \in Z(\mathfrak{g})$ if and only if its coordinate vector $(a_1, a_2, \cdots, a_n)$ is a solution to the homogeneous system of equations whose matrix M is obtained by reshaping the array C as follows. We describe C as consisting of n sheets $C^{(1)}, C^{(2)}, \cdots, C^{(n)}$, each sheet being an $n \times n$ matrix. Specifically, $C^{(k)}$ is the $n \times n$ matrix $[c_{ij}^k]$. Using this notation M is the partitioned matrix

$$M = \begin{bmatrix} C^{(1)} \\ \hline C^{(2)} \\ \hline \vdots \\ \hline C^{(n)} \end{bmatrix}$$

We now obtain a basis for $Z(\mathfrak{g})$ by finding a basis for the solution space of M. If $Z(\mathfrak{g}) = 0$, we stop. If $Z(\mathfrak{g}) \neq 0$, we continue to (b).

(b) New Basis of $\mathfrak{g}$. If $\dim Z(\mathfrak{g}) = p$, we change the current basis for $\mathfrak{g}$ so that the last p vectors in the new basis form a basis for $Z(\mathfrak{g})$.

(c) Factoring Out $Z(\mathfrak{g})$. The array giving the structure constants of $\mathfrak{g}/Z(\mathfrak{g})$ is the $(n-p) \times (n-p) \times (n-p)$ array obtained from the array C by choosing the first n-p indices for i, j, and k. We compute the multiplication table and print it to identify $\mathfrak{g}/Z(\mathfrak{g})$ and then return to step (a).

The algorithm described above uses the following standard algorithms from linear algebra:

(a) Find the inverse of a matrix
(b) Find a basis for the solution space of a homogeneous system of equations.
(c) Transform a given matrix to reduced row echelon form.
(d) Transform a matrix to reduced row form.

3. Construction of Nilpotent Lie Algebras as Central Extensions

The problem of finding a central extension $\tilde{\mathfrak{g}} = \mathfrak{g}(B)$ of a given nilpotent Lie algebra $\mathfrak{g}$ consists of two phases: the selection of a suitable bilinear form $B: \mathfrak{g} \times \mathfrak{g} \to F^k$, and the computation of $\mathfrak{g}(B)$ and its associated algebraic objects. We assume that $\mathfrak{g}$ itself has been obtained iteratively as a sequence of central extensions of nilpotent Lie algebras of lower dimension. Thus, for the given nilpotent Lie algebra $\mathfrak{g}$ we know the following algebraic objects associated with $\mathfrak{g}$:

$Z(\mathfrak{g})$, the center of $\mathfrak{g}$

$C_1(\mathfrak{g})$, the space of all closed skew symmetric bilinear forms $B: \mathfrak{g} \times \mathfrak{g} \to F$

$E_1(\mathfrak{g})$, the space of all exact forms in $C_1(\mathfrak{g})$

$C_1(\mathfrak{g})/E_1(\mathfrak{g})$

$\mathrm{Aut}(\mathfrak{g})$, the automorphism group of $\mathfrak{g}$

$\mathrm{Der}(\mathfrak{g})$, the derivation group of $\mathfrak{g}$ (this group, which is not needed for the computation of $\mathfrak{g}(B)$, is useful in many other situations; its computation is very similar to the computation of $\mathrm{Aut}(\mathfrak{g})$).

For the sake of simplicity, we first turn to the second phase. Suppose that an acceptable B has been obtained. This vector valued bilinear form can be represented by a $k \times \frac{1}{2}n(n-1)$ matrix, where each row represents one component of B expressed as a linear combination of the standard basis for $\Lambda^2\mathfrak{g}$. Each of these rows can then be shaped into an $n \times n$ skew-symmetric matrix.

(a) Structure Constant Array $\tilde{C}$ of $\mathfrak{g}(B)$. To the structure constant array C for $\mathfrak{g}$, adjoin k sheets, each of which comes from one row of B shaped into an $n \times n$ matrix. To each of these n+k sheets, adjoin k rows and k columns of zeros. This construction of $\tilde{C}$ reflects the fact that the center of $\mathfrak{g}(B)$ is isomorphic to F^k and is spanned by $\{e_{n+1}, e_{n+2}, \cdots, e_{n+k}\}$.

(b) $C_1(\mathfrak{g}(B))$. The space $C_1(\mathfrak{g})$ of closed forms on $\mathfrak{g}$ is a subspace of $\Lambda^2(\mathfrak{g})$. Thus we may represent a basis for $C_1(\mathfrak{g})$ as a $\dim C_1(\mathfrak{g}) \times \frac{1}{2}n(n-1)$ matrix, each row of which gives the coordinates of a basis element of $C_1(\mathfrak{g})$ with respect to the standard basis for $\Lambda^2\mathfrak{g}$. Likewise we may represent a basis for $C_1(\mathfrak{g}(B))$ as a $\dim C_1(\mathfrak{g}(B)) \times \frac{1}{2}(n+k)(n+k-1)$ matrix. The $(n+k) \times (n+k)$ skew-symmetric matrix $[\beta_{ij}]$ corresponds to a closed form of $C_1(\mathfrak{g}(B))$ if and only if

$$\sum_{r=1}^{n+k} (\tilde{c}_{ij}^r \beta_{rt} + \tilde{c}_{ti}^r \beta_{rj} + \tilde{c}_{jt}^r \beta_{ri}) = 0 \qquad (1)$$

for $1 \le i < j < t \le n$

and

$$\sum_{r=1}^{n+k} c_{ij}^r \beta_{rt} = 0 \qquad (2)$$

for $t = n+1, \cdots, n+k$ and i and j such that $[e_i, e_j] \neq 0$.

Moreover, if the $n \times n$ matrix M represents a closed form on $\mathfrak{g}$, then it is easy to show that the $(n+k) \times (n+k)$ matrix

$$\begin{bmatrix} M & 0 \\ 0 & 0 \end{bmatrix}$$

represents a closed form on $\mathfrak{g}(B)$. Thus, $\dim C_1(\mathfrak{g}) \le \dim C_1(\mathfrak{g}(B))$. A basis for $C_1(\mathfrak{g}(B))$ is obtained by finding a basis for the homogeneous system given by (1) and (2).

(c) $E_1(\mathfrak{g}(B))$. The space of exact forms is spanned by the set of all B_{rs} such that $[e_r,e_s] \neq 0$. For convenience, we partition the indexing set of $\Lambda^2\mathfrak{g}(B)$ into S_0 and S_1 where

$$S_0 = \{(r,s) \mid [e_r,e_s] = 0\}$$

$$S_1 = \{(r,s) \mid [e_r,e_s] \neq 0\} \quad .$$

Thus S_1 indexes the basis for E_1.

(d) $C_1(\mathfrak{g}(B))/E_1(\mathfrak{g}(B))$. Instead of finding an actual basis for C_1/E_1 we use elements of C_1 which represent equivalence classes that form a basis for C_1/E_1. In the previously obtained matrix representing a basis for $C_1(\mathfrak{g}(B))$, set the entries in the columns whose indices belong to S_1 equal to zero. The nonzero rows in the resulting matrix form a basis for C_1/E_1.

(e) $\mathrm{Aut}(\mathfrak{g}(B))$. It follows from (1) that $\alpha \in \mathrm{Aut}(\mathfrak{g}(B))$ if and only if

$$\alpha = \begin{bmatrix} \alpha_1 & 0 \\ \alpha_3 & \alpha_4 \end{bmatrix} \tag{3}$$

where $\alpha_1 \in \mathrm{Aut}(\mathfrak{g})$, $\alpha_3 \in \mathrm{Hom}(\mathfrak{g}, F^k)$, $\alpha_4 \in GL(k,F)$ and

$$\alpha_3 \circ [\cdot,\cdot] + \alpha_4 \circ B = B \circ (\alpha_1 \times \alpha_1). \tag{4}$$

In (3), α_1 is known while α_3 and α_4 are arbitrary. Of course, the entries in α_1,α_3, and α_4 may by symbolic. Equation (4) is an equation in F^k and thus must be checked for each component by evaluating it on all possible pairs of basis vectors of $\mathfrak{g}$. If $(i,j) \in S_1(\mathfrak{g})$ then $B(e_i,e_j) = 0$ since B_{ij} is exact and therefore the coefficient of B_{ij} in the representation of B with respect to the standard basis for $\Lambda^2\mathfrak{g}$ must be zero. In this case, (4) reduces to

$$\alpha_3([e_i,e_j]) = B(\alpha_1(e_i),\alpha_1(e_j)) \tag{5}$$

The vector representing $[e_i,e_j]$ can be constructed from the entries of MULTAB which start with I,J. The arguments of B are the ith and jth columns of the matrix representing α_1. Similarly if $(i,j) \in S_0(\mathfrak{g})$, then (4) reduces to

$$\alpha_4(B(e_i,e_j)) = B(\alpha_1(e_i),\alpha_1(e_j)) \tag{6}$$

Since α_3 is initially an arbitrary matrix, (5) represents expressions for the arbitrary elements of α_3 in terms of the coefficients of B and the entries of α_1. Likewise equation (6) represents such expressions for α_4 if $B(e_i,e_j) \neq 0$. Otherwise it may induce additional relations on the entries of α_1.

A symbolic computation language such as Reduce can be used to generate all the identities coming from equations (5) and (6). In general, however, the identities are too complicated to solve under program control. Instead, an interactive approach has been taken.

(f) $\mathrm{Der}(\mathfrak{g}(B))$. Proceeding as in the case of $\mathrm{Aut}(\mathfrak{g}(B))$, it is easy to show that $\delta \in \mathrm{Der}(\mathfrak{g}(B))$ if and only if

$$\delta = \begin{bmatrix} \delta_1 & 0 \\ \delta_3 & \delta_4 \end{bmatrix}$$

where $\delta_1 \in \mathrm{Der}(\mathfrak{g})$, $\delta_3 \in \mathrm{Hom}(\mathfrak{g}, F^k)$, $\delta_4 \in gl(k,F)$ and

$$\delta_3 \circ [\cdot,\cdot] + \delta_4 \circ B = B \circ (\delta_1 \times \mathrm{id}) + B \circ (\mathrm{id} \times \delta_1) \tag{7}$$

Equation (7) then can be used to generate identities which are solved to obtain conditions on the entries in δ_1, δ_3, and δ_4.

We next turn to the first phase, the problem of finding a suitable bilinear form. To construct a single k-dimensional extension of a given n-dimensional Lie algebra $\mathfrak{g}$, it suffices to find a bilinear form $B \in C_k(\mathfrak{g})$ such that $\mathrm{rad}(B) \cap Z(\mathfrak{g}) = 0$. The problem of finding all k-dimensional extensions of $\mathfrak{g}$ is much more difficult, and the computer is used in an interactive manner. For this problem recall that S_k is the set of all k-dimensional subspaces of C_1/E_1 and that $\mathrm{Aut}(\mathfrak{g})$ acts on S_k. A Z-orbit is one in which every subspace in the orbit contains only E-equivalence classes of bilinear forms such that $\mathrm{rad}(B) \cap Z(\mathfrak{g}) = 0$. Now the set of indecomposable k-dimensional extensions of $\mathfrak{g}$ is in one-to-one correspondence with the set of Z-orbits of S_k under the action of $\mathrm{Aut}(\mathfrak{g})$.

To find all the Z-orbits we must compute the images of various subspaces $W \in S_k$ under the action of an arbitrary $\alpha \in \mathrm{Aut}(\mathfrak{g})$ and then by choosing a canonical representative for each orbit, verify that all the subspaces in S_k have been accounted for.

(g) Radical of B Let $B \in C_k(\mathfrak{g})$. The radical of B is defined by

$$\mathrm{rad}(B) = \{x \in \mathfrak{g} \mid B(x,y) = 0 \text{ for all } y \in \mathfrak{g}\}$$

If $B = (B^{(1)},B^{(2)},\cdots,B^{(k)})$, then it follows

from [1] that

$$\mathrm{rad}(B) = \bigcap_{i=1}^{k} \mathrm{rad}(B^{(i)}).$$

Thus, it suffices to have an algorithm to compute the radical of a field-valued form combined with an algorithm to compute the intersection of several subspaces of a vector space.

Let $B = \sum_{r<s} b_{rs}B_{rs} \in C_1(\mathfrak{g})$. Note that the property which defines rad(B) is linear in y. Thus, it suffices to work only with a basis for $\mathfrak{g}$. Let $x = \sum_{i=1}^{n} a_i e_i \in \mathrm{rad}(B)$. Then for $j = 1,2,\cdots,n$ we have

$$0 = B(x,e_j) = B\left(\sum_{i=1}^{n} a_i e_i, e_j\right)$$

$$= \sum_{i=1}^{n} \sum_{r<s} a_i b_{rs} B_{rs}(e_i,e_j)$$

which yields

$$\sum_{i=1}^{n} b_{ij} a_i = 0 \qquad j = 1,2,\cdots,n \quad (5)$$

A basis for rad(B) is obtained by finding a basis for the solution space of the homogeneous system (5) in the unknowns $a_1, a_2, \cdots, a_n$.

(h) Computation of Z-orbit. The elements of S_k are k-dimensional subspaces of C_1/E_1 and thus can be represented by $k \times \dim C_1/E_1$ matrices. In fact, these matrices can be chosen to be in reduced row echelon form since the elementary row operations used in this transformation correspond to changes of basis for a given subspace W. To decompose S_k into Z-orbits we choose a subspace $W \in S_k$, represent it by a $k \times \dim C_1/E_1$ matrix MW in reduced row echelon form, and interpret the ith row of MW as a skew-symmetric form $B^{(i)} \in C_1/E_1$.

We can represent $\alpha \in \mathrm{Aut}(\mathfrak{g})$ by the n×n matrix ALPHA, whose entries will generally be symbolic and also will satisfy certain identities. For $i = 1,2,\cdots,k$ we convert the ith row of MW to a $1 \times \frac{1}{2}n(n-1)$ matrix whose entries represent the components of $B^{(i)}$ with respect to the fixed basis for $\Lambda^2\mathfrak{g}$, and then this matrix is in turn transformed to an n×n skew-symmetric matrix MSI. We now compute

$$(\mathrm{ALPHA})^t(\mathrm{MSI})(\mathrm{ALPHA})$$

The resulting skew-symmetric matrix is then reshaped back to a $1 \times \frac{1}{2}n(n-1)$ matrix and then to a $1 \times \dim(C_1/E_1)$ matrix. Adjoining all the $1 \times \dim(C_1/E_1)$ matrices for $B^{(1)}, B^{(2)}, \cdots, B^{(k)}$ yields a $k \times \dim(C_1/E_1)$ matrix which can then be transformed to reduced row echelon form. The resulting matrix WALPHA represents the image of W under an arbitrary element $\alpha \in \mathrm{Aut}(\mathfrak{g})$ and its entries are expressed in terms of the entries of ALPHA. At this point we identify all the reduced row echelon form matrices which can be formed from WALPHA by choosing special values for the entries of ALPHA. The subspaces corresponding to those reduced row echelon form matrices comprise one Z-orbit of S_k.

This procedure for computing Z-orbits is repeated until the Z-orbits of all $k \times \dim C_1/E_1$ reduced row echelon form matrices have been identified.

REFERENCES

[1] Beck, R.E. and Bernard Kolman, "Construction of Nilpotent Lie Algebras over Arbitrary Fields," Proceedings SYMSAC '81, Paul S. Wang (editor), ACM, New York, 1981.

Computing an Invariant Subring of k[X,Y]

Rosalind Neuman
Washington University

1. Introduction and notations

We are interested in computing the invariant subring (also referred to as fixed ring) which is obtained as a result of the action of a certain finite group of k-linear automorphisms on the polynomial ring in two variables with coefficients in a field of characteristic p.

Let $GA_2(k)$ denote the group of k-linear ring automorphisms of the polynomial ring $k[X,Y]$ where k is a field having characteristic p. For $\alpha \in GA_2(k)$ we write $\alpha = (F_1,F_2)$, $F_1,F_2 \in k[X,Y]$, to mean $\alpha(X) = F_1$ and $\alpha(Y) = F_2$. If $\beta = (G_1,G_2)$ is also an element of $GA_2(k)$, then the law of composition gives $\alpha\beta = (G_1(F_1,F_2),G_2(F_1,F_2))$.

$GA_2(k)$ contains the following two subgroups. The affine subgroup denoted Af_2 is the group of automorphisms of the form

$$(a_1X+b_1Y+c_1, a_2X+b_2Y+c_2) \quad \text{where} \quad \begin{bmatrix} a_1 & a_2 \\ b_1 & b_2 \end{bmatrix} \in GL_2(k).$$

By E_2 we mean the subgroup of automorphisms of the form $(vX+f(Y), wY+c)$ with $v,w \in k^*$, $c \in k$, and $f(Y) \in k[Y]$. (k^* is the group of non-zero elements of k). As is well known [1, Theorem 3.3], $GA_2(k)$ is the amalgamated product of its subgroups Af_2 and E_2 along their intersection which is the lower triangular subgroup of Af_2. It follows that if G is a finite subgroup of $GA_2(k)$, G is conjugate to a subgroup of Af_2 or E_2 [3, §1.3, Corollary 1].

We denote by $k[X,Y]^G$ the invariant subring of any subgroup G of $GA_2(k)$, that is, $k[X,Y]^G = \{F \in k[X,Y] \mid \alpha(F) = F \text{ for every } \alpha \in G\}$. For a finite subgroup $G \subset Af_2(k)$, $k[X,Y]^G$ has been the object of considerable study [4,5]. This paper is concerned with $k[X,Y]^G$ for a finite subgroup $G \subset E_2$ where the order of G is a power of p. The main theorem in this work asserts that for such a G, $k[X,Y]^G$ is always a polynomial ring. Furthermore, a precise method for computing $k[X,Y]^G$ is indicated within the proof of the theorem.

2. A fixed subring of k[X,Y]

Lemma. Let $\alpha = (vX+f(Y), wY+c) \in E_2$. Then $v=w=1$ if and only if the order of α, denoted $|\alpha|$, is a power of p.

Proof. If $|\alpha| = p^s$ for some $s = 0,1,2,\ldots$, we then have $v^{p^s} = w^{p^s} = 1$ and hence $v=w=1$.

Assume now that $v=w=1$. If $c=0$ then $\alpha^p = (X+pf(Y),Y) = (X,Y)$ and therefore $|\alpha| = p$. If $c \neq 0$, then α^p is of the form $(X+g(Y),Y)$ and so $\alpha^{p^2} = (X,Y)$ giving us the desired result.

Corollary. Let G be a finite subgroup of E_2. Then $|G| = p^s$ if and only if $v=w=1$ for every $\alpha \in G$.

Proof. Follows immediately from the lemma.

Definition. Let $\alpha \in E_2$. We will call α a pseudo-reflection if α can be conjugated to the form $\alpha = (vX+f(Y),Y)$ or $\alpha = (X,wY+c)$. This is the same as saying that after a suitable change of variables, α fixes a variable.

Proposition 1. Let $\alpha \in E_2$ with $v = w = 1$. Then α is a pseudo-reflection if and only if the order of α is p.

Proof. Clearly if α is a pseudo-reflection, α has order p. So suppose the order of α is p. Say $\alpha = (X + f(Y), Y + c)$ where $c \neq 0$ and $f(Y) = \sum_i a_i Y^i \neq 0$, $a_i \in k$. If we conjugate α with $\gamma = (X + g(Y), Y)$ where $g(Y) = -[a_j/(j+1)c]Y^{j+1}$, we eliminate the Y^j term in $f(Y)$ where $j + 1 \not\equiv 0(p)$ without affecting the higher degree terms of f. By repeating this process several times we may assume that $f(Y) = Y^{p-1} f_1(Y^p)$ where $f_1 \in k[Y^p]$. A brief calculation gives:

$$\alpha^p = (X + \sum_{j=0}^{p-1} (Y + jc)^{p-1} f_1(Y^p + jc^p), Y)$$

$$= (X - c^{p-1} f_1(Y^p) + f_2(Y^p) + f_3(Y), Y)$$

where $f_1, f_2 \in k[Y^p]$, $\deg f_2(Y^p) \lneq \deg f_1(Y^p)$ and if $f_3(Y) = \sum_j b_j Y^j$, for $b_j \neq 0$, $p \nmid j$. Since $\alpha^p = (X, Y)$, we obtain $f_1(Y^p) = 0$ and therefore α may be conjugated to $\alpha = (X, Y + c)$, i.e., α is a pseudo-reflection.

Proposition 2. Suppose $\alpha \in E_2$ such that α fixes a variable. Then $k[X,Y]^{\alpha} = k[\prod_{j=0}^{p-1}(X + jf(Y)), Y]$ if $\alpha = (X + f(Y), Y)$ or $k[X,Y]^{\alpha} = k[X, \prod_{j=0}^{p-1}(Y + jc)]$ if $\alpha = (X, Y + c)$.

Proof. See [2].

Theorem. Let G be a finite p-group in E_2. Then $k[X,Y]^G = k[U,V]$ where U and V are algebraically independent.

Proof. Since G is a p-group there exists a composition series: $\{1\} = G_0 \subset G_1 \subset \dots G_n = G$ where $G_{i-1} \Delta G_i$ and $|G_i/G_{i-1}| = p$, $i = 1,2,\dots,n$. Furthermore G_i is generated by β_i and G_{i-1} where β_i is any element in G_i such that $\bar{\beta}_i$ is a generator of G_i/G_{i-1}. We may assume that $k[X,Y]^{G_{i-1}} = k[U_{i-1}, V_{i-1}]$ with $U_{i-1}, V_{i-1} \in k[X,Y]$ algebraically independent. Since $G_{i-1} \Delta G_i$, G_i acts on $k[U_{i-1}, V_{i-1}]$. We also have that $\beta_i^p \in G_{i-1}$, that is β_i has order p on $k[U_{i-1}, V_{i-1}]$. By proposition 1, β_i is a pseudo-reflection on $k[U_{i-1}, V_{i-1}]$ and so, by proposition 2, $k[U_{i-1}, V_{i-1}]^{G_i} = k[U_i, V_i]$ with $U_i, V_i \in k[X,Y]$ algebraically independent. Continuing in this fashion we have that

$$k[X,Y]^G = k[U_{n-1}, V_{n-1}]^{G_n} = k[U,V]$$

with U and V algebraically independent as desired.

The proof of the theorem contains a precise method for computing the fixed ring of any finite p-group in E_2. The following example illustrates this method. We make note of the fact that for any $G \subset GA_2(k)$ and $\rho \in GA_2(k)$, $k[X,Y]^G = \rho[k[X,Y]^{\rho^{-1}G\rho}]$.

Example. Let k be a field of characteristic 2 and $Z \neq 0$ any transcendental element in k. Let G be the group generated by $\alpha = (X + Y^2 + Y, Y + 1)$ and $\beta = (X, Y + Z)$. Then $|G| = 2^3$. G consists of the following nontrivial elements:

$$\alpha = (X + Y^2 + Y, Y + 1),$$

$$\beta = (X, Y + Z),$$

$$\alpha\beta = (X + Y^2 + Y, Y + Z + 1),$$

$$(\alpha\beta)^2 = (X + Z + Z^2, Y),$$

$$(\alpha\beta)^3 = (X + Y^2 + Y + Z^2 + Z, Y + Z + 1),$$

$$(\alpha\beta)\alpha = (X + Z^2 + Z, Y + Z),$$

$$(\alpha\beta)^2\alpha = (X + Y^2 + Y + Z^2 + Z, Y + 1).$$

Let $\gamma = (\alpha\beta)^2$. We may use the following composition series to find $k[X,Y]^G$:

$$\{1\} \subset G_1 \subset G_2 \subset G_3 = G$$

where G_1 is generated by γ, G_2 is generated by $\alpha\beta$, and G_3 is generated by G_2 and β.

By proposition 2, $k[X,Y]^{G_1} = k[U_1, V_1]$ where $U_1 = X(X + Z + Z^2)$ and $V_1 = Y$.

We now have G_2 acting on $k[U_1, V_1]$ with

$$(\alpha\beta)U_1 = U_1 + V_1^4 + (1 + Z + Z^2)V_1^2 + (Z + Z^2)V_1$$

and

$$(\alpha\beta)V_1 = V_1 + Z + 1.$$

As indicated in the proof of proposition 1, if we conjugate $\alpha\beta$ by

$$\rho_1 = (U_1 + (Z+1)^{-1}V_1^5 + (Z+1)^{-1}(1+Z+Z^2)V_1^3 + (Z+Z^2)V_1, V_1)$$

we obtain

$$\rho_1^{-1}(\alpha\beta)\rho_1 = (U_1, V_1 + Z + 1) .$$

Hence by proposition 2 $k[U_1,V_1]^{G_2} = k[U_2,V_2]$ where $U_2 = \rho_1(U_1)$ and $V_2 = V_1(V_1 + Z + 1)$.

$G = G_3$ acts on $k[U_2,V_2]$ with

$$\beta(U_2) = U_2 + Z(Z+1)^{-1}(V_2^2 + ZV_2) + Z^2$$

and

$$\beta(V_2) = V_2 + Z .$$

Conjugate β by

$$\rho_2 = (U_2 + (Z+1)^{-1}V_2^3 + (Z+1)^{-1}ZV_2, V_2) .$$

Then $\rho_2^{-1}\beta\rho_2 = (U_2, V_2 + Z)$ and hence $k[U_2,V_2]^G = k[U_3,V_3]$ where $U_3 = \rho_2 U_2$ and $V_3 = V_2(V_2 + Z)$. But since $k[U_2,V_2]^G = k[X,Y]^G$ we have obtained $k[X,Y]^G = k[U_3,V_3]$ where

$$U_3 = X^2 + (Z+Z^2)X + (Z+1)^{-1}Y^6 + Z(Z+1)^{-1}Y^5 + (Z+1)Y^4 + Z^3(Z+1)^{-1}Y^3 + (Z+1)^{-1}ZY^2 + Z^2Y$$

$$V_3 = Y^4 + (Z^2 + Z + 1)Y^2 + (Z^2 + Z)Y .$$

In [4, Theorem 1] Serre proves that if $G \subset Af_2(k)$ is finite and if $(\text{card}\, G, p) = 1$ then $k[X,Y]^G$ is a polynomial ring if and only if G is generated by the pseudo-reflections it contains. The theorem in this paper shows that if $(\text{card}\, G, p) \neq 1$, $k[X,Y]^G$ may be a polynomial ring even though $G \subset Af_2(k)$ is not generated by pseudo-reflections. For example, let G be the group generated by $\alpha = (X+Y, Y+1)$ over a field k of characteristic 2. $G \subset Af_2(k) \cap E_2$ has order 4; therefore α is not a pseudo-reflection. Using the method developed in the theorem, we find that the fixed ring of G is the polynomial ring $k[U,V]$ where $U = X^2 + X + Y^3 + Y$ and $V = Y^2 + Y$.

References

[1] M. Nagata, On automorphism group of k[X,Y], Lectures in Mathematics, Kyoto University, Kinokuniya Bookstore, Tokyo.

[2] P. Russell, Simple galois extensions of two-dimensional affine rational domains, to appear.

[3] J.P. Serre, Arbres, Amalgames, SL_2, Société Mathématique de France, 1977.

[4] ______, Groupes finis d'automorphismes d'anneaux locaux réguliers, Colloque d'Algèbre (Paris, 1967) No. 8, 11 p. (preprint).

[5] G.C. Shephard and J.A. Todd, Finite unitary reflection groups, Canadian Journal of Mathematics, Vol. 6, 1954, p. 274-304.

Double Cosets and Searching Small Groups

Gregory Butler*

Department of Computer Science
Concordia University
Montreal H3G 1M8

* This work was partially supported by the Natural Sciences and Engineering Research Council of Canada.

Introduction

Double cosets are an important concept of group theory. Although the desirability of algorithms to compute double cosets has been recognized, there has not appeared any algorithm in the literature. The algorithm which we present is a variant of Dimino's algorithm for computing a list of elements of a small group. (By "small" we mean groups of order less than 10^4, whose list of elements we can explicitly store.)

The paper focusses on the problem of searching a small group for elements with a given property. For the record we present Dimino's algorithm and a general algorithm for searching a small group. These two algorithms are not original. We analyse the search algorithm and discuss the role of double cosets in searching. The use of double cosets in the search algorithm does not appear to lead to an improvement over the use of right cosets.

The algorithms will be described in both standard mathematical notation, and in a natural combination of PASCAL and English.

Dimino's Algorithm

Let G be group given by generators $s_1, s_2, \ldots, s_m$. We must be able to multiply elements of G and to determine whether two elements of G are equal. This usually means that we are given a "concrete" representation of G either as permutations or matrices. The first problem we will address is that of generating a list of all the elements of G. The algorithm we give is due to Lou Dimino of Bell Labs and dates from 1971.

Consider the special case where $G = \langle H, s_m \rangle$ and we have a list of elements of $H = \langle s_1, s_2, \ldots, s_{m-1} \rangle$. The group G is a disjoint union

©1981 ACM O-89791-047-8/81-0800-0182 $00.75

$$G = Hy_1 \cup Hy_2 \cup \ldots \cup Hy_N$$

of right cosets of H, where

$$y_1 = \text{identity},$$

$$y_2 = s_m,$$

and for $i \geq 3$

$$y_i = y_j z$$

for some $j < i$ and for some $z \in \{s_1, s_2, \ldots, s_m\}$.

Given y_i the elements of $Hy_i = \{hy_i | h \in H\}$ are enumerated from the list of elements of H. This requires $|H|$ multiplications. For each j and for each $z \in \{s_1, \ldots, s_m\}$ we must form $y_j z$, and determine if it belongs to the union of cosets already computed. If it does not, then $y_j z$ defines the next coset representative y_i. This requires Nm multiplications and searches. The total number of multiplications is therefore

$$H\ (N-1) + Nm - (N-1).$$

The last term compensates for counting the formation of $y_1, y_2, \ldots, y_N$ in both of the previous terms.

A special case which should be noted is the following. If s_m normalizes H then we may choose

$$y_i = s_m^{i-1}.$$

This avoids the formation of the products $y_j z$ at the cost of forming the conjugates

$$s_m^{-1} s_j s_m, \quad \text{for } i = 1, 2, \ldots, m-1,$$

and determining if each conjugate belongs to H.

For a complete analysis of the algorithm we introduce some notation. Let $H_i = \langle s_1, s_2, \ldots, s_i \rangle$. So H_0 is the identity subgroup. Let N_i denote the index of H_{i-1} in H_i. Thus $|G| = N_1 N_2 \ldots N_m$. Summing the costs of computing each $H_i = \langle H_{i-1}, s_i \rangle$ gives a total of

(1) $|G| - 1 + \sum_{i=1}^{m} (N_i + 1)(i-1)$ multiplications, and

$$m + \sum_{i=1}^{m} [(N_i - 1)i + i - 1] \quad \text{searches}$$

for the worst case where the test "s_i normalizes H_{i-1}" always fails.

Note that in the case of a two generator group (where m = 2) the number of multiplications is minimized by taking the generator of greater order as s_1.

For example, the permutations a =(1,2) and b = (1,2,3,4) generate the symmetric group of degree 4. If s_1 = a and s_2 = b then the algorithm requires 36 multiplications and 25 searches. If s_1 = b and s_2 = a then the algorithm requires only 30 multiplications and 15 searches.

In order to efficiently perform the searches the list of elements may be "hash" addressed. See Knuth [2] for more details of this technique.

Searching a small group

The algorithm of this section is a refinement of the most direct approach: look at each element of the group. We assume that the group G is given as a list of elements. Furthermore we are to find all the elements of G with a given property P. There are two restrictions on P. First we must know how to decide, given an element y of G, whether y has property P. Second the set of elements

$$G(P) = \{y \in G \mid y \text{ has property } P\}$$

must be a subgroup of G. Some typical examples are the computation of the centralizer of an element, the normalizer of a subgroup, or the centre of G.

The refinements of the direct approach which are used in the algorithm are best described by viewing the direct algorithm as a discarding process. In the direct method we have an (implicit) set Γ of the elements of G. Each element y is considered in turn. If y has property P then it belongs to G(P) and is discarded from Γ. If y does not have property P then it is discarded from Γ.

The first refinement follows from the fact that G(P) is a subgroup. Let H be the subgroup of G(P) generated by the elements y which the algorithm has already considered and which have property P. Each element y of an explicit set Γ is considered in turn. If y has property P then H becomes $\langle H,y \rangle$ and H is discarded from Γ. Dimino's algorithm will compute $\langle H,y \rangle$. For this additional cost we may avoid many tests for property P.

The second refinement again follows from the fact that G(P) is a subgroup. If y does not have property P then we discard a set D(y) of elements all of which do not have property P. Some obvious candidates for D(y) are $\{y\}$, $\{y,y^{-1}\}$, the group $\langle y \rangle$ generated by y, the right coset Hy, and the double coset HyH. It is easily seen that in these cases if y does not belong to G(P) then $D(y) \cap G(P)$ is empty. That is, no element of D(y) has property P. Algorithms currently use D(y) = Hy. Later we will analyse these cases with particular attention to the cases D(y) = Hy and D(y) = HyH.

The algorithm for the case D(y) = Hy is given below. The modifications required in order to alter the choice of D(y) are minor. Usually the algorithm is supplied with an initial subgroup H_0 of G(P). This may be the identity subgroup, but is often larger. For example, when computing the centralizer of an element z in G then $H_0 = \langle z \rangle$. When computing the normalizer of a subgroup K of G then H_0 = K.

The set Γ may be efficiently implemented as a characteristic function of a subset of the list of elements of G. This requires one binary digit for each element of G.

Double cosets

Let H and K be subgroups of G and let y be an element of G. The double coset HyK is the set of elements

$$\{hyk \mid h \in H, k \in K\}.$$

It is an elementary fact that a double coset HyK is a disjoint union of right cosets of H. In fact

$$HyK = Hy_1 \cup Hy_2 \cup \ldots \cup Hy_N$$

where

$$y_1 = y,$$

and for $i \geq 2$,

$$y_i = y_j z$$

for some j < i and for some generator z of K.

Note the similarity with the setting for Dimino's algorithm. In the special case where $G = \langle H,s_m \rangle$ we can view Dimino's algorithm as a computation of the double coset Hs_mG.

We can modify Dimino's algorithm to compute the double coset HyK. The information we require is a list of the elements of H (in order to enumerate Hy_i) and a set of generators $\{k_1,k_2,\ldots,k_\ell\}$ of K. The algorithm will list the elements of HyK.

An analysis of the algorithm yields a cost of

(2) $|HyK| + N\ell - N$ multiplications, and

$N\ell$ searches.

We remark that if y normalizes H then HyH = Hy. In this case the above algorithm enumerates Hy at less cost than testing whether y normalizes H and forming Hy directly. Therefore the special case is not incorporated into algorithm DOUBLE-COSET even though it is used in algorithm DIMINO.

Analysis and comparison of search algorithms

This section will concentrate on the second refinement to the direct search algorithm and examine the choice of the discard set D(y). But first we should say something about the first refinement. In discussing either of the refinements it is necessary to bear in mind the cost of deciding whether an element has property P. This cost will be denoted c(P). Some typical value are

c(P) = 2 multiplications
    + 1 comparison of elements,
        for centralizer of an element,
    = 3r multiplications
    + search time,
        for normalizer of $K = <k_1,...,k_r>$,
    = 2m multiplications
    + m comparison of elements,
        for centre of $G = <s_1,s_2,...,s_m>$.

Introducing the first refinement saves $|G(P)| - t$ tests (where t generators for G(P) are found by the search algorithm) at the cost of Dimino's algorithm for G(P). Equation (1) is maximized when N_m is $|G|/2^{m-1}$ and each other N_i is 2. This gives upper bounds of

$$\left(1 + \frac{t-1}{2^{t-1}}\right)|G(P)| + \frac{3}{2}(t-1)(t-2) - 1$$

multiplications, and

$$\frac{t}{2^{t-1}}|G(P)| + t^2 \qquad \text{searches}$$

for the cost of Dimino's algorithm for G(P).

When considering the second refinement the cost of computing D(y) relative to the number of tests it saves is also important. For some choices of D(y) however the issue is complicated by the fact that an element may be discarded more than once. For example, the algorithm may consider y_1 and y_2 but we may have $D(y_1) \cap D(y_2)$ nonempty. An example of this is the case $D(y) = <y>$, where an element may be discarded, and later its square root may be discarded. We will call this phenomena "multiple rejection".

In the case of single or double cosets, multiple rejection can only occur if the subgroup H is extended. The reason for this is that the cosets for a fixed H are disjoint. As the subgroup is extended each time a generator is found, each element can be rejected a maximum of t + 1 times. The average seems much lower than this.

For simplicity let us ignore multiple rejection. In the cases where $D(y) = \{y\}$, $\{y,y^{-1}\}$ or $<y>$ the cost of computing D(y) is small, but the number of elements discarded is also small. Typically the group G will not have elements of very large order so the best we can hope for is to reduce the number of tests by one or two orders of magnitude. However the subgroup H may be much larger than this, and taking D(y) = Hy will discard |H| elements, at the small cost of one multiplication per element. Therefore we should take D(y) = Hy in preference to the first three cases. The next decision is not so clearcut. Taking D(y) = HyH will enable the algorithm to discard more elements but the cost (as given in (2)) is now greater than one multiplication per element. The advantage, if any, of double cosets will be in cases where the cost c(P) of testing is high, as it does significantly reduce the number of tests.

Of course multiple rejection clouds the picture, and increases the cost of discarding an element, as an element may be discarded several times. In both cases, D(y) = Hy and HyH, the bound on the number of times an element is discarded is t + 1. To obtain some feel for the interplay of the three factors, size of discard set, cost of computing discard set, and multiple rejection we did an experiment using CAYLEY [1] to simulate the algorithms and to monitor the above factors.

The group G of degree 12 generated by (1,6,7)(2,5,8,3,4,9)(11,12), and (1,3)(4,9,12)(5,8,10,6,7,11) has order 648. Dimino's algorithm computed its list of elements. For each of the first fifty elements we used algorithm SEARCH to compute its centralizer. This was done using both single and double cosets. For no element did the centralizer require more than four generators. The following averages over these fifty runs are listed for comparison. (All costs are in terms of multiplications.)

Table I: Average of all 50 cases.

	Hy	HyH
|G(P)| = 18.2		
number of generators t = 1.96		
number of tests	88.4	21.9
rejection rate per element	1.18	1.24
cost per discard	1.00	1.18
cost per element discarded	1.18	1.47
total cost (including tests)	950	998

The first thing to note is that discarding double cosets greatly reduces the number of tests. Perhaps surprisingly the rejection rate for each element is higher for double cosets than for single cosets, so the cost for each element discarded will also be higher. In this example where the cost c(P) is only 2 multiplications, there is only a 5% difference in performance.

It is felt that the cases where $G(P) = H_0$ were very similar for both single and double cosets, and therefore hid the true difference in performance. In Table II we present the data from the 26 cases (of the above 50) where the algorithm extends H_0 in order to compute G(P). Note that the differences are more marked. There is now a 15% difference in total cost.

Table II: Average of 26 cases where $G(P) \neq H_0$.

	Hy	HyH
$\|G(P)\|$ = 28.4		
number of generators t = 2.8		
number of tests	85.0	26.8
rejection rate per element	1.35	1.48
cost per discard	1.00	1.20
cost per element discarded	1.35	1.77
total cost (including tests)	1058	1206

Conclusion

Although the findings indicate that discarding right cosets is "better" than discarding double cosets in the search algorithm, I feel that the use of double cosets should be further investigated. A more efficient method of enumerating a double coset could reverse the above situation. The use of double cosets could also be better in cases where it is costly to determine whether an element has property P, as is indicated by the large reduction in the number of tests when discarding double cosets.

As the determination of the distribution of groups is a difficult open problem, a complete analysis of the search algorithm appears intractable at this time. In particular, only an intuitive understanding of the phenomena of multiple rejection is possible.

Bibliography

[1] John J. Cannon, Software tools for group theory, Proc. AMS Symp. Pure Math. 37 (1980) 495-502.

[2] Donald E. Knuth, The Art of Computer Programming, Volume 3, Addison-Wesley, 1973.


```
Algorithm   DIMINO ( group G )
{  Given a set {s_1,s_2,...,s_m} of generators of G
   compute a list elements of the elements of G.  }
begin    { DIMINO }
  order := 1;  elements [ 1 ] := identity;
  for  i := 1 to  m  do
    if not s_i in elements then
    begin  { extend by i-th generator.  It is not redundant }
      last-gen := s_i;
      if  s_i^-1 s_j s_i in elements for all j < i
        then  first-gen := s_i;
        else  first-gen := s_1;
      old-order := order; coset-start := old-order + 1;
      append s_i to elements;
      append elements [ 2.. old-order ] * s_i to elements;
```

```
      order := order + old-order;
      repeat  { for each coset representative }
        coset-rep := elements [ coset-start ];
        for each generator gen in { first-gen,...,last gen } do
        begin
          elt := coset-rep * gen;
          if not elt in elements then
          begin
           append elt to elements;
           append elements [ 2..old-orders ] * elt to element;
           order := order + old-order;
          end;
        end;
        coset-start := coset-start + old-order;
      until coset-start > order;
    end;
end; { DIMINO }
```

```
Algorithm   SEARCH ( group G, property P, initial subgroup H0 )
{  Search G for the subgroup G(P) of elements with property
   P.  The subgroup H0 is the initial approximation to G(P). }
begin  { SEARCH }
  H := H0;
  elements-to-search := G \ H0;
  while elements-to-search not empty do
  begin
    choose elt from elements-to-search
    if elt has property P
    then { extend H }
      H := <H,elt>;
      elements-to-search := elements-to-search \ H;
    else { discard D(elt) = H * elt }
      elements-to-search := elements-to-search \ (H * elt);
  end;
end;    { SEARCH . G(P) is H }
```

```
Algorithm   DOUBLE-COSET ( groups H,K, element y )
{  Given a list of the elements of H and generators of K,
   the algorithm produces a list double-coset-list of
   the elements of the double coset HyK. }
 begin  { DOUBLE-COSET }
   coset-start := 1; double-coset-list [ 1 ] := y;
   append (H \ { identity })* y to double-coset-list;
   length := |H|;
```

```
repeat  { for each right coset of H in the double coset }
  coset-rep := double-coset-list [ coset-start ];
  for each generator gen of K do
  begin
    elt := coset-rep * gen;
    if not elt in double-coset-list then
    begin  { adjoin a new right coset of H }
      append elt to double-coset-list;
      append (H ∖ { identity })* elt to double-coset-list;
      length := length + |H|;
    end;
  end;
  coset-start := coset-start + |H|;
until  coset-start > length;
end;  { DOUBLE-COSET }
```

A Generalized Class of Polynomials That Are Hard to Factor

by

Erich Kaltofen,* Rensselaer Polytechnic Institute, Troy, New York 12180

David R. Musser, Computer Science Branch, General Electric Research & Development Center, Schenectady, New York 12345

B. David Saunders,* Rensselaer Polytechnic Institute, Troy, New York 12180

Abstract

A class of univariate polynomials is defined which make the Berlekamp-Hensel factorization algorithm take an exponential amount of time. The class contains as subclasses the Swinnerton-Dyer polynomials discussed by Berlekamp and a subset of the cyclotomic polynomials. Aside from shedding light on the complexity of polynomial factorization this class is also useful in testing implementations of the Berlekamp-Hensel and related algorithms.

1. Introduction and Summary of Results

This paper generalizes a class of univariate polynomials with integral coefficients attributed to H.P.F. Swinnerton-Dyer by E.R. Berlekamp [BERL70, p733]. We use Galois theoretical methods to prove their properties of interest.

These polynomials are of particular interest for the Berlekamp-Hensel factorization algorithm [KNUD81, p433], which determines factors modulo p and lifts them to find the integral factors of a polynomial. Because the polynomials in the class we will define are irreducible over the integers but have a large number of factors modulo p for every prime p, the Berlekamp-Hensel algorithm behaves badly on them: In determining their irreducibility it needs a number of operations that is exponential in the degree and coefficient lengths of the polynomials. This was also true for the Swinnerton-Dyer polynomials, but the class we define is much larger and contains subclasses that, unlike the Swinnewton-Dyer polynomials would make an alternative algorithm (see Section 4) remain super-polynomial in its computing time.

Let n be a positive integer and let r be an integer ≥ 2. By ζ_r we denote $\exp(2\pi i/r)$, the first primitive r-th root of unity. Let $p_1,\ldots,p_n$ be n distinct positive prime numbers. By $f_{r;p_1,\ldots,p_n}(x)$ we denote the monic univariate polynomial in x whose roots are $\zeta_r^{i_1} \sqrt[r]{p_1} + \cdots + \zeta_r^{i_n} \sqrt[r]{p_n}$ with $1 \leq i_1,\ldots,i_n \leq r$.

All $f_{r;p_1,\ldots,p_n}$ have integral coefficients and are irreducible polynomials of degree r^n over the integers. If r is a prime number the following will be shown: If the coefficients of $f_{r;p_1,\ldots,p_n}$ are projected into a field of residues modulo any prime number q, henceforth denoted by Z_q, the image polynomials $f_{r;p_1,\ldots,p_n}$ (mod q) factor into irreducible polynomials over Z_q which have degree at most r.

If r = 2 this construction gives a slightly simpler version of the Swinnerton-Dyer polynomials which treat $\sqrt{-1}$ as an additional prime number. But our Galois theoretical proofs can be easily extended to yield this special case.

The condition of r being a prime number is not crucial for the unpleasant running time behavior for the factorization of these polynomials. For composite r the degrees of the irreducible factors in the modular domain are then bounded by r^2 (we will actually prove a somewhat better bound).

A modified version of these polyno-

---

*This work has been supported by the National Science Foundation under Grant # MCS-7909158.

mials is also presented because of its closely related properties: By $f^*_{r;p_1,\ldots,p_n}$ we denote the polynomial whose roots are $\zeta_r^{i_0} + \zeta_r^{i_1}\sqrt[r]{p_1} + \cdots + \zeta_r^{i_n}\sqrt[r]{p_n}$ where $1 \le i_0, i_1, \ldots, i_n \le r$ and $(i_0, r) = 1$, meaning that the greatest common divisor of i_0 and r is 1.

Again all $f^*_{r;p_1,\ldots,p_n}$ are integer polynomials which factor modulo any prime q into polynomials whose degrees are bounded as for $f_{r;p_1,\ldots,p_n}$. If r is 2, 4, 6 or an odd integer, $f^*_{r;p_1,\ldots,p_n}$ is also irreducible over the integers. Otherwise these polynomials may be reducible but we can guarantee that all factors over the integers are of degree at least $2r^n$.

If n=0, $f^*_{r;\emptyset}$ are the cyclotomic polynomials $\Psi_r(x)$. We will show that for certain composite r the maximum degree of factors in any residue field implies a super-polynomial running time for the Berlekamp-Hensel factorization algorithm. This fact is discussed in [MUSD75, p302]. D. Knuth [KNUD81, p437] uses Berlekamp's algorithm to prove the modular factorization property for Ψ_8.

In section 2 we introduce the number theory and Galois theory required to prove our main theorems in section 3. In section 4 we give examples of these polynomials and additionally analyze the computing time required to factor them. We conclude in section 5 with a recap of the Galois theoretical considerations underlying our results.

Notation: By Z we denote the integers and by Q the rational numbers.

2. Number Theoretical and Galois Theoretical Background

Let r be an integer ≥ 2. By U_r we denote the set of residues modulo r which are relatively prime to r. This set forms a group under multiplication modulo r and there exists a minimal non-negative integer $\lambda(r)$ such that for each $s \in U_r$: $s^{\lambda(r)} \equiv 1 \pmod r$. $\lambda(r)$ is called the minimum universal exponent modulo r. It is known (see [KNUD81, p19]) that

$$\lambda(2) = 1,\ \lambda(4) = 2,\ \lambda(2^\alpha) = 2^{\alpha-2} \text{ for } \alpha \ge 3$$

$$\lambda(2^{\alpha_0} q_1^{\alpha_1} \cdots q_n^{\alpha_n}) = \mathrm{lcm}(\lambda(2^{\alpha_0}), \phi(q_1^{\alpha_1}), \ldots, \phi(q_n^{\alpha_n}))$$

where the q_i are distinct prime numbers, ϕ is Euler's totient function and lcm means the least common multiple. Let p_i be the i-th consecutive prime number. As a consequence of Tchebycheff's theorem $p_i < 2^i$ for all $i > 1$ [SIEW64, p138]. Actually p_i is of order $O(i \log i)$ [HAWR79, p10] but in the following lemma we only need the previous estimate:

<u>Lemma 1</u>: Let j be a fixed integer ≥ 2. Then there are infinitely many positive integers m (namely the product of the first k odd prime numbers with k sufficiently large) such that

$$\phi(m)/\lambda(m) > (\log_2(\phi(m)))^j.$$

<u>Proof</u>: Let $m = p_2 \cdots p_k$. Then $\phi(m) = (p_2-1)\cdots(p_k-1) < 2^{k(k+1)/2}$ by the above estimate for p_i. Therefore $(\log_2(\phi(m)))^j < k^{2j}$. Also $\lambda(m) = \mathrm{lcm}(p_2-1, \ldots, p_k-1) < 2(p_2-1)/2 \cdots (p_k-1)/2 = 2^{2-k}\phi(m)$. Hence $\phi(m)/\lambda(m) > 2^{k-2} > k^{2j}$ for k chosen large enough. Therefore for all sufficiently large k: $\phi(m)/\lambda(m) > (\log_2(\phi(m)))^j$. □

In the proof of theorem 2 below we will make use of the fact that for every prime number r and for all $s \in U_r - \{1\}$: $(s^{r-1}-1)/(s-1)$ is a multiple of r. This follows from the Euler theorem ($a^{\phi(b)} \equiv 1 \pmod b$ for $(a,b) = 1$) and the fact that r is a prime number. In order to treat composite r we generalize this matter:

<u>Lemma 2</u>: Let r be a positive composite integer. By $\eta(r)$ we denote the minimum exponent such that for each $s \in U_r - \{1\}$: $(s^{\eta(r)}-1)/(s-1)$ is divisible by r. Then $\eta(r) \le r\,\lambda(r)$. In fact, $\eta(r) \le d\lambda(r)$ where $d = \mathrm{lcm}(\{(s-1,r) : s \in \bar{U}_r - \{1\}\})$.

<u>Proof</u>: Since for any s, $(s-1,r)$ divides r so must d and therefore $d \le r$. We claim that $(s^{d\lambda(r)}-1)/(s-1)$ is a multiple of r: To prove this we first factor $s^{d\lambda(r)}-1$ as

$$(s^{\lambda(r)}-1)\,(s^{(d-1)\lambda(r)} + s^{(d-2)\lambda(r)} + \cdots + 1).$$

Now the left factor is a multiple of r. It is therefore sufficient to show that the right factor is a multiple of d since that means it can absorb any factor of r in s-1 (by definition of d). But $s^{k\lambda(r)} \equiv (s^{\lambda(r)})^k \equiv 1 \pmod d$ for $0 \le k \le d-1$ since d divides r and thus $(s^{(d-1)\lambda(r)} + \cdots + 1) \equiv d \cdot 1 \equiv 0 \pmod d$, as required. Therefore

$\eta(r) \le d\lambda(r) \le r\lambda(r)$. □

At this point the question arises what the actual value of $\eta(r)$ is. We have not found an explicit formula. Theorem 2 could be improved if we knew that the η value of a divisor of r is bounded by $\eta(r)$. (This is true for the function λ.) The following table indicates for some r the values of $\eta(r)$ and its bounds from lemma 2:

Table 1:

r	4	9	15	22	27	35	3.7^2
$\eta(r)$	2	18	60	10	54	420	294
$d\lambda(r)$	4	18	60	20	162	420	6174
$r\lambda(r)$	8	54	60	220	486	420	6174

We will need some well known properties of the cyclotomic polynomials later and shall mention them now: Let r be an integer ≥ 2 and let ζ_r be a primitive r-th root of unity. There always exist $\phi(r)$ distinct primitive r-th roots of unity in an extension field of Q or Z_q provided that q is a prime number not dividing r. They are the powers of ζ_r whose exponents are relatively prime to r. Then

$$\Psi_r(x) = \prod_{\substack{i=1\\(i,r)=1}}^{r} (x-\zeta_r^{\,i}) = \prod_{d|r} (x^d-1)^{\mu(r/d)}$$

denotes the r-th cyclotomic polynomial which has all integer coefficients (or their residues modulo q if the ground field is Z_q). By $d|r$ we mean that d is a divisor of r and μ denotes the Moebius function. (See [VdWA53, p112].)

If $\zeta_r = \exp(2\pi i/r)$ (i.e. the ground field is Q) then Ψ_r is irreducible over Z [VdWA53, p162].

Lemma 3: Let q be a prime number and let m and r be positive integers such that r is relatively prime to q. Then

$$\Psi_{rq^m}(x) \equiv \Psi_r(x)^{\phi(q^m)} \pmod q.$$

Proof: First we notice the fact that for any integral polynomial f and any integer $i > 0$: $f(x^{q^i}) \equiv f(x)^{q^i} \pmod q$. Then by using the formulas for the cyclotomic polynomials with the Moebius function given above the stated congruence can be easily shown. □

By the Galois group of a polynomial we mean the automorphism group of its splitting field over the field of its coefficients. Then the Galois group of Ψ_r over Q is isomorphic to U_r under multiplication modulo r [VdWA53, p162].

Let f and g be two monic polynomials whose coefficients lie in some integral domain R. Let α_i, $1 \le i \le \deg(f)$ and β_j, $1 \le j \le \deg(g)$ denote their roots respectively. Since the polynomial

$$\prod_{i=1}^{\deg(f)} \prod_{j=1}^{\deg(g)} (x-\alpha_i-\beta_j)$$

is symmetric in both the α_i and the β_j it follows from the fundamental theorem of symmetric functions [VdWA53, p79] that its coefficients also lie in R. There is a resultant method which makes it possible to compute this polynomial by operations in R:

Lemma 4: Let R be an integral domain and let f and g be monic polynomials in R[x]. Then

$$(-1)^{\deg(f)\,\deg(g)}\ \mathrm{res}_y(f(x-y),g(y))$$

is a monic polynomial in R[x] of degree $\deg(f)\cdot\deg(g)$ whose roots are $\alpha_i+\beta_j$ where α_i $(1 \le i \le \deg(f))$ are the roots of f and β_j $(1 \le j \le \deg(g))$ are the roots of g. By res_y we mean the resultant with respect to the indeterminate y (see [VdWA53, p84]).

Proof: See [LOOS73]. □

The next two lemmas will help explain why our polynomials split into so many factors modulo any prime number. First we show what happens to the Galois group when an integral polynomial is projected to a polynomial over a residue field.

Lemma 5: Let f be a monic separable polynomial in Z[x] and let $\bar f \in Z_q[x]$ be its natural projection modulo a prime number q. If $\bar f$ is separable (over Z_q) the Galois group of $\bar f$ over Z_q is a subgroup (as a permutation group on the suitably arranged roots) of the Galois group of f over Q.

Proof: See [VdWA53, p190]. □

The above lemma has been generalized by [ZASH79] for the case in which $\bar f$ is not separable. But, $\bar f$ is inseparable only if q divides the discriminant of f, and one must avoid those primes in order to be able to perform the Hensel factor lifting algorithm.

Lemma 6: Let $\bar f \in Z_q[x]$, q being prime. Assume that all elements of the Galois group $\bar f$ (as permutations on the distinct roots of $\bar f$) are written as products of disjoint cycles. Then $\bar f$ does not contain an irreducible factor of degree higher than the length of the longest cycle.

Proof: This follows immediately from the

statement made about the generating element of the Galois group of $\overline{f}$ in [VdWA53, p191] or likewise from theorem 14 in [GAAL73, p230]. □

The next two lemmas constitute the key for our irreducibility proofs. By [K:F] we denote the degree of a field K over a subfield F and by $F(\theta_1,\ldots,\theta_n)$ we denote the field F extended by the elements $\theta_1,\ldots,\theta_n$.

<u>Lemma 7</u>: Let r be an integer ≥ 2, ζ_r a primitive r-th root of unity, and let $p_1,\ldots,p_n$ be distinct positive primes:

a) $[Q(\sqrt[r]{p_1},\ldots,\sqrt[r]{p_n}):Q] = r^n$.

b) If $r \geq 3$ then $2r^n \leq [Q(\zeta_r,\sqrt[r]{p_1},\ldots,\sqrt[r]{p_n}):Q] \leq \phi(r)r^n$.

c) If r is odd or 2, 4, or 6 then

$$[Q(\zeta_r,\sqrt[r]{p_1},\ldots,\sqrt[r]{p_n}):Q] = \phi(r)r^n.$$

<u>Proof</u>: Partial statements of this lemma are usually shown by Kummerian theory [LANG71, p218]. Elementary proofs of part a) appear in [BESI40] and [RICI74], of part c) for odd r in [CAVB68, p50] and [RICI74]. Part b) follows immediately from part a) and the fact that for $r \geq 3$ every ξ_r is a non-real algebraic number of degree $\phi(r)$. If r = 2 part c) is the same as part a). For r=4 or 6 we combine part b) and the fact that both $\phi(4)$ and $\phi(6)$ are 2. □

Notice that part c) may not hold for even $r \geq 8$ depending on what primes $p_1,\ldots,p_n$ are chosen. Counterexamples may be constructed using the fact that $\sqrt{2} \in Q(\zeta_8)$ or $\sqrt{5} \in Q(\zeta_{10})$ etc.

<u>Lemma 8</u>: Let r be an integer ≥ 2, ζ_r a primitive r-th root of unity, and let $p_1,\ldots,p_n$ be distinct prime numbers. Then $\sqrt[r]{p_n}$ is not an element of the field $Q(\zeta_r,\sqrt[r]{p_1},\ldots,\sqrt[r]{p_{n-1}})$.

<u>Proof</u>: If r is 2 the fact follows from part a) of lemma 8. By F_k we denote the field $Q(\zeta_r,\sqrt[r]{p_1},\ldots,\sqrt[r]{p_k})$ with $1 \leq k \leq n$. Now assume that r is ≥ 3 and $\sqrt[r]{p_n} \in F_{n-1}$ which implies that $F_n = F_{n-1}$. Applying part b) of lemma 8 we get $2r^n \leq [F_n:Q] = [F_{n-1}:Q] \leq \phi(r)r^{n-1}$, which is impossible. □

3. <u>Main Results</u>

<u>Theorem 1</u>: Let r be an integer ≥ 2 and let $p_1,\ldots p_n$ be distinct prime numbers. Then $f_{r;p_1,\ldots,p_n}$ and $f^*_{r;p_1,\ldots,p_n}$ have integer coefficients and the following irreducibility conditions hold:

a) $f_{r;p_1,\ldots,p_n}$ is irreducible over the integers and each irreducible factor of $f^*_{r;p_1,\ldots,p_n}$ over the integers with $r \geq 3$ has degree at least $2r^n$.

b) If r = 2, 4, 6 or odd then $f^*_{r;p_1,\ldots,p_n}$ is irreducible.

<u>Proof</u>: Using lemma 4 inductively we see that the coefficients of $f_{r;p_1,\ldots,p_n}$ and $f^*_{r;p_1,\ldots,p_n}$ are integers and that their degrees are r^n and $\phi(r)r^n$ respectively. (Notice that Ψ_r has integer coefficients as mentioned in section 1.) First we prove by induction that $\sqrt[r]{p_1}+\cdots+\sqrt[r]{p_n}$ is a primitive element of $Q(\sqrt[r]{p_1},\ldots,\sqrt[r]{p_n})$. We make use of the construction of a primitive element given in [VdWA53, p126]: Let $\alpha_1 = \sqrt[r]{p_1}+\cdots+\sqrt[r]{p_{n-1}}$ and $\alpha_2, \ldots, \alpha_{r^{n-1}}$ be the remaining roots of $f_{r;p_1,\ldots,p_{n-1}}$. By the induction hypothesis $Q(\alpha_1) = Q(\sqrt[r]{p_1},\ldots,\sqrt[r]{p_{n-1}})$. The minimal polynomial of α_1 is of degree $[Q(\alpha_1):Q]$ which is r^{n-1} by lemma 7. Therefore $f_{r;p_1,\ldots,p_{n-1}}$ is the minimal polynomial. Let $\beta_1 = \sqrt[r]{p_n}, \beta_2,\ldots,\beta_r$ be the roots of x^r-p_n which is irreducible by Eisenstein's criterion. Then $\alpha_1+\beta_1$ is a primitive element of $Q(\alpha_1,\beta_1) = Q(\sqrt[r]{p_1},\ldots,\sqrt[r]{p_n})$ provided that $\alpha_1+\beta_1 \neq \alpha_i+\beta_j$ for $1 \leq i \leq r^{n-1}$ and $1 < j \leq r$. For the sake of contradiction assume that this condition cannot be achieved, namely there exist an i and a $j > 1$ such that $\alpha_1-\alpha_i = \beta_1-\beta_j$. Since $\beta_j = \zeta_r^k \sqrt[r]{p_n}$ for some positive $k < r$ it follows that $\alpha_1-\alpha_i = \sqrt[r]{p_n}(1-\zeta_r^k)$ and therefore $\sqrt[r]{p_n} = (\alpha_1-\alpha_i) / (1-\zeta_r^k)$ which is an element of $Q(\zeta_r, \sqrt[r]{p_1}, \ldots, \sqrt[r]{p_{n-1}})$, in contradiction to lemma 8. Noticing that Ψ_r is irreducible we can prove in exactly the same way that $\zeta_r + \sqrt[r]{p_1}+\cdots+\sqrt[r]{p_n}$ is a primitive element of $Q(\zeta_r, \sqrt[r]{p_1}, \ldots, \sqrt[r]{p_n})$. However, the α_i will be the roots of an irreducible factor of $f^*_{r;p_1,\ldots,p_{n-1}}$. We now conclude that the minimal polyno-

mials of these primitive elements are of the same degree as the field extensions obtained by adjoining them to the rationals which we know by lemma 7, part a) and c). Therefore $f_{r;p_1,\ldots,p_n}$ and, in the case that $r = 2, 4, 6$ or an odd integer, $f^*_{r;p_1,\ldots,p_n}$ are these minimal polynomials and hence must be irreducible. All irreducible factors of $f^*_{r;p_1,\ldots,p_n}$ have degree at least $2r^n$ because all roots are primitive elements by the argument above and the lower bound of the corresponding field extension is known from lemma 7b). □

Theorem 2: Let r be an integer ≥ 2 and let $p_1,\ldots,p_n$ be prime numbers. For any prime number q the following factorization properties hold for the projected polynomials $f_{r;p_1,\ldots,p_n}$ (mod q) and $f^*_{r;p_1,\ldots,p_n}$ (mod q):
a) The maximum degree of any irreducible factor of both polynomials over the residue field modulo q is at most $r\lambda(r)$. Special case: If r is a prime number the maximum degree is r.
b) If $n = 0$ then the maximum degree of an irreducible factor of $f^*_{r;\emptyset}$ (mod q) $= \Psi_r$ (mod q) is $\lambda(r)$.

Proof:
a) We first show that the length of the longest cycle in any permutation of the Galois group of $f_{r;p_1,\ldots,p_n}$ or $f^*_{r;p_1,\ldots,p_n}$ is at most $\max(r,\eta(r))$, where $\eta(r)$ is as defined in lemma 2. Let σ be an automorphism on $Q(\zeta_r, \sqrt[r]{p_1}, \ldots, \sqrt[r]{p_n})$. As such it has to map the roots of the polynomials Ψ_r and x^r-p_i into roots of the same polynomials. In particular $\sigma(\zeta_r) = \zeta_r^{s_\sigma}$ where ζ_r is a primitive r-th root of unity and s_σ is relatively prime to r. Also $\sigma(\sqrt[r]{p_i}) = \zeta_r^{m_i}\sqrt[r]{p_i}$, where the m_i depend also on σ $(1 \leq i \leq n)$. We now distinguish two cases:

Case 1: $s_\sigma = 1$.
Applying σ r times we get $\sigma^{(r)}(\sqrt[r]{p_i}) = \sqrt[r]{p_i}$ for all $1 \leq i \leq n$ and therefore $\sigma^{(r)}$ maps each root of $f_{r;p_1,\ldots,p_n}$ and $f^*_{r;p_1,\ldots,p_n}$ onto itself which is to say that the permutation corresponding to σ has cycles of length at most r.

Case 2: $s_\sigma > 1$.
By lemma 2 we know that both $s_\sigma^{\eta(r)} \equiv 1$ (mod r) and $(s_\sigma^{\eta(r)}-1) / (s_\sigma-1) \equiv 0$ (mod r). A short computation shows that then $\sigma^{(\eta(r))}(\zeta_r) = \zeta_r$ and $\sigma^{(\eta(r))}(\sqrt[r]{p_i}) = \sqrt[r]{p_i}$ for all $1 \leq i \leq n$. Therefore the cycle lengths of the permutation corresponding to σ are at most $\eta(r)$. Cases 1 and 2 together prove the statement made initially. If the image polynomials are separable we are finished by virtue of the lemmas 2, 5 and 6. But we can repeat the above arguments for automorphisms on the splitting field of $f_{r;p_1,\ldots,p_n}$ (mod q) itself because as we mentioned before the properties of r-th roots of unity carry over for ground fields of characteristic q, provided that q does not divide r.

Finally let q^m be the highest power of q dividing r. By using the identity introduced in the proof of lemma 3 and by using lemma 3 itself we can determine the multiplicities of the roots of Ψ_r (mod q) and x^r-p_i (mod q) (which lie in some Galois field). Therefore $f_{r;p_1,\ldots,p_n} \equiv (f_{r/q^m;p_1,\ldots,p_n})^{q^{mn}}$ (mod q) and $f^*_{r;p_1,\ldots,p_n} \equiv (f^*_{r/q^m;p_1,\ldots,p_n})^{\phi(q^m)q^{mn}}$ (mod q). It follows from the formula for λ given at the beginning of section 2 that $\lambda(r/q^m)$ divides $\lambda(r)$. Then by lemma 2 and the already proven theorem for the case that q does not divide r we conclude that the maximum degree in this case is $r/q^m\,\lambda(r/q^m) < r\lambda(r)$. If r is a prime number the above proof together with the remark made above lemma 2 actually gives the degree bound r.

b) If Ψ_r (mod q) is separable we know its Galois group to be a subgroup of U_r under multiplication modulo r. (This by lemma 5 but one may verify it directly.) The definition of λ and lemma 6 then lead to the statement. If Ψ_r (mod q) is inseparable q necessarily divides r. Again putting together the above, lemma 3 and the fact that $\lambda(r/q^m)$ divides $\lambda(r)$ proves the theorem for this case. □

In many cases the above arguments reveal that the bound $r\lambda(r)$ is too pessimistic: If the image polynomial is separable or more generally if q does not divide r we have proven that the bound $\max(r,\eta(r))$ suffices which may be considerably smaller than $r\lambda(r)$ (see table 1 below lemma 2). If q divides r a bound is $r\lambda(r)/q$.

4. Computational Considerations

One may use lemma 4 in connection with a method to compute cyclotomic polynomials [KNUD81, p440] to actually generate sample polynomials. In producing some of the following examples we used the computer algebra systems MACSYMA and SAC2.

Table 2:

n=0:

(1) $f^*_{8;\emptyset}(x) = \Psi_8(x) = x^4+1$, $\lambda(x)=2$ [KNUD81, p437].

(2) $f^*_{12;\emptyset}(x) = \Psi_{12}(x) = x^4+x^2+1$, $\lambda(12)=2$ [VdWA53, p115].

n=1:

(3) $f^*_{8;2}(x) = (x^{16}+4x^{12}-16x^{11}+80x^9+2x^8+160x^7+128x^6-160x^5+28x^4-48x^3+128x^2-16x+1)(x^{16}+4x^{12}+16x^{11}-80x^9+2x^8-160x^7+128x^6+160x^5+28x^4+48x^3+128x^2+16x+1)$.

n=2:

(4) $f_{2;2,3}(x) = x^4-10x^2+1$ [GAAL73, p233].

(5) $f_{3;2,3}(x) = x^9-15x^6-87x^3-125$.

(6) $f_{4;2,3}(x) = x^{16}-20x^{12}+666x^8-3860x^4+1$.

(7) $f^*_{3;2,3}(x) = x^{18}+9x^{17}+45x^{16}+126x^{15}+189x^{14}+27x^{13}-540x^{12}-1215x^{11}+1377x^{10}+15444x^9+46899x^8+90153x^7+133893x^6+125388x^5+29160x^4-32076x^3+26244x^2-8748x+2916$.

n=3:

(8) $f_{2;2,3,5} = x^8-40x^6+352x^4-960x^2+576$.

(9) $f_{2;-1,2,3} = x^8-16x^6+88x^4+192x^2+144$ [KNUD81, p625].

Table 2 illustrates very well our results: All but polynomial (3) are irreducible over the integers. Since $\sqrt{2} \in \mathbb{Q}(\zeta_8)$ we also know that (3) must be composite. All the polynomials (1)-(9) factor in any modular field into polynomials of smaller degrees and make excellent test cases for implementations of the Berlekamp-Hensel factorization algorithm. E.g. polynomial (7) factors

mod 7: $(x^3+x^2+4x+3)\ (x^3+2x^2+5x+5)\ (x^3+2x^2+4x+2)\ (x^3+x^2+3x+5)\ (x^3+2x^2+2x+3)\ (x^3+x^2+x+2)$

mod 1979: $(x^2+1823x+1632)\ (x^2+85x+6)\ (x^2+828x+749)\ (x^2+1069x+6)\ (x^2+1069x+749)\ (x^2+1069x+1632)\ (x^2+1069x+878)\ (x^2+85x+1744)\ (x^2+828x+1744)$.

The Berlekamp-Hensel factorization algorithm contains the following "bottleneck" [KNUD81, p434]: If f is a polynomial of degree k and splits in a chosen residue field into j irreducible factors then one must perform at least $2^{j-1}-1$ trial divisions to prove its irreducibility over the integers.

If we assume that r is kept fixed and that $p_1,\ldots,p_n$ are the first n prime numbers we obtain for $f_{r;p_1,\ldots,p_n}$ that $k=r^n$ and $j \geq r^n/r\lambda(r) \geq r^{n-2}$. Thus at least $O(2^k)$ factor combinations must be tested. As we will show below the lengths of the coefficients under these assumptions are of magnitude $O(k \log\log(k))$. Hence the worst case time complexity of the Berlekamp-Hensel algorithm is indeed an exponential function in the degree and coefficient lengths of its inputs. Since the degrees of all irreducible factors of $f_{r;p_1,\ldots,p_n}$ are independent of n the modifications of this algorithm suggested in [KNUD81, p434] do not eliminate the exponential running time behavior.

However, additional possibilities have arisen: New probabilistic algorithms factor polynomials over large modular fields sufficiently fast (see [RABM80]). Therefore the Hensel factor lifting step can be avoided and inseparable modular factorizations used. It may then happen that there occur only a few distinct factors with high multiplicities from which, for example, certain irreducibility criteria can be inferred. (Cf. exercise 19 in [KNUD81, p626].)

Unfortunately, these considerations are not applicable for the cyclotomic polynomials Ψ_m with $m=p_2\cdots p_k$ as in lemma 1 such that $\phi(m)/\lambda(m) > (\log \phi(m))^4$. Then $\Psi_m \pmod q$ is inseparable only if $q \mid m$ i.e. $q=p_i$ for some i. Generally q is still too small to justify the omission of the Hensel lifting procedure. Using lemma 3 and the estimate for p_i above lemma 1 we can also show that $\Psi_m \pmod q$ has at least $O((\log \phi(m))^2)$ distinct factors which would make the factor combination step superpolynomial.

Finally we establish certain bounds for the coefficients of our polynomials when the primes p_i are as small as possible. For a polynomial f, let norm(f) denote the sum of the absolute values of the coefficients of f.

<u>Theorem 3</u>: Let r be a fixed integer ≥ 2 and let $p_1,\ldots,p_n$ be the first n primes.

a) $\log(\mathrm{norm}(f)) = O(\deg(f) \log\log(\deg(f)))$ for $f = f_{r;p_1,\ldots,p_n}$ and for $f = f^*_{r;p_1,\ldots,p_n}$.

b) $\log_2(\mathrm{norm}(\Psi_m)) \leq \phi(m)$ for $m \geq 1$.

<u>Proof</u>: Given $f(x) = a_0 + a_1x + \cdots + a_{k-1}x^{k-1} + x^k \in Z[x]$, let B denote the maximum of the absolute values of the roots of f. Then, since the coefficients are the

elementary symmetric functions of the roots, it follows that $|a_i| \leq \binom{k}{i} B^{k-i}$ for $0 \leq i \leq k$. Therefore $\text{norm}(f) \leq (B+1)^k$.

a) For $f = f_{r;p_1,\ldots,p_n}$ the maximum absolute value of the roots is $B = \sqrt[r]{p_1} + \cdots + \sqrt[r]{p_n}$ and for $f^* = f^*_{r;p_1,\ldots,p_n}$ it is $B^* = 1 + B$. As noted above p_i is of order $O(i \log(i))$, so that B and B^* are of order $O(n^2)$. Since r is fixed n is of order $O(\log \deg(f))$ and $O(\log \deg(f^*))$. Taking the logarithm of the previous inequality for the norm immediately establishes part a).

b) Every root of Ψ_m has absolute value 1 and hence $\text{norm}(\Psi_m) \leq (1+1)^{\phi(m)}$. □

5. Remarks

By way of summary, we attempt to abstract from our arguments about our class of polynomials an "intuitive" explanation of their unusual factorization behavior. Their irreducibility over the integers seems quite plausible because a set of roots of powers of primes is linearly independent over the field of rational numbers. However, their Galois groups have very short cycle lengths and hence a modular projection will preserve very little structure of these groups. Therefore it seems that the extreme "compression" of the Galois groups by taking the remainders is to be blamed for the abundance of factors we then get.

The average behavior of the Berlekamp-Hensel algorithm is quite well understood [COLG79]. But the density theorems used for its analysis degenerate in our case. Peter Weinberger's irreducibility criterion is based on the correctness of the Generalized Riemann Hypothesis and may constitute an algorithm for testing irreducibility in polynomial time (see [KNUD81, p632]). However, no polynomial time algorithm is known which computes the factors of reducible polynomials such as $f^*_{8;2,3,\ldots,p_n}$.

REFERENCES

[BERL70] E. R. Berlekamp, Factoring polynomials over large finite fields, Math. Comput. 24 (1970), pp 713-735.

[BESI40] A. S. Besicovitch, On the linear independence of fractional powers of integers, J. London Math. Soc. 15 (1940), pp 1-3.

[CAVB68] B. F. Caviness, On Canonical Forms and Simplification, Carnegie-Mellon Univ., Ph.D. Thesis 1968.

[COLG79] G. E. Collins, Factoring univariate integral polynomials in polynomial average time, Lecture Notes in Comp. Sci. 72 (1979), Springer Verlag, pp 317-29.

[GAAL73] L. Gaal, Classical Galois Theory with Examples, Chelsea Publ. Co., New York 1973.

[HAWR79] G. H. Hardy and E. M. Wright, An Introduction to the Theory of Numbers, Fifth Ed., Oxford Univ. Press, Oxford 1979.

[KNUD81] D. E. Knuth, The Art of Computer Programming, Vol. 2, Sec. Ed., Addison-Wesley, Reading MA. 1981.

[LANG71] S. Lang, Algebra, Addison-Wesley, Reading MA. 1971.

[LOOS73] R. G. K. Loos, A constructive approach to algebraic numbers, Unpublished Article, May 1973.

[MUSD75] D. R. Musser, Multivariate polynomial factorization, J. of ACM 22 (1975), pp 291-308.

[RABM80] M. O. Rabin, Probabilistic algorithms in finite fields, SIAM J. Comput. 9 (1980), pp 273-280.

[RICI74] I. Richards, An application of Galois theory to elementary arithmetic, Adv. in Math. 13 (1974), pp 268-273.

[SIEW64] W. Sierpinski, Elementary Theory of Numbers, Polish Sci. Publ., Warszawa 1964.

[VdWA53] B. L. Van der Waerden, Modern Algebra, Vol. 1, Ungar Publ. Co., New York 1953.

[ZASH79] H. Zassenhaus, On the Van der Waerden criterion for the group of an equation, Lecture Notes in Comp. Sci. 72 (1979), Springer Verlag, pp 95-107.

Some Inequalities About Univariate Polynomials

Maurice MIGNOTTE
Strasbourg

Plane

I - Bounds for the height of divisors of polynomials.
   1. Classical result
   2. Other results
   3. Examples

II - Isolating roots of polynomials.
   1. Classical result
   2. Other results
   3. Examples

Abstract - This paper deals with the two following topics : bounds for the heights of divisors of polynomials, minimal distance between distinct roots of integral univariate polynomials. In each case we recall the best known results, we give some new inequalities and, constructing suitable examples, we show that these inequalities are not "too bad".

Introduction.

In many fundamental algorithms on polynomials in Symbolic Algebraic Computation one has to solve the two following problems : factorize an univariate polynomial F , isolate its roots. For the first problem current methods need the knowledge of an upper bound for the height of the possible divisors of F . To isolate the distinct roots of F one has to find a lower bound for the minimal distance between distinct roots of F . This paper is devoted to these two problems.

©1981 ACM O-89791-047-8/81-0800-0195 $00.75

Thus we are interested by inequalities on these two subjects ; of course by sharp inequalities as simple and general as possible. For each of these two questions we choose the same presentation : firstly we recall the best "classical" inequalities, secondly we give some new inequalities which -in some cases- are sharper than the classical ones, then we close each section by the construction of examples which show that the classical inequalities are not "too bad" but indeed reasonably sharp. In our opinion these examples constitue the main interest of this paper, because they go against the common opinion that the classical inequalities we are considering are much too crude. Of course in many cases these inequalities give bounds much worse that the exact bounds but our examples show that this is not always true !

In the sequel we shall use the following notation. The polynomial P we consider has complex coefficients and is given by the formula

$$P(X) = a_d X^d + \dots + a_0 , \quad a_d \neq 0 ,$$

d is called the degree of P , the number

$$H(P) := \mathrm{Max}\{|a_i| \ ; \ 0 \leq i \leq d\}$$

is called the height of P .

The roots of P (in the complex field) are denoted by $z_1, \dots, z_d$, so that

$$P(X) = a_d(X-z_1)\dots(X-z_d).$$

We also put

$$\|P\| = (|a_0|^2+\dots+|a_d|^2)^{1/2},$$
$$L(P) = |a_0| +\dots+|a_d|,$$

the length of P.

Notice that

$$H(P) \le \|P\| \le L(P).$$

I - Bounds for the height of divisors of polynomials.

1. Classical result

Suppose that Q is any divisor of P of degree q, say

$$Q = b_q X^q + \dots, \quad b_q \ne 0.$$

We want to obtain an upper bound for the height of Q. It is clear that this upper bound must depend on b_q and on the height of P.

The best known general result seems to be the following.

THEOREM 1.. If Q given above is a divisor of P then

$$L(Q) \le |b_q/a_d|.2^q\|P\|.$$

$ This theorem was first proved by Landau [2], and rediscovered several times. An elementary algebraic simple proof can be found in [4].

The result is an easy consequence of the fundamental inequality

$$|a_d| \prod_{j=1}^{d} \operatorname{Max}(1, |z_j|) \le \|P\|. \; \$$$

COROLLARY.. If P and Q are integral polynomials such that Q divides P then

$$L(Q) \le 2^q \|P\|,$$

where q is the degree of Q.

2. Other results

Sometimes the following inequality gives a better result than theorem 1.

THEOREM 2.. Let P and Q be polynomials as in theorem 1. Then

$$L(Q) \le |b_q/a_d|.d^{d-q}L(P).$$

$ By Güting [1] (lemma H), this result is true when $d-q = 1$. The general case follows by an obvious induction on $d-q$. $

Remark : For $q = d-2$ and $q = d-3$ Güting proved the stronger inequality

$$L(Q) \le |b_q/a_d| . \binom{d}{q} L(P).$$

This inequality is not proved for $q < d-3$. But, of course this stronger result can be used to improve slightly theorem 2. For example the following is true

$$L(Q) \le \sqrt[3]{6}\,|b_q/a_d|\, d(d-1)\dots(d-q+1)6^{-q/3}\, L(P).$$

3. Examples

As we noticed in the introduction many people consider theorem 1 as too crude. We shall show that this theorem is not so bad. To do this we have to construct examples where P has small coefficients whereas Q has big ones.

Firstly we choose Q:

$$Q = (X-1)^q$$

so that

$$L(Q) = 2^q.$$

Now we have to find an integral polynomial P which is a multiple of Q and has small coefficients (P non zero !). Suppose that d is the degree of P and that $H(P) = 1$. We have to solve the following problem : prove that there exists such a polynomial P of degree not too big (with respect to q). The proof uses the well-known pigeon-hole principle.

Suppose that q is a given integer and that d is an integer to be specified later. The condition "Q divides P" is equivalent to the following system of linear equations

$$\sum_{i=0}^{d} \binom{i}{t} a_i = 0 \ , \quad t = 0, 1, \ldots, q-1 \ ,$$

where $a_0, \ldots, a_d$ are the (unknown) coefficients of P.

Let E be the set of polynomials of degree at most d and coefficients 0 or 1. The map

$$E \to \mathbb{N}^q$$

$$(c_d X^d + \ldots + c_0) \mapsto \left(\binom{0}{t} c_0 + \ldots + \binom{d}{t} c_d\right)_{0 \le t < q}$$

sends the set E of 2^{d+1} elements into a set at most $(d+1)^{1+2+\ldots+q}$ elements. For a suitable positive constant c_1, the condition

$$d = [c_1 q^2 \operatorname{Log} q]$$

implies

$$2^{d+1} > (d+1)^{q(q+1)/2}$$

and then two distinct polynomials P_1 and P_2 of E have the same image and the polynomial $P = P_1 - P_2$ satisfies

- $\deg(P) \le d$,
- $H(P) = 1$,
- Q divides P .

This construction implies the following result.

THEOREM 3.. For any positive integer q there exist integral non zero polynomials P and Q such that $\deg(Q) = q$, Q divides P and

$$L(Q) \ge c_2 \, 2^q (q^2 \operatorname{Log} q)^{-1/2} \, \|P\| \ ,$$

where c_2 $(= 1/c_1)$ is a positive absolute constant.

Remark : It can be proved that the polynomial P we constructed must satisfy

$$\deg(P) \ge c_3 \, q^2 / \operatorname{Log} q \ ,$$

where c_3 is an absolute constant.

II - Isolating roots of polynomials.

1. Classical result

We want to obtain a lower bound for

$$\operatorname{sep}(P) := \min_{z_i \ne z_j} |z_i - z_j| \ .$$

For reasons of simplicity we consider only polynomials with simple zeroes (i.e. square free polynomials).

The best known result seems to be (see [5]) the following.

THEOREM 4.. Let P be a square-free polynomial of discriminant D. Then

$$\operatorname{sep}(P) > \sqrt{3}\, d^{-(d+2)/2} |D|^{1/2} \|P\|^{1-d} \ .$$

\$ In [3] Mahler proved the inequality

$$\operatorname{sep}(P) > \sqrt{3}\, d^{-(d+2)/2} |D|^{1/2} M(P)^{1-d}$$

where

$$M(P) = |a_d| \prod_{j=1}^{d} \operatorname{Max}(1, |z_j|) \ .$$

To conclude use the upper bound of $M(P)$ given in the proof of theorem 1 . \$

COROLLARY.. When P is a square-free integral polynomial $\operatorname{sep}(P)$ verifies

$$\operatorname{sep}(P) > \sqrt{3}\, d^{-(d+2)/2} \|P\|^{1-d} .$$

Remark : The restriction to square-free polynomials in the integral case is not a serious drawback because it is easy to obtain the square-free factorization of an integral polynomial. For results in the general case see [1] .

2. Other results

As was noticed in [1], improvements of theorem 4 can be otained when the roots of P satisfy some suitable hypotheses. Variants of these results and others can be obtained from the following theorem.

THEOREM 5.. Let S be a subset of $\{(i, j) ; 1 \le i < j \le d\}$. Define S' as $\{j ; (i, j) \in S\}$. Suppose that the roots of P satisfy $|z_1| \le \dots \le |z_d|$.

Then, with the notations of theorem 4,

$$\prod_{(i,j)\in S} |z_i - z_j| \ge |D|^{1/2}\, 2^{k-d(d-1)/2} \prod_{j\in S'} |z_j| .$$
$$|a_d^{1-d} z_1^{1-d} z_2^{2-d} \dots z_{d-1}^{-1}|$$

where $k = \mathrm{Card}(S)$.

\$ This is inequality (6) of [6]. The proof consists in an easy estimation of D. \$

COROLLARY.. Suppose that P has real coefficients. Then

(i) if z_i and z_j are non real and $z_i \ne \overline{z}_j$,

$$|z_i - z_j| \ge |D|^{1/4}\, 2^{1-d(d-1)/4} |z_j|$$
$$(|a_d^{1-d} z_1^{1-d} \dots z_{d-1}^{-1}|)^{1/2}$$
$$\ge |D|^{1/4}\, 2^{1-d(d-1)/4} M(P)^{(1-d)/2} ,$$

(ii) if z_i is real and z_j is not,

$$|z_i - z_j| \ge (|D|^{1/2}\, 2^{2-(d-1)d/2}$$
$$|z_j^3 a_d^{1-d} z_1^{1-d} \dots z_{d-1}^{-1}|)^{1/3}$$
$$\ge (|D|^{1/2}\, 2^{2-d(d-1)/2} M(P)^{1-d})^{1/3} .$$

\$ In the first case take $S = \{(i, j), (i', j')\}$ where $z_{i'} = \overline{z_i}$ and $z_{j'} = \overline{z_j}$. In the second case $z_{j'} = \overline{z_j}$ and $S = \{(i, j), (i, j'), (j, j')\}$. \$

3. Examples

We are mainly interested in the dependence of sep(P) with respect to H(P), the degree of P being fixed. This is the reason we introduce the quantity

$$\ell(P) = \frac{\mathrm{Log}(\mathrm{sep}(P))}{\mathrm{Log}(H(P))} .$$

It is natural to define

$$L(d) = \limsup \ell(P)$$

where P runs over integral polynomials of degree d, and

$$L_0^*(d) = \limsup \ell(P)$$

where P runs over integral monic irreductible polynomials of degree d. Of course

$$L_0^*(d) \le L(d) ,$$

and theorem 4 implies

$$L(d) \le d-1 .$$

In [5], using a theorem of W. M. Schmidt, we proved (non constructively) that

$$L(d) \ge (d+1)/2 .$$

In [6], using the previous inequality, we gave a complicated proof of the lower bound

$$L_0^*(d) \ge c_4\, d^{1/3} ,$$

and conjectured

$$L_0^*(d) \ge c_5\, d .$$

Here we prove this conjecture.

Let $d \ge 3$ be a fixed integer and $a \ge 3$ be some integer. Consider the following monic integral polynomial

$$P = X^d - 2(aX-1)^2 .$$

Eisenstein's criterion shows that P is irreductible over the integers (consider the prime number 2).

We show that P has two real roots close to $1/a$. Clearly

$$P(1/a) > 0$$

and if

$$h = a^{-(d+2)/2}$$

then

$$P(1/a \pm h) < 2a^{-d} - 2a^2 h^2 = 0 .$$

This proves that P has two real roots in the interval $(1/a-h, 1/a+h)$. Thus

$$\mathrm{sep}(P) < 2h = 2a^{-(d+2)/2} .$$

This implies the following result.

THEOREM 6.. For each integer $d \geq 3$ there exist infinitely many integral monic irreducible polynomials P which satisfy

$$\mathrm{sep}(P) < 2^d H(P)^{-(d+2)/4}$$

and $\deg(P) = d$.

COROLLARY.. For any integer $d \geq 3$

$$L_0^*(d) \geq (d+2)/4 .$$

<u>Remark</u> : The exact value of $L_0^*(d)$ is unknown for each $d \geq 3$. Trivially $L_0^*(2) = 0$.

REFERENCES

[1] GÜTING R. - Polynomials with multiple zeroes, Mathematika 14, 1967, p. 181-196.

[2] LANDAU E. - Sur quelques théorèmes de M. Petrovic relatifs aux zéros des fonctions analytiques, Bull. Soc. France 33, 1905, p. 251-261.

[3] MAHLER K. - An inequality for the discriminant of a polynomial, Michigan Math. J. 11, 1964, p. 257-262.

[4] MIGNOTTE M. - An inequality about factors of polynomials, Math. of Comp. 28, 1974, p. 1153-1157.

[5] MIGNOTTE M. - Sur la complexité de certains algorithmes où intervient la séparation des racines d'un polynôme, R. A. I. R. O. Informatique Théorique 10, 1976, p. 51-55.

[6] MIGNOTTE M. PAYAFAR M. - Distance entre les racines d'un polynôme, R. A. I. R. O. Analyse Numérique 13, 1979, p. 181-192.

Maurice E. MIGNOTTE
Université Louis Pasteur
Centre de Calcul
7, rue René Descartes
67084 STRASBOURG Cédex
FRANCE

Factorization over Finitely Generated Fields

James H. Davenport*

Emmanuel College
Cambridge
England
CB2 3AP

Barry M. Trager

Mathematical Sciences Department
IBM Thomas J. Watson Research Center
P.O. Box 218
Yorktown Heights
NY 10598

Abstract. This paper considers the problem of factoring polynomials over a variety of domains. We first describe the current methods of factoring polynomials over the integers, and extend them to the integers mod p. We then consider the problem of factoring over algebraic domains. Having produced several negative results, showing that, if the domain is not properly specified, then the problem is insoluble, we then show that, for a properly specified finitely generated extension of the rationals or the integers mod p, the problem is soluble. We conclude by discussing the problems of factoring over algebraic closures.

1. Introduction.

The problem of factoring polynomials is one that has received great attention in computer algebra. Not only is factoring of great inherent interest and utility, but it is also required by a great many other algorithms, e.g. integration. In fact, when it comes to algebraic numbers, factorisation is a prerequisite for ensuring unique representations - the only way we can discover how to represent $6^{1/2}$ in terms of $2^{1/2}$ and $3^{1/2}$ is to observe that x^2-6 factors over the extension of the rationals generated by $2^{1/2}$ and $3^{1/2}$. While factoring over the integers, or over algebraic extensions, is the most common requirement, recently factoring over other domains has been required. Indeed the investigations that lead to this paper were prompted by the authors' interests in integration, which leads to requirements to factor over algebraic extensions of polynomial domains, over algebraic extensions of finite fields, and over algebraic closures of such objects.

---

* Many of the discussions leading to this paper took place while the first author was a post-doctoral fellow at the IBM Thomas J. Watson Research Center.

2. Factoring over polynomial domains.

One of the major advances in the field was the development of effective methods for factoring univariate polynomials with integer coefficients (originally due to Zassenhaus[1969], and implemented, inter alia, by Musser[1971]). This process has subsequently been refined in a variety of ways (see, for example, Wang[1978] and [Moore & Norman,1981], and the recent ideas of Zassenhaus[1981]), but the fundamental principles are the same. We first outline the univariate method since the multivariate process relies on it.

1) The problem is reduced to factoring square-free polynomials. This is done by square-free decomposition, to which there are many references, e.g. [Yun,1977].
2) A 'suitable' prime p is chosen.
3) The coefficients of the polynomial to be factored are reduced mod p.
4) This polynomial is factored (over the integers mod p) by Berlekamp's[1967] algorithm.

5) This factorisation is 'lifted' to one mod p^n, for suitably large n;

6) This is examined to yield the factorisation over the integers. This is not necessarily a trivial process, since one factor over the integers may be represented by several factors mod p^n, and we need some way of combining factors mod p^n into factors over the integers. Collins[1979] discusses the average complexity of this process.

Since Berlekamp's method* will factorise polynomials over the integers mod p, this means that the problem of factoring univariate polynomials over the prime fields (viz. Q and the fields of integers modulo p) was solved.

3. Multivariate Polynomials.

The next problem to be considered is the factorisation of multivariate polynomials. The process (due originally, in the case of characteristic 0, to Wang and Rothschild[1975]) is in fact very similar to that for factoring univariate polynomials over the integers, and proceeds as follows:

1) Ensure the polynomial is square-free (doing square-free decompositions over the integers mod p is slightly tricky, since, for example $(x^p-1)' = 0$, but it can be done - see algorithm square-free-decompose in Davenport[1981] for one way).
2) Find values for all but one of the variables such that the result of substituting them in leaves one with a square-free polynomial. Since there are only finitely many values of any variable which do not change the square-free nature of a polynomial, this can always be done over the integers: the rare case when it cannot be done over the integers modulo p is discussed below.
3) Factor the resulting univariate polynomial by the methods discussed above.
4) 'Lift' this factorisation back to a multivariate one by methods analogous to those used in the univariate case. This can be done either variable by variable or by lifting all the variables at once [Wang, 1978], and Zippel[1979] has some interesting remarks on the interaction of this process with the sparsity of the original polynomial.

There is a potentially serious problem here, inasmuch as the factorisation in step (3) may yield many factors, several of which correspond to one multivariate factor. In the worst case, one may need to try all combinations of the results of extending the univariate factors before finding the multivariate factors (or asserting that the multivariate polynomial is irreducible). We shall not discuss this problem, often known as the "combinatorial explosion", further here, except to note that a variety of methods have been proposed, e.g. in [Wang,1978], to minimise the cost.

There remains the possibility that step (2) above cannot be completed over a finite field - for example no value of y leaves x(x+1)(x+y) square-free over the integers modulo 2. The solution to this is to make the ground field larger, by taking an algebraic extension of it, and to admit values of y from this larger ground field. This leaves us with a univariate polynomial over an algebraic extension of the integers modulo p, which can be factored by the methods of Berlekamp[1970]. This can then be 'lifted' to a multivariate factorisation exactly as above, so that we now have a multivariate factorisation of the original polynomial over the larger ground field. This can be converted to one over the original ground field by considering the norm of each factor - more precisely, having lifted our factorisation from $k[\theta][x]$ to $k[\theta][x][y]$, we then consider this as a factorisation over $k[x][y][\theta]$, and take norms with respect to the extension by θ, as described by Trager[1976]. Of course, in practice this case is extremely rare, but nevertheless, as we have seen, it does not pose any theoretical embarrassment, though it is likely to cost a great deal in computer time, since our ground field is now an extension of the

---

* Though, if p is large, an alternative method [Berlekamp,1970] is better.

integers modulo p, rather than being just the integers modulo p.

Hence we can factor multivariate polynomials over any prime field.

Furthermore, since K[x,y] = K[x][y], we can factor over polynomial extensions of prime fields. As a result of Gauss's Lemma [van der Waerden, 1949, p. 73], this extends to rational function extensions, since, once we have cleared denominators, factoring in K[x][y] is the same as factoring in K(x)[y]. Hence we can factor polynomials over any finitely generated, purely transcendental extension of any prime field, and, since any polynomial can only involve a finite number of items, we can drop the restriction "finitely generated" in the above.

4. A Negative Result.

This therefore leaves algebraic fields as the next major problem. However, it is certainly not possible to solve the factorisation problem (i.e. produce an algorithm that will factorise any polynomial) for infinitely generated algebraic extensions of the integers, as is shown by the following example (due to Fröhlich & Shepherdson [1956]).

Let f be a function from the natural numbers into themselves whose image is a recursively* enumerable but not recursive set (such functions exist [Kleene, 1938]). Then let K be the field $Q(p_{f(1)}^{1/2}, p_{f(2)}^{1/2}, \ldots)$, where p_i is the i-th prime. Then consider attempting to factorise the polynomial x^2-p_n. This has one factor if p_n is not a square in K, i.e. n is none of the f(i), and two factors if n is one of the f(i). Hence any algorithm to factor polynomials, even of this very simple kind, over K would enable us to solve the question "is n one of the f(i)", which is known to be insoluble.

---

*For the benefit of those not familiar with recursive function theory, this means that we can compute f(n) for any natural number n, but there can be no procedure for deciding if a given natural number m lies in the range of f, i.e. whether or not m=f(n) for some n.

In case the above example be thought too abstract, here is a simple illustration of the fact that the ability to factorise polynomials (even quadratics) over infinitely presented fields is deeper than might be thought. Define g(n) to be 1 if 2n is the sum of two primes, and -1 otherwise. g is clearly a computable function. Let L be $Q(g(1)^{1/2}, g(2)^{1/2}, \ldots)$ and consider the factorisation of x^2+1 over L. It has one factor if L = Q, i.e. the Goldbach conjecture (that every even number is the some of two primes) is true, and two factors if the conjecture is false. This is more embarrassing, in some ways, because L is definitely finitely generated (being either Q or Q[i]), so a straight-forward restriction to being finitely-generated will not help here: we must insist, in some way, that the field be explicitly finitely generated.

While the above could be regarded as pedantry ("after all, who would actually state a problem like that"), it has an interesting consequence. Classical algebra texts (e.g. van der Waerden[1949]) show that, in the presence of a descending chain condition on divisors (which is nearly always present), the existence of greatest common divisors is equivalent to unique factorisation. Now it is certainly easy to construct greatest common divisors in the domains discussed above, while we cannot construct factorisations. This is a formalisation of the widely-held belief among computer algebraists that factorisation is "inherently" more complicated than g.c.d. computations.

5. Algebraic Extensions

So let us now assume that we have a field L = $k(t_1, \ldots, t_n, s_1, \ldots, s_m)$, which we shall also write as $K(s_1, \ldots, s_m)$, where the t_i are all transcendental over $k(t_1, \ldots, t_{i-1})$, and the s_i are algebraic, with given minimal polynomial, over $K(s_1, \ldots, s_{i-1})$. This choice of order, placing all the algebraics after all the transcendentals, is not a real limitation, because the only constraint on the position of an algebraic is that it should come after everything in its minimal polynomial. Suppose that we wish to factor a (potentially multivariate) polynomial over L. If L is of characteristic 0 (i.e. if k is the rational numbers), there is no intrinsic problem - Trager's [1976] algorithm sqfr-norm can be used to reduce the problem to factoring a (generally much larger[+]) polynomial in the same variables over K, which we know how to do. Having factored this polynomial, we can recover the factors of the original polynomial over L, as in his algorithm alg-factor.

Life is not so simple if L has finite characteristic. There are two reasons for this:

a) (only applicable if n=0) There may not be enough elements of L, because sqfr-norm searches through L looking for a substitution which will produce a square-free norm, and it is possible (but only in a finite number of cases) for a substitution not to yield a square-free norm;

b) The whole theory of algebraic field extensions is much more complicated in the case of characteristic p, because an algebraic extension can now be inseparable (see van der Waerden[1949] Section 38). $K[\theta]$ is said to be an inseparable extension of K if the minimal polynomial of θ is irreducible over K, but has multiple roots in $K[\theta]$. To see how this can happen, consider $K=k[x]$, where k is the finite field with p elements, and x is transcendental over k, and let θ be defined by $\theta^p=x$. Then the minimal polynomial for θ, viz. y^p-x, is irreducible over K, but over $K[\theta]$ it factors into $(y-\theta)^p$.

These two problems require different solutions. Problem (a) is solved by observing that this can only occur if L is finite. This implies that n=0, and that L is an algebraic extension of a field with p elements. In that case we can reduce the problem to a univariate one, as described in section 3 (including, if necessary, the ground field extension process), and then factor the univariate polynomial thus produced by the method of Berlekamp[1970].

Problem (b) requires a rather different approach. We first observe (see van der Waerden[1949] for details) that an element can only be inseparable if its minimal polynomial is a polynomial in x^p. In fact we can split an inseparable extension into several parts - a separable extension followed by one or more purely inseparable extensions, viz. those generated by an element with minimum polynomial of the form y^p-z. It is a fact (often attributed to Krull[1953], but actually proved by Endler[1952]) that one can dispense with the inseparable extension by a change of generating elements. This is not hard to see in any special case*, and Endler provides an algorithm for performing the change of representation. Of course, once one has a separable representation, the problem polynomial can be transformed into that representation, and the factorisation problem solved there, and the resulting factors transformed back.

6. Algebraic Closures.

We have shown how one can factor polynomials over an explicitly finitely generated field, and this often what is required. However, one sometimes wants to factor over algebraic closures. A good example of this is integration, where the ground field has to be considered to be algebraically closed (see Risch[1969], where it is shown that $1/(x^2-2)$ is only integrable if the ground field contains $2^{1/2}$). The previous work is not of any direct help here, since algebraic closures are infinitely generated.

+ Even if the minimal polynomials for the s_i contain none of the t_i, the degree of the polynomial to be factored over the integers has been multiplied by the degrees of all the minimal polynomials.

---

* For example, if k is a field of p elements, and we consider $k[x][y]$, where x is transcendental over k, and y satisfies $y^p=x$, then this field is isomorphic to $k[y]$, under the mapping $y \to y$ & $x \to y^p$.

This problem was discussed by Risch[1969, p.178], who used the Kronecker trick of mapping $x_i \rightarrow t^{d^{i-1}}$, where d is larger than any integer occurring in the problem, to reduce the problem to a univariate one. Further details, and a cost analysis, are presented by Trager[1981, chapter 3].

So, let us suppose that we wish to factor a polynomial in $x_1, \ldots, x_n$ over the field K, where K is an explicitly finitely generated extension of a prime field*.

1) Ensure the polynomial is square-free, as in step (1) of Section 3.
2) Reduce the problem to a univariate one, as in step (2) of Section 3. Call this univariate polynomial f(x), defined in K[x].
3) Factor this over K[x], as in Sections 3 and 5. So $f(x) = f_1(x)f_2(x)\ldots f_m(x)$.
4) Until f is the product of linear factors, extend K by a root of one of the non-linear factors, and re-factor all the factors over this larger field.
5) Lift this back to a multivariate factorisation in $x_1, \ldots, x_n$, as in step (4) of Section (3). Note that the "combinatorial explosion" mentioned there is quite likely to occur in this case, because we have ensured that our original factorisation is into linear factors, all combinations of which will need to be tried before one can assert that the original polynomial is irreducible over the algebraic closure (if indeed it is).

----

* The reader may object that we need only consider K to be a transcendental extension of a prime field, since the algebraic closure of an algebraic extension of K is the same as the algebraic closure of K. This is perfectly true mathematically, but is a pitfall computationally, since algebraic numbers only have a unique representation after one has chosen a basis. As an example, consider factoring x^2+4 over Q[i]. It factors as $(x+2i)(x-2i)$. But if we try to factor it over Q, we find that is does not factor over Q, so we introduce a new algebraic number θ defined by $\theta^2=-4$, and factor the polynomial as $(x+\theta)(x-\theta)$. If we try to use this as an extension of Q[i] we are in trouble, since i and θ are not independent.

7. Conclusions

We have shown that, assuming the problem is stated in a suitably explicit form, we can factor multivariate polynomials over any finitely generated extension of a prime field, or over the algebraic closure of such a field. In practice, this means that any 'reasonable' factorisation problem is solved in principle.

This is not to say that factorisation is a dead area. Many problems of implementation remain, and there is much that has to be done before all the algorithms described in this paper can be made available to the user.

Also, while these algorithms all work, and while parts of them are in use and proving relatively efficient, other parts are very expensive. In this context one thinks particularly of the "combinatorial explosion" of section 3, and Trager's[1976] algorithm sqfr-norm of Section 5. It is the authors' intuitive feeling that, in general, the exponential nature of sqfr-norm is inherent in the problem, but they have no proof. Indeed, there is remarkably little known about the complexity of factoring as a whole, though Yun[1977] deals with the square-free part of factorisation algorithms, and Collins[1979] studies the "average" complexity of the univariate factor-combining process.

Two particular areas that the authors feel deserve attention are:

1) The ideas of Zassenhaus[1981]. Can they be adapted to eliminate the combinatorial explosion of section 3.
2) The ideas of Wang[1976] and Weinberger and Rothschild[1976]. These deal with the factoring of multivariate polynomials over algebraic number fields (whereas, of course, our section 5 can deal with arbitrary extensions), but they are often more efficient when they apply. It is obviously possible to achieve some compromise between them and sqfr-norm - it would be interesting to investigate the details.

8. References

Berlekamp,1967
Berlekamp,E.R., Factoring Polynomials over Finite Fields. Bell System Tech. J. 46(1967) pp. 1853-1859.

Berlekamp,1970
Berlekamp,E.R., Factoring Polynomials over Large Finite Fields. Math. Comp. 24 (1970) pp. 713-735.

Collins,1979
Collins,G.E., Factoring univariate integral polynomials in polynomial average time. Proc. EUROSAM 79 [Springer Lecture Notes in Computer Science 72, Springer-Verlag, Berlin-Heidelberg-New York, 1979], pp. 317-329.

Davenport,1981
Davenport,J.H., On the Integration of Algebraic Functions. Springer Lecture Notes in Computer Science 102, Springer-Verlag, Berlin-Heidelberg-New York, 1981.

Endler,1952
Endler,O., Konstruktion von allgemeinen Normalkörpen in beschränkt vielen Schritten. Hausdorff-Gedächtnis-Preisarbeit der Math.-Nat.-Fakultät Bonn 1952.

Fröhlich & Shepherdson,1956
Fröhlich,A. & Shepherdson,J.C., Effective Procedures in Field Theory. Phil. Trans. Roy. Soc. Ser. A 248(1955-6) pp. 407-432.

Kleene,1938
Kleene,S.C., On Notation of Ordinal Numbers. Journal of Symbolic Logic 3(1938) pp.150-155.

Krull,1953
Krull,W., Über Polynomzerlegung mit endlich vielen Schritten. Math. Z. 59(1953) pp. 57-60.

Moore & Norman,1981
Moore,P.M.A. & Norman,A.C., Implementing a polynomial factorization and GCD package. Proc. SYMSAC 81.

Musser,1971
Musser,D.R., Algorithms for Polynomial Factorization. Ph.D. Thesis, Computer Science Department, University of Wisconsin (Madison) 1971.

Risch,1969
Risch,R.H., The Problem of Integration in Finite Terms. Trans. A.M.S. 139(1969) pp. 167-189 (MR 38 #5759).

Trager,1976
Trager,B.M., Algebraic Factoring and Rational Function Integration. Proc. SYMSAC 76, pp. 219-226.

Trager,1981
Trager,B.M., Integration of Algebraic Functions. Ph.D. Thesis, M.I.T. Dept. of EE&CS, expected date Sept. 1981.

van der Waerden,1949
van der Waerden,B.L., Modern Algebra, Frederick Ungar, New York, 1949 (Translated from Moderne Algebra 2nd. ed.).

Wang,1976
Wang,P.S., Factoring Multivariate Polynomials over Algebraic Number Fields. Math. Comp. 30(1976) pp. 324-336.

Wang,1978
Wang,P.S., An Improved Multivariable Polynomial Factorising Algorithm. Math. Comp. 32(1978) pp. 1215-1231.

Wang & Rothschild,1975
Wang,P.S. & Rothschild,L.P., Factoring Multi-Variate Polynomials over the Integers. Math. Comp. 29(1975) pp. 935-950.

Weinberger & Rothschild,1976
Weinberger,P.J. & Rothschild,L.P., Factoring Polynomials over Algebraic Number Fields. ACM Transactions on Mathematical Software 2(1976) pp. 335-350.

Yun,1977
Yun,D.Y.Y., On the Equivalence of Polynomial Gcd and Squarefree Factorization Algorithms. Proc. 1977 MACSYMA Users' Conference [NASA publication CP-2012, National Technical Information Service, Springfield, Virginia], pp. 65-70.

Zassenhaus,1969
Zassenhaus,H., On Hensel Factorization. I. J. Number Theory 1(1969) pp. 291-311. MR 39 #4120.

Zassenhaus,1981
Zassenhaus,H., Polynomial time factoring of integer polynomials. Presentation to Symbolic Mathematics Committee of SHARE Europe Association, Antwerp, 15th. Jan 1981. [An earlier version of this, entitled "On a problem of Collins", was presented at the AMS Summer Meeting in Ann Arbor, Aug. 1980.]

Zippel,1979
Zippel,R.E., Probabilistic Algorithms for Sparse Polynomials. Proc. EUROSAM 79 [Springer Lecture Notes in Computer Science 72, Springer-Verlag, Berlin-Heidelberg-New York, 1979], pp. 216-226.

ON SOLVING SYSTEMS OF ALGEBRAIC EQUATIONS VIA IDEAL BASES AND ELIMINATION THEORY

by

Michael E. Pohst
University of Cologne
Weyertal 86 - 90
5 Cologne 41, West Germany

David Y. Y. Yun*
IBM Watson Research Center
P. O. Box 218
Yorktown Heights, NY 10598

(Extended Abstract)

1. INTRODUCTION

The determination of solutions of a system of algebraic equations is still a problem for which an efficient solution does not exist. In the last few years several authors have suggested new or refined methods, but none of them seems to be satisfactory. In this paper we are mainly concerned with exploring the use of Buchberger's algorithm for finding Groebner ideal bases [2] and combine/compare it with the more familiar methods of polynomial remainder sequences (pseudo-division) and of variable elimination (resultants) [4].

Buchberger's algorithm determines a certain system of generators for a polynomial ideal, called a "Groebner basis". Apart from other properties, a Groebner basis gives an easy mechanism for deciding whether another given polynomial is contained in the ideal. Buchberger [1] suggested the use of a Groebner basis algorithm to solve systems of algebraic equations, but he did not take up this idea in his subsequent reports. In 1977 Trinks [3] applied Buchberger's algorithm to solve a system of six equations in six variables over $\mathbb{Q}$. We will use this example as a basis for discussion and comparison of various methods in Section 4.

©1981 ACM 0-89791-047-8/81-0800-0206 $00.75

One of the first conclusions we reached is that Buchberger's algorithm is not as suitable as expected, for example, by Trinks [3]. We therefore suggest a combined method - using resultants, pseudo-division, and S-polynomials - which worked extremely well for our examples. Since this method seems to work best when guided by some mathematical intuition, using it interactively is highly recommended. However, we give a set of criteria (R1-R4), which allows a systematic exploration of the solution mechanism in a non-interactive mode. Our experience stems mainly from sessions with SCRATCHPAD -- an interactive computer algebra system developed at the IBM Watson Research Center. We present an algorithm and give insights derived from such experience.

2. BUCHBERGER'S METHOD

The general problem can be stated as follows. Let $\mathbf{R}$ be a commutative domain with 1 and $f_1,...,f_k \in \mathbf{R}[x_1,...,x_n]$. Then we search for elements $\alpha_1,...,\alpha_n$ in a suitable extension $\overline{\mathbf{R}}$ of $\mathbf{R}$ such that

$$f_i(\alpha_1,...,\alpha_n) = 0 \quad (i = 1,...,k). \qquad (1)$$

In general $\mathbf{R}$ must have additional properties. For example, Buchberger's algorithm requires $\mathbf{R}$ to be Noetherian with two additional assumptions [3]. However, we are currently only interested in solving (1) in rings $\mathbf{R}$ which are fields or principal ideal domains.

* Partially supported by NSF Grant No. MCS79-09158

Every $f \in \mathbf{R}[x_1,...,x_n]$ can be written as

$$f(x_1,...,x_n) = \sum_{finite} a_{(i_1,...,i_n)}\, x_1^{i_1}...x_n^{i_n} \qquad (2)$$

where $a_{(i_1,...,i_n)} \in \mathbf{R}-\{0\}$, for distinct $(i_1,...,i_n) \in (\mathbb{Z}^{\geq 0})^n$. Each term $a_{(i_1,...,i_n)} x_1^{i_1}...x_n^{i_n} = a_{\underline{i}}\underline{x}^{\underline{i}}$ in (2) is called a monomial. We order these monomials with respect to an ordering of their exponent vectors $\underline{i}$ and $\underline{j}$. For the lexicographic ordering, we denote $\underline{i} <_L \underline{j}$ if and only if there exists $k \in \{1,...,n\}$ such that $i_k < j_k$ and $i_l = j_l$ $(l = 1,...,k-1)$. For the total-degree ordering (used by Buchberger), $\underline{i} <_T \underline{j}$ if and only if $\sum_{l=1}^{n} i_l \geq \sum_{l=1}^{n} j_l$. The highest monomial, $a_{\underline{i}}\underline{x}^{\underline{i}}$, with respect to a particular ordering is called the "headterm", $h(f)$, and its coefficient, $a_{\underline{i}}$, the "leading coefficient" of f, $lc(f)$. The "total degree" of f is $\delta(f) = \sum_{l=1}^{n} i_l$ for the headterm. (For $n=1$ the total degree coincides with the degree of f.) When the lexicographic ordering is used, $f \in \mathbf{R}[x_1,...,x_n]$ is considered a polynomial in the (main) variable x_1 with coefficients in $\mathbf{R}[x_2,...,x_n]$. The degree of f with respect to the variable x_j, $\delta_{x_j}(f)$, is given by $\max\{i_j \mid a_{\underline{i}} \neq 0,\ j = 1,...,n\}$. We say that the monomial $a_{\underline{i}}\underline{x}^{\underline{i}}$ divides $b_{\underline{j}}\underline{x}^{\underline{j}}$, if $i_l \leq j_l$ $(l = 1,...,n)$ for their exponent vectors.

Buchberger's algorithm is based on the construction of polynomials with headterms of low order in a "finitely generated ideal" $\mathcal{A} = <f_1,...,f_k>$ of $\mathbf{R}[x_1,...,x_n]$. To obtain such elements, Buchberger defines "S-polynomials" for pairs f_i, $f_j \in \mathcal{A}$, which is simply a cross-multiply-and-subtract, or one step of the standard polynomial division:

$$S(f_i,\ f_j) \leftarrow \frac{h(f_j)f_i - h(f_i)f_j}{\gcd(h(f_i),h(f_j))} \qquad (3)$$

Obviously $S(f_i,\ f_j) \in \mathcal{A}$. Further we define a reduction of a polynomial f modulo a finite set of polynomials $\mathcal{F} = \{f_1,...,f_k\}$. We call "$f$ reduced modulo $\mathcal{F}$" if no headterm $h(f_i)$ divides $h(f)$ for i = 1,...,k. Otherwise let $i \in \{1,...,k\}$ with $h(f_i) \mid h(f)$. Then we replace f by $S(f,\ f_i)$. In this way we obtain an $\mathcal{F}$-reduced polynomial in a finite number of steps.

<u>Algorithm A (Buchberger)</u>

Input: $\mathcal{F} = \{f_1,...,f_k\} \in \mathbf{R}[x_1,...,x_n]$ and $B = \{(i,j) \in \mathbb{Z}^2 \mid 1 \leq i < j \leq k\}$.

Output: A Groebner basis for the ideal $<f_1,...,f_k>$ stored in $\mathcal{F}$.

Step 1. If B is empty then return $\mathcal{F}$ else take the next pair $(i,j) \in B$ and set $B \leftarrow B - \{(i,j)\}$.

Step 2. Set $f \leftarrow S(f_i,\ f_j)$ and compute $\bar{f} =$ the reduction of f modulo $\mathcal{F}$.

Step 3. If $\bar{f} = 0$ then goto 1 else $k \leftarrow k+1$, $f_k \leftarrow \bar{f}$, $\mathcal{F} \leftarrow \mathcal{F} \cup lcbf_k\}$, $B \leftarrow B \cup \{(i,k) \mid i = 1,...,k-1\}$ and goto 1 .

The great advantage of this algorithm is that it is so simple. On the other hand, no realistic bound for the number of necessary steps is known in case $n \geq 3$ (when $n=2$, Buchberger established a polynomial bound [2]). The advantage of a Groebner basis for the determination of the common zeros of $f_1,...,f_k$ strongly depends on the ordering of the variables [3]. An elimination of variables is only obtained if we choose a lexicographic ordering of the variables.

The algorithm can be speeded up substantially by a criterion which determines whether the reduction of an S-polynomial will yield zero without actually computing it [2]. Namely, in Step 2 those S-polynomials, $S(f_i,\ f_j)$, are not computed for which there is an $l \in \{1,...,k\}$ different from i and j, such that the ordered pairs of i and l, and of j and l are not in B and $h(f_l)$ divides $\mathrm{lcm}(h(f_i),\ h(f_j))$. Of course this criterion depends on the information that has already been obtained by the computation of other S-polynomials. Therefore, the set $\mathcal{F}$ of polynomials rapidly increases at the beginning of the algorithm and the criterion effectively reduces the amount of computation only at a later stage. A further disadvantage is the fast growth of the size of coefficients [3].

3. A NEW COMBINED APPROACH

It is our aim to consider what the algorithm does for separating and eliminating variables. If we choose Buchberger's ordering, no separation or elimination takes place and the obtained Groebner basis has no value for the computation of the zeros of a polynomial ideal (see the example in [2]). If we choose a lexicographic ordering $x_1>x_2>...>x_n$ instead, the variables are eliminated in the course of computations of a Groebner basis. So in the Groebner basis, there may be a polynomial depending on x_n only. But to achieve this, we must not compute a Groebner basis in the normal sense. Rather, we eliminate variables by successively forming xSx-polynomials first for those polynomials which still depend on x_1, then for those still depending on x_2, and so on. Thus we are led to the following approach.

Description of an Algorithm

Given a set of polynomials, $\mathcal{P} = \{P_1, P_2, ..., P_k\}$, contained in $\mathbf{R}[x_1,...,x_n]$, distribute them initially in n subsets $\mathcal{P}_i$ of $\mathcal{P}$ for $i=1,...,n$, where

$$\mathcal{P}_i = \{P_j \in \mathcal{P} \mid P_j \in \mathbf{R}[x_i,...,x_n] - \mathbf{R}[x_{i+1},...,x_n]\}.$$

Namely these are subsets of $\mathcal{P}$ which contain polynomials with x_i as the main variable. (Of course, some of these subsets may be empty at the beginning. In fact, in many cases $\mathcal{P}_1 = \mathcal{P}$.) Let k_i be the number of elements in $\mathcal{P}_i$ and $\ell_i = \sum_{j=1}^{i} k_j$. The basic idea of our algorithm is to iterate on $i = 1,2,...,n$ until there is at least one element in each $\mathcal{P}_i$, if possible. There are three potential methods to generate elements of $\mathcal{P}_j$ from those of $\mathcal{P}_i$ for $j>i$. These three methods are S-polynomial, pseudo-division, and resultants. The use of S-polynomials for polynomial ideals has already been mentioned. The use of the other two methods and the relationship among them will be discussed shortly. (In fact, the advantages of using resultants for this purpose should be clear.) In general, if one of the $\mathcal{P}_i$'s contains more than one element, then each of these three methods can be applied to obtain an element for some $\mathcal{P}_j$, $j>i$. Thus we proceed on i until $\mathcal{P}_i = \phi$ and backtrack to attempt adding an element to $\mathcal{P}_i$ from some $\mathcal{P}_\ell$ for $\ell<i$.

The most obvious of the three methods is by the computation of resultants. Its theory and history goes well into the 19th century. We include a modern statement of the "Fundamental Theorem of Resultants" [4]:

Theorem: Let **F** be an algebraically closed field. The resultant $\mathbf{r}(x_1,...,x_{\nu-1})$ of two polynomials $p(x_1,...,x_\nu) = \sum_{i=0}^{m} p_i(x_1,...,x_{\nu-1})x_\nu^i$, $q(x_1,...,x_\nu) = \sum_{i=0}^{n} q_i(x_1,...,x_{\nu-1})x_\nu^i$ from $\mathbf{F}[x_1,...,x_\nu]$ with positive degrees m and n in x_ν, is a polynomial with one less variable. If $(a_1,...,a_\nu)$ is a common root of p and q, then $\mathbf{r}(a_1,...,a_{\nu-1}) = 0$. Conversely, if $\mathbf{r}(a_1,...,a_{\nu-1}) = 0$, then at least one of the following holds:

(a) $p_0(a_1,...,a_{\nu-1}) = ... = p_m(a_1,...,a_{\nu-1}) = 0$

(b) $q_0(a_1,...,a_{\nu-1}) = ... = q_n(a_1,...,a_{\nu-1}) = 0$

(c) $p_m(a_1,...,a_{\nu-1}) = q_n(a_1,...,a_{\nu-1}) = 0$

(d) for some a_ν in **F**, $(a_1,...,a_\nu)$ is a common root of p and q.

One immediate consequence of the Theorem is that the resultant of two polynomials from $\mathcal{P}_i$ usually yields a polynomial for $\mathcal{P}_{i+1}$.

The next method involves "pseudo-division": given $A(x)$ and $B(x) \neq 0$ in $\mathbf{R}[x]$, where $\mathbf{R}$ is an integral domain, let β be the leading coefficient of B and $\alpha = \beta^{\delta(A)-\delta(B)+1}$ then $Q(x)$ and $R(x)$ (known as the "pseudo-quotient" and "pseudo-remainder" respectivelly) in $\mathbf{R}[x]$ can be found such that $\alpha A(x) = Q(x)B(x) + R(x)$ and $\delta(R)<\delta(B)$. If we apply pseudo-division to two polynomials of

$\mathscr{P}_i$ with respect to x_i, then we either obtain an element of $\mathscr{P}_j$, $j>i$, or a polynomial of $\mathscr{P}_i$ with lower degree. In case of the latter, we include this element in $\mathscr{P}_i$ and then choose another pair of polynomials to continue the process for generation of elements for $\mathscr{P}_j$, $j>i$. At this point, any of the three methods can be used. It is well known that the computation of the resultant of two polynomials can be done through a sequence of pseudo-divisions. However, this means staying with a fixed sequence of remainders starting with a chosen pair of polynomials, until the remainder is independent of the variable. The process we suggest of choosing another pair from $\mathscr{P}_i$ offers the opportunity of continuing with different elements. The advantage of this added flexibility is that the results are often smaller polynomials. Of course, the underlying assumption is that there are more than two polynomials to work with.

The last of the three methods is by forming S-polynomials and a reduction modulo $\mathscr{P}_i$. Unless the result is in **R**, we obtain a polynomial of lower headterm. We include it in $\mathscr{P}_i$ and again choose a new pair to continue. It is easily seen that pseudo-division consists of a sequence of S-polynomial computations with the "divisor" fixed. Thus the possibility of choosing a new pair of polynomials to form the S-polynomial allows greater flexibility, in much the same way as the use of pseudo-division with different polynomials being more flexible than the computation of the resultant of a fixed pair.

For choosing a pair of elements from $\mathscr{P}_i$ to apply one of the three methods, a fundamental rule is always to take polynomials whose sum of the degrees is the smallest. Namely, the pair (k,ℓ) with the smallest $\deg_{x_i}(f_k) + \deg_{x_i}(f_\ell)$ will be taken. When there are several pairs achieving the minimum degree sum, a pair satisfying the first applicable rule (described as follows) for choosing one of three methods will be taken.

(R1) If one polynomial has a headterm of total degree 1, then make a substitution for that variable of the headterm.

(R2) If both polynomials have degree ≤ 3, then compute the resultant eliminating x_i.

(R3) If one polynomial has small degree (2 or 3) and the other larger (by a degree difference ≥ 2), then apply pseudo-division.

(R4) Otherwise (i.e. both with large degrees) compute the S-polynomial (and reduce it modulo $\mathscr{P}_i$).

This method of using S-polynomials also offers a termination condition for the algorithm. If each $\mathscr{P}_i$ is a Groebner basis for itself then forming any S-polynomial will not yield any new element for any $\mathscr{P}_j$, $j\geq i$. Termination is, thus, forced upon us. In this case, however, there is a side benefit, i.e. the union of all $\mathscr{P}_i$'s is a Groebner basis for the original set $\mathscr{P}$ of polynomials, considered as generators of an ideal.

<u>Algorithm B (Eliminating Variables)</u>

Input: A set $\mathscr{P} = \{P_1,...,P_k\}$ of polynomials in $\mathbf{R}[x_1,...,x_n]$.

Output: There are three possibilities:

(i) $P_1,..,P_k$ have no common zeros.

(ii) The result is a set of polynomials involving successively less variables, $\{x_n,...,x_j\}$ for $j = n,n-1,...,1$. The common zeros of $\mathscr{P}$ can therefore be determined successively from this "triangular system" by "back-substitution" (cf. [4]).

(iii) There exists an index $j\in\{1,...,n\}$ such that we do not obtain a polynomial with x_j as the main variable (degenerate case). What we obtain are Groebner-bases for $\langle\mathscr{P}\rangle \cap \mathbf{R}[x_i,...,x_n]$ $(i = 1,...,n)$, namely $\mathscr{P}_n \cup ... \cup \mathscr{P}_i$ (defined in step 1 below).

Step 1. [Initialization] Decompose $\mathscr{P}$ into subsets $\mathscr{P}_1,...,\mathscr{P}_n$ such that for $i=1,...,n$, $\mathscr{P}_i =$

$\{P_j \in \mathcal{P} \mid P_j \in \mathbf{R}\ [x_i,\ldots,x_n] - \mathbf{R}[x_{i+1},\ldots,x_n]\}$, by $B_i \leftarrow \{(\ell,j) \mid 1 \leq \ell < j \leq k$

f sub scriptl , f sub j memberof $\mathcal{P}$ sub i .rcb , $k_i \leftarrow |\mathcal{P}_i|$, $C_i \leftarrow B_i$, $m \leftarrow 1$, $m_0 \leftarrow 0$. [m_0 is the largest non-negative integer for which $C_i = \phi$ throughout all further steps.]

Step 2. [Existence of polynomial in $\mathbf{R}[x_m,\ldots,x_n]$ $-\mathbf{R}[x_{m+1},\ldots,x_n]$] If $k_m \geq 1$ then goto 7 else $\ell \leftarrow m-1$.

Step 3. [Degenerate case ?] If $\ell \leq m_0$ then goto 6.

Step 4. [Generate new polynomials from $\mathcal{P}_\ell, \ell < m$] If $C_\ell = \phi$, decrease ℓ by 1 and goto 3. Otherwise there are two cases:

(i) For $B_\ell = \phi$, choose $(i,j) \in C_\ell$ and compute $\bar{f} = S(f_i,f_j)$, $C_\ell \leftarrow C_\ell - \{(i,j)\}$.

(ii) For $B_\ell \neq \phi$, choose $(i,j) \in B_\ell$ according to rules (R1) - (R4) and compute $\bar{f}$ correspondingly. Set $B_\ell \leftarrow B_\ell - \{(i,j)\}$ and, if (R4) was applied, $C_\ell \leftarrow C_\ell - \{(i,j)\}$.

Step 5. [Insert new polynomial $\bar{f}$ into suitable $\mathcal{P}_g$] For $g = \ell, \ell+1,\ldots,n$ reduce $\bar{f}$ modulo $\mathcal{P}_g$ and check.

(i) If $\bar{f} \in \mathbf{R} - \{0\}$, the polynomials $P_1,\ldots,P_k$ cannot have any common zeros. Return this result and terminate.

(ii) If $\bar{f} \equiv 0$, then goto Step 4.

(iii) If $\bar{f} \in \mathbf{R}[x_g,\ldots,x_n] - \mathbf{R}[x_{g+1},\ldots,x_n]$, then $k \leftarrow k+1$, $f_k \leftarrow \bar{f}$, $B_g \leftarrow \{(i,k) \mid f_i \in \mathcal{P}_g\} \cup B_g$, $C_g \leftarrow \{(i,k) \mid f_i \in \mathcal{P}_g\} \cup C_g$, $\mathcal{P}_g \leftarrow \mathcal{P}_g \cup \{f_k\}$. If $g = m$ then goto 7 else replace ℓ by g and goto 4.

Step 6. [Degenerate case] Increase m until $m \geq n$ or $k_m > 0$. In the first case return the sets $\mathcal{P}_1,\ldots,\mathcal{P}_n$ and terminate. In the second case replace m_0 by $m-1$ and goto 7.

Step 7. [Increase m] $m \leftarrow m+1$. If $m \leq n$ then goto 2 else terminate and return the non-empty sets $\mathcal{P}_1,\ldots,\mathcal{P}_n$.

4. EXAMPLES AND COMPARISONS

Let us first consider the example of Trinks [3], whose task is to determine common zeros of

$$F_1 = 45p + 35s - 165b - 36,$$
$$F_2 = 35p + 40z + 25t - 27s,$$
$$F_3 = 15w + 25ps + 30z - 18t - 165b^2,$$
$$F_4 = -9w + 15pt + 20zs,$$
$$F_5 = wp + 2zt - 11b^3,$$
$$F_6 = 99w - 11\,sb + 3b^2$$

in $\mathbb{Q}(\sqrt{-11})$. The order of the variables is $w>p>z>t>s>b$. It took Trinks' program 11 minutes on a UNIVAC 1108 to compute a Groebner basis for $\mathcal{A} = \langle f_1,\ldots,f_6 \rangle$ in $\mathbb{Q}[w, p, z, t, s, b]$. A polynomial in the variable b alone of degree 10 results. It has coefficients involving numbers up to 160 digits. It seems (though he does not state explicitly) that he does not divide the polynomials by their contents (the gcd of all the coefficients). This polyomial factors over $\mathbb{Q}$ into two factors of degrees 8 and 2, respectively. Trinks is only interested in the one of degree 2, namely,

$$F_7 \leftarrow b^2 + \frac{33}{50}\,b + \frac{2673}{10000}$$

which is reducible over $\mathbb{Q}(\sqrt{-11})$. Trinks' computation of the corresponding zeros of $F_1,\ldots,F_6$, i.e. of the corresponding values for w, p, z, t, s, took another 3 minutes.

The same example was attempted on SCRATCHPAD, guided by intuition and each step done interactively. After a successful interactive session, Algorithm B was applied directly. The computing time for F_7 on IBM 370/168 in such a batch mode was only a few seconds. Moreover, the coefficients of the tenth degree polynomial in the variable b alone had no more than 60 digits. Taking out the content, the coefficients shrunk to only 16 digits. We carried out the computation only with integral coefficients, until the elimination process was completed.

The same example was also attempted on MACSYMA using Spear's algorithm [5] for finding ideal basis, which is also based on a lexical ordering of the variables. The polynomials and variables were entered in their natural order as appeared in Trinks' paper. After approximately 15 minutes terminal time it was able to give a Groebner basis whose elements involved successively less variables. Our attempts of changing the order of the variables to coerce the program to give the desired F_7 failed twice each after more than 40 minutes of wait at the terminal.

5. CONCLUSIONS

This work started out as a simple attempt to understand the difficulties in computational methods for ideal bases. The slightly different objective of finding a basis, which can also be useful for solving a system of polynomial equations, led us to some simplifications and unifications that were previously somewhat mysterious. However, the problem of finding common zeros of polynomial equations are still some ways from being satifactorily solved. Polynomial systems often result in a large number of solutions, which can grow exponentially according to the number and degrees of the polynomials. As we have seen, some methods can also be very sensitive to the ordering of the variables. Our experience, therefore, reaffirms the importance of interaction through guidance by intuition or the need to easily combine and test heuristic and algorithmic approaches in programs.(cf. [4]).

REFERENCES

[1] Buchberger, B.: "Ein algorithmisches Kriterium fur die Loesbarkeit eines algebraischen Gleichungssystems", *Aequationes Math. 4,* (1969), 374-383.

[2] Buchberger, B.: "Some properties of Groebner-bases for polynomial ideals", *ACM SIGSAM Bulletin No. 40,* (1976), 19-24.

[3] Trinks, W.: "Ueber B. Buchbergers Verfahren, Systeme algebraischer Gleichungen zu loesen", *Journal of Number Theory,* Vol. 10, No. 4, Nov. 1978, 475-488.

[4] Yun, D. Y. Y.: "On algorithms for solving systems of polynomial equations", *ACM SIGSAM Bulletin No. 27,* Sept. 1973, pp. 9-15.

[5] Spear, D. A.: "A Constructive approach to commutative ring theory", *Proceedings 1977 MACSYMA Users' Conference,* NASA CP-2012, July 1977, pp. 369-376.

A p-adic Algorithm for Univariate Partial Fractions

Paul S. Wang*
Kent State University
Department of Mathematical Sciences
Kent, Ohio 44242

1. Introduction

Partial fractions is an important algebraic operation with many applications in applied mathematics, physics and engineering [3], [11]. It is also an important operation in any computer symbolic and algebraic system. Among other things, it is used in the integration algorithm [7], [8].

Let $R(x)$ be a rational function over the integers, $\mathbb{Z}$, with the denominator completely factored.

$$(1.1) \qquad R(x) = C(x)/\prod_{i=1}^{t} D_i(x)^{e_i}, \quad e_i > 0,$$

where $C(x)$ and $D_i(x)$ are pairwise relatively prime polynomials with integer coefficients and each $D_i(x)$ is irreducible over $\mathbb{Z}$. Without loss of generality, let us also assume that $C(x)$ and $D_i(x)$ are primitive and

$$\deg(C) < \sum_{i=1}^{t} e_i \cdot \deg(D_i) = n.$$

There exist unique polynomials $B_{i,j}(x)$, over the field of rational numbers, $\mathbb{R}$, such that

$$(1.2) \qquad R(x) = \sum_{i=1}^{t} \sum_{j=1}^{e_i} B_{i,j}(x)/D_i(x)^j,$$

with $\deg(B_{i,j}) < \deg(D_i)$ for all i, j. The partial fraction expansion (PFE) of $R(x)$ is to obtain the polynomials $B_{i,j}(x)$ in equation (1.2) given the $C(x)$, $D_i(x)$ and e_i.

If the denominator of $R(x)$ is not given in a factored form, it can be factored over $\mathbb{Z}$ first [9], [10]. This factoring process is quite involved and is usually not considered a part of the PFE process.

To calculate the PFE of $R(x)$, we first compute what's known as the incomplete PFE of $R(x)$,

$$(1.3) \qquad R(x) = \sum_{i=1}^{t} A_i(x)/Q_i(x), \quad Q_i(x) = D_i(x)^{e_i},$$

by finding the unique polynomials $A_i(x)$ over $\mathbb{R}$ with $\deg(A_i) < \deg(Q_i)$. For each i, the $B_{i,j}$ can then be computed from $A_i(x)$ by repeated division. Classical methods for the incomplete PFE involve solving systems of linear equations. More recent methods (e.g., [5]) involve polynomial divisions and polynomial remainder sequence (PRS) operations.

*Research supported in part by the National Science Foundation and the Department of Energy.

The PRS is used in solving for the unique polynomials α and β in

(1.4) $\alpha F_1 + \beta F_2 = 1$, $\deg(\alpha) < \deg(F_2)$, $\deg(\beta) < \deg(F_1)$,

given relatively prime polynomials $F_1(x)$ and $F_2(x)$. For each A_i there is a different equation (1.4). Therefore it is solved t times. If the PRS is carried out with rational number coefficents, the cost of integer gcd's is excessive. One way to avoid rational number arithmetic is to use pseudo-division [4] and the reduced PRS algorithm [1], [2]. Through these techniques, the algorithm for PFE is made more efficient. However, the size of the integer coefficients grows large in the pseudo division and reduced PRS. Often the size of the integer coefficients of the intermediate results are much larger than those in the final answers. Asymptotic timing analysis by Kung and Tong [5] indicates that the number of coefficient arithmetic operations for this algorithm is $O(n \log^2 n)$. But the coefficients involved in the arithmetic operations grow in size considerably, even for practical problems. The actual computing time for, say, dividing small integers is far less than that for large integers. Thus, the PFE algorithm can be made more efficient if the size of the coefficients can be better controlled.

A p-adic algorithm for the incomplete PFE (1.3) is presented. This algorithm solves the given PFE problem modulo a suitably selected prime, p, first. Then the solution mod(p) is lifted p-adically to give the desired solution over $\mathbb{R}$. Efficiency in this approach comes from controlling coefficient size and the parallel computation of all the $A_i(x)$.

There are three phases to this p-adic PFE algorithm.

Phase I: For given $C(x)$, $Q_i(x)$ and a suitably selected integer prime, p, not exceeding single-precision, the algorithm computes, for all i, $A_i^{(1)}$ satisfying

$$A_i^{(1)} = A_i(x) \bmod (p).$$

Thus all arithmetic in the PRS is in the field $\mathbb{Z}_p$, avoiding coefficient growth. (See Section 2).

Phase II: The $A_i^{(1)}$ are lifted by a p-adic construction algorithm to $A_i^{(j)}$ such that

$$A_i^{(j)} = A_i(x) \bmod (p^j), \; j \geq 1.$$

(See Section 3).

Phase III: For sufficiently large p^j, the rational coefficients of each $A_i(x)$ can be determined from the corresponding coefficients of $A_i^{(j)}$. (See Section 4).

Programs for this algorithm have been written and implemented in the VAXIMA system [6] for testing. Machine examples with timings using both the p-adic algorithm and the PRS type algorithm are included in the appendix.

2. Solution in $\mathbb{Z}_p$

Let

$$F_i(x) = \prod_{\substack{j=1 \\ j \neq i}}^{t} Q_j(x).$$

It is clear from (1.3) that $A_i(x)$ uniquely satisfy

(2.1) $\sum_{i=1}^{t} A_i F_i = C(x)$, $\deg(A_i) < \deg(Q_i)$.

Let $p \in \mathbb{Z}$ be a prime such that (I) p does not divide the leading coefficient of $C(x)$ or $Q_i(x)$ for any i and (II) $Q_i(x) \bmod (p)$ stay pairwise relatively prime over $\mathbb{Z}_p$.

Lemma 1. If p is any prime satisfying the conditions (I) and (II), then p does not divide the denominator of any rational number coefficient of $A_i(x)$ for any i.

Proof. Let m be the integer least common multiple of all denominators of coefficients in $A_i(x)$ for all i. From (2.1), we have

$$\sum_{i=1}^{t} mA_i(x)\ F_i(x) = m\ C(x)$$

where each mA_i has all integer coefficients. Suppose the lemma is false. Then

$$\sum_{i=1}^{t} mA_i\ F_i \equiv 0 \bmod(p).$$

But not all $mA_i \equiv 0 \bmod(p)$. Since $\deg(A_i) < \deg(Q_i)$, this means that the $Q_i(x)$ are not relatively prime over $\mathbb{Z}_p$, a contradiction.

In our algorithm, a prime, p, is selected randomly just below the word size which satisfies condition (I). This prime is used as though it satisfies (II) also. If it does not satisfy (II) an error will occur later in the algorithm which enables us to select another prime. This approach saves time because condition (II) is almost always satisfied without being checked first.

As stated before, we want to obtain polynomials $A_i^{(j)}$ such that

$$(2.2)\quad A_i^{(j)}(x) = A_i(x) \bmod(p^j),\ j \geq 1.$$

Presented in this section is an algorithm to compute $A_i^{(1)}(x)$ over $\mathbb{Z}_p$ which satisfies

$$(2.3)\quad \sum_{i=1}^{t} A_i^{(1)}(x)\ F_i(x) \equiv C(x) \bmod(p).$$

We first obtain polynomials $\alpha_i(x)$ such that

$$(2.4)\quad \sum_{i=1}^{t} \alpha_i(x)\ F_i(x) \equiv 1 \bmod(p),\ \deg(\alpha_i) < \deg(Q_i).$$

The $\alpha_i(x)$ exist uniquely if $Q_i(x)$ are relatively prime over $\mathbb{Z}_p$. Let us denote by $G_i(x)$ the product $Q_{i+1} \cdots Q_{i+t}$, i.e.,

$$G_i(x) = \prod_{j=i+1}^{t} Q_j(x),\ i = 1, 2, \cdots, t-1.$$

Note that $G_1 = F_1$. By means of PRS we can find $\alpha_i(x)$ as a result of the following computations,

$$\beta_{i+1}Q_i + \alpha_i G_i \equiv \beta_i(x) \bmod(p),\ i = 1, 2, \cdots, t-1,$$

where $\beta_1(x) = 1$, $\alpha_t(x) = \beta_t(x)$ and $\deg(\alpha_i) < \deg(Q_i)$. It can readily be deduced that the $\alpha_i(x)$ satisfy (2.4). The polynomials $\alpha_i(x)$ are useful both here and later in the p-adic lifting phase. Thus they are computed and stored for repeated reference.

If the prime, p, violates condition (II), our attempt to solve (2.4) will result in an error. In this case the algorithm goes back and selects another prime p.

We shall denote by $C^{(1)}(x)$ and $Q_i^{(1)}(x)$ respectively $C(x)$ and $Q_i(x)$ modulo (p). For any given right-hand side of (2.3) as long as $\deg(C) < n$, the following algorithm is used to obtain the $A_i^{(1)}(x)$ from the $\alpha_i(x)$. It is called with $C^{(1)}(x)$ as an argument. The α_i, p and $Q_i^{(1)}$ are used as non-local variables.

ALGORITHM ALPHA(c)/*All arithmetic operations are in $\mathbb{Z}_p$*/

A0. Local $a_1, \cdots, a_t$

A1. For i:= 1 To t Do

$$a_i^{(1)}(x) := c\alpha_i(x) \bmod(Q_i(x))$$

A2. Return $(a_1^{(1)}, \cdots, a_t^{(1)})$

We see clearly that

$$(A_1^{(1)}, \cdots, A_t^{(1)}) = \text{ALPHA}(C^{(1)}(x)).$$

3. p-adic Construction

Using the method described in the previous section, the $A_i^{(1)}(x)$ can be obtained. From each $A_i^{(1)}$, a sequence of polynomials $A_i^{(j)}$ will be constructed, for $j = 2, 3, \cdots$, such that

$$(3.1) \quad \sum_{i=1}^{t} A_i^{(j)} F_i(x) \equiv C(x) \bmod (p^j).$$

It can be shown that the $A_i^{(j)}$ are unique. Therefore,

$$A_i^{(j)}(x) = A_i(x) \bmod (p^j).$$

We present a p-adic algorithm for constructing $A_i^{(j+1)}$ from $A_i^{(j)}$ for any $j \geq 1$. Let

$$(3.2) \quad W_j(x) = C(x) - \sum_{i=1}^{t} A_i^{(j)}(x) F_i(x) \text{ over } \mathbb{Z}.$$

By induction, suppose $A_i^{(j)}$ and $W_j(x)$ are obtained for some $j \geq 1$. If $W_j(x)$ is identically zero, then the $A_i^{(j)}(x)$ are the desired $A_i(x)$ and the incomplete PFE is finished. Otherwise, we compute

$$\tilde{W} = W_j(x) \bmod (p^{j+1})/p^j.$$

The division is exact over $\mathbb{Z}$ since $W_j(x) = 0 \bmod (p^j)$. If $\tilde{W}$ is zero then $A_i^{(j+1)} = A_i^{(j)}$ and $W_{j+1}(x) = W_j(x)$ and the induction is complete. If $\tilde{W} \neq 0$, we proceed to compute polynomials $H_i(x)$ using ALGORITHM ALPHA.

$$(H_1, \cdots, H_t) = \text{ALPHA}(\tilde{W}).$$

Needless to say, the $H_i(x)$ satisfy

$$\sum_{i=1}^{t} H_i(x) F_i(x) \equiv \tilde{W}(x) \bmod (p).$$

Now $A_i^{(j+1)}$ is obtained by "correcting" the $A_i^{(j)}$ as follows.

$$A_i^{(j+1)}(x) = A_i^{(j)}(x) + p^j H_i(x) \bmod (p^{j+1}),$$

$$i = 1, 2, \cdots, t.$$

And $W_{j+1}(x)$ is given by

$$W_{j+1}(x) = W_j(x) - \sum_{i=1}^{t} H_i(x) F_i(x).$$

It is easy to show that $A_i^{(j+1)}$ and W_{j+1} thus obtained satisfy (3.1) and (3.2) with j replaced by j+1.

Now we can obtain $A_i^{(j)}(x)$ for any $j \geq 1$. How big should j become before we stop p-adic lifting? Given $R(x)$, there exist a positive integer S sufficiently large such that for any coefficient a/b in any $A_i(x)$, we have $S > 2a^2$ and $S > 2b^2$. The p-adic process will finish when $p^j > S$. The final $A_i^{(j)}(x)$, are then used to obtain the $A_i(x)$ by recovering the rational coefficients from their $\bmod(p^j)$ images. Ideally, we should compute the smallest S for the given problem then stop as soon as $p^j > S$. However, it is difficult to make S a tight bound. Fortunately, we don't really need to compute S. Instead, we use a heuristic value for S which is trivially computed. Our algorithm proceeds and checks the final partial fractions obtained over $\mathbb{Z}$. If the answer does not check, the algorithm goes back to p-adic lift some more. Let c_{max} be the maximum coefficient, in absolute value, of the $D_i(x)$ and d_{max} be the maximum degree of the $Q_i(x)$. The following heuristic value for S is used in the program.

$$S = c_{max} ** (d_{Max} + t).$$

4. Recovering Rational Coefficients

Let a and b be relatively prime integers with $b > 0$. Let a/b be a rational coefficient of any $A_i(x)$. If p is a prime satisfying conditions (I) and (II) as set forth in Section 2, then p does not divide b by Lemma 1. Let us denote p^j by m. Furthermore, let us assume that we know the integer c such that

(4.1) $a/b \equiv c \bmod(m)$.

Lemma 2. Given c and m in (4.1), the integers a and b can be uniquely determined if $|a| < \sqrt{m/2}$ and $|b| < \sqrt{m/2}$.

Proof. Suppose that there exists a'/b' such that $|a'| < \sqrt{m/2}$, $|b'| < \sqrt{m/2}$, $\gcd(a', b') = 1$ and

$$a/b \equiv a'/b' \bmod(m)$$

then we have

$$ab' \equiv ba' \bmod(m),$$

which implies that for some integer i

(4.2) $ab' - ba' = i \cdot m$.

But we have $|ab'| < m/2$ and $|ba'| < m/2$ which implies

(4.3) $-m < (ab' - ba') < m$.

It follows from (4.2) and (4.3) that $i = 0$. Therefore, $a = a'$ and $b = b'$.

For given c and m the following algorithm based on the Euclidean algorithm for integer greatest common divisor can be used to calculate a and b.

ALGORITHM RATCONVERT (c, m, m2) /* m2 is $\sqrt{m/2}$ */

R1. u:= (1, 0, m), v:= (0, 1, c)

R2. WHILE m2 $\le v_3$ DO

R3. {q:= u_3/v_3, r:= u - qv, u:= v, v:= r}

R4. IF abs(v_2) $\ge$ m2 THEN error()

R5. a:= sign(v_2)v_3, b:= abs(v_2)

R6. return((a,b)).

In algorithm RATCONVERT the following relations always hold,

$$u_1 m + u_2 c = u_3$$

and

$$v_1 m + v_2 c = v_3.$$

After the WHILE loop we have

$$v_3 < \sqrt{m/2}.$$

If $|v_2| < \sqrt{m/2}$ also, then we have found v_3/v_2 such that

$$v_3/v_2 \equiv c \bmod(m).$$

When the error condition in step R4 is satisfied, it means that our heuristic S value is not large enough. Thus, the algorithm goes back to perform a few more iterations of p-adic construction. If all coefficients are successfully recovered, then we use (2.1) to check the final answer.

Appendix

Contained here are ten examples of incomplete partial fraction expansion. The timing is done on the MACSYMA system [6] which runs on a DEC KL-10 with a memory cycle time of about two microseconds. The examples were done using the p-adic algorithm described in this paper and the PRS type algorithm (PRS) already in MACSYMA. The timings in Table 1 do not include the factoring of the denominator.

As one can see in Table 1, the p-adic algorithm is often faster but not always as compared to the PRS algorithm. Only further investigation can reveal the contributing factors to this behaviour.

Table 1

Timing Comparisons

Problem	p-adic	PRS
1	23 msec	33 msec
2	40 msec	48 msec
3	72 msec	129 msec
4	76 msec	121 msec
5	76 msec	766 msec
6	71 msec	105 msec
7	77 msec	722 msec
8	117 msec	259 msec
9	157 msec	116 msec
10	1133 msec	1024 msec

The ten problems are listed below by giving the numerator $C(x)$ and the factors $Q_i(x)$ of the denominator of $R(x)$. Thus

$$R(x) = C(x)/\pi Q_i(x).$$

The $Q_i(x)$ are given in the form of a list as in

$$Q_1(x),\ Q_2(x),\ Q_3(x).$$

1) $Q_i(x)$: $x-1,\ x+1$

$C(x)$: 1

2) $Q_i(x)$: $(x+2)^2,\ (x-2)^2$

$C(x)$: 1

3) $Q_i(x)$: $4x^2-12x-9,\ x^2+15,\ 7x^2-11x+8$

$C(x)$: $11x^4+48x^3-557x^2-2070x+3$

4) $Q_i(x)$: $(3x^2+1)^3,\ (3x^2-1)^3$

$C(x)$: $x^{10}-1$

5) $Q_i(x)$: $x^3+2,\ 3x^3-2,\ 5x^3-x+7$

$C(x)$: $47x^6-42x^5+350x^4-96x^3-68x^2+788x-260$

6) $Q_i(x)$: $3x^4-2,\ 2x^4+3,\ (x^2+1)^2$

$C(x)$: 1

7) $Q_i(x)$: $2x^2+3x+1,\ 3x^2-2x-1,\ x^2+x+1$

$C(x)$: x^5

8) $Q_i(x)$: $48x^4-96x^3+72x^2-24x-29,$

$16x^4-32x^3+24x^2-8x+25,$

$16x^4-32x^3+56x^2-40x+25$

$C(x)$: 1

9) $Q_i(x)$: $2x^2-12x-19,\ x^2+15,\ 3x^2-11x+8$

$C(x)$: 6

10) $Q_i(x)$: $(12x^3+5x-13)^2,\ 4x^2+15x-9,\ (x+2)^3$

$C(x)$: $3x^4-28.$

References

1. W. S. Brown, "The Subresultant PRS Algorithm", TUMS 4, 1978, pp. 237-249.

2. G. E. Collins, "Subresultants and Reduced Polynomial Remainder Sequences", JACM, Vol. 14, Jan. 1967, pp. 128-142.

3. P. Henrici, Applied and Computational Complex Analysis, Vol. 1, Wiley-Intersicence, N.Y. 1974, Chapter 7.

4. D. E. Knuth, "The Art of Computer Programming", Vol. 2: Seminumerical Algorithms, Addison-Wesley, Reading, Mass., 1969.

5. H. T. Kung and D. M. Tong, "Fast Algorithms for Partial Fraction Decomposition", Department of Computer Science report, Carnegie-Mellon University, Jan. 1976.

6. MACSYMA Reference Manual, "The MATHLAB Group", Laboratory for Computer Science, MIT, Cambridge, Mass., Dec. 1977.

7. J. Moses, "Symbolic Integration: The Stormy Decade", Trans. ACM, Vol. 139, May 1969, pp. 167-189.

8. R. Risch, "The Problem of Integration in Finite Terms", Bulletin of AMS, May 1970, pp. 605-608.

9. P. Wang, "An Improved Multivariate Polynomial Factoring Algorithm", Math Comp, Vol. 32, No. 144, Oct. 1978, pp. 1215-1231.

10. P. Wang, "Parallel p-adic Construction in the Univariate Polynomial Factoring Algorithm", Proceedings, Second MACSYMA USERS CONFERENCE, Washington, D.C., June, 1979, pp. 310-317.

11. L. Weinberg, Network Analysis and Synthesis, McGraw-Hill, N.Y., 1962.

Use of VLSI in Algebraic Computation: Some Suggestions

H.T. Kung

Department of Computer Science
Carnegie-Mellon University
Pittsburgh, Pennsylvania 15213

This paper reviews issues in the design of special-purpose VLSI chips in general, and suggests VLSI designs for polynomial multiplication and division, which are basic functional modules in algebraic computation.

1. Introduction

Current large-scale integration (LSI) technology allows tens of thousands of transistors to fit on a single chip; the emerging very-large-scale integration (VLSI) technology promises an increase of this number by at least one or two orders of magnitude in the next decade. Some of the recently announced 32-bit microprocessor chips already contain over 100,000 transistors. VLSI will undoubtedly improve the cost-effectiveness of the traditional, general-purpose computer in terms of its component cost and reliability, and its power and size. A more striking impact of VLSI however is related to the expectation that VLSI will make it practical to employ special-purpose systems in many application areas. This paper explores issues in building special-purpose VLSI systems, and suggests VLSI designs for polynomial multiplication and division that could result in high-speed, yet cost-effective building blocks for algebraic computation.

The author is currently on leave from CMU with ESL, a subsidiary of TRW. Most of the research reported here was carried out at CMU and was supported in part by Office of Naval Research under contracts N00014-76-C-0370, NR 044-422 and N00014-80-C-0236, NR 048-659, and in part by the National Science Foundation under grant MCS 78-236-76.

2. Issues in Special-Purpose VLSI Designs

2.1. Understanding the Application

The first step in the development of a special-purpose VLSI system is to define the task. Ideally we hope to identify some system bottleneck to which a relatively inexpensive VLSI solution can improve the overall system in a substantial way. A typical area that could benefit from VLSI solutions is composed of those low level, compute-intensive, operations which are repetitive and well-understood. For example, in signal processing such an operation could be the inner product operation,

$$\sum_{i=1}^{n} a_i \cdot b_i.$$

The speed and accuracy of a signal processor is often dominated by the calculation of this operation. Thus it is worthwhile to build special-purpose chips for performing the inner product operation. Indeed, several commercially available chips have been built for this purpose, and found to be extremely cost-effective. Examples include the TRW multiplier-accumulator chips [11] and the NEC signal processor chip [9]. Another important operation that is frequently performed in signal and image processing is the FFT computation. Again, special-purpose FFT chips such as the AMI S2814A chip have proven to be useful. Choosing a proper task to be implemented in chips is most fundamental to the success of the special-purpose approach; it requires a good understanding of computational needs in a given application area.

2.2. Simple and Regular Design

One of the most genuine concerns about the special-purpose approach is its cost-effectiveness. The cost of a special-purpose system must be low enough to justify its limited applicability. For VLSI chips the dominant costs are typically on design and testing. If a design can truly be decomposed into few types of simple modules which are used repetitively with simple interfaces, then great savings in design and testing costs can be achieved [8]. In addition, the resulting modularity makes the design easily adjustable to various design goals. Thus it is important that VLSI designs be simple and regular.

2.3. Concurrency and Communication

VLSI components based on commercially proven technologies such as MOS are likely not much faster than their TTL counter parts of 10 years ago (see, for example, [10]). The power of VLSI comes from its potential of providing a large number of simultaneously executable processing elements, rather than fast components. Thus the performance of a VLSI chip is closely tied to the degree of concurrency in the underlying algorithm. When a large number of processing elements work simultaneously, their co-ordination and communication becomes non-trivial. The communication problem is especially significant for VLSI where routing costs dominate the area and power required to implement a circuit in silicon. The issue here is to design algorithms that support high degrees of concurrency, and in the mean time employ only simple and regular communications and controls so that their chip implementations would be easy and efficient.

2.4. Balancing Computation with I/O

When a special-purpose device is attached to a host from which it gets input data and to which it output results, I/O costs play an important role in the overall system performance. The ultimate performance goal of a special-purpose system is (and *should be no more than*) that of achieving a computation rate that balances the available I/O bandwidth with the host.

The pin bandwidth limitation often forms the bottleneck for performance. The bottleneck can be greatly resolved if multiple computations are performed per I/O access. However, the repetitive use of a data item requires it to be stored inside the chip for a sufficient length of time. Thus the I/O cost is related not only to the available pin bandwidth, but also to the available memory internal to the chip. Tradeoffs between the memory size and the I/O bandwidth requirements must be carefully studied [4].

When a computation task is larger than what a single chip can handle, some multi-chip configurations will be adopted. In this case, chips communicate not only with the host but also among themselves. How to partition a task among multiple chips to achieve a balance between computation and inter-chip communication presents another challenge to a chip designer.

3. VLSI Designs for Polynomial Multiplication and Division

This section sketches VLSI designs of polynomial multiplication and division. We shall see that these designs possess the desirable design properties discussed in the preceding section.

3.1. Polynomial Multiplication

Given polynomials $A(t) = \sum_{i=0}^{m} a_i t^i$ and $X(t) = \sum_{i=0}^{n} x_i t^i$, we want to compute their product $Y(t) = \sum_{i=0}^{m+n} y_i t^i$. It is well known that this is a convolution problem, that is, y_i's are defined by

$$y_i = \sum_{k=0}^{i} a_k \cdot x_{i-k},\ i = 0, \ldots, m+n,$$

where $a_i = 0$ for $i > m$ and $x_i = 0$ for $i > n$. Suppose that m is a fixed constant and n is arbitrarily large. Then a VLSI array of m processors called cells can compute all the y_i's in $O(n)$ time, independent of m. The processor array and its cell definition are depicted in Figure 3-1 for the case $m = 3$.

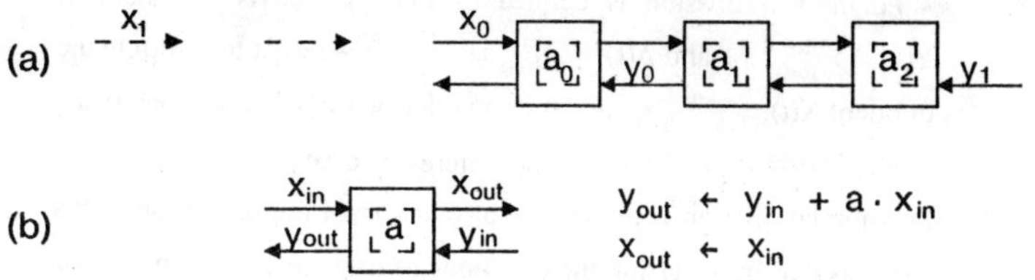

Figure 3-1: (a) Processor array for polynomial multiplication, and (b) the definition of the basic cell.

The coefficients of A(t) are downloaded to the cells, one at each cell, before the computation starts. During the computation, the y_i's (initialized as zeros before entering the array) and the x_i's move synchronously in opposite directions. When a y meets an x at a cell, the product of x and the particular coefficient of A(t) that was preloaded to that cell is accumulated to the y. To ensure that each y_i is able to meet the three consecutive x_i's it depends on, the x_i's on the x data-stream are separated by two cycle times and so are the y_i's on the y data-stream. As Figure 3-1 shows, during cycle one a_0x_0 is accumulated to y_0 at the left-most cell, and during cycle two a_1x_0 is accumulated to y_1 at the middle cell. During cycle three a_0x_1 and a_2x_0 are accumulated to y_1 and y_2 at the left-most and right-most cells, respectively. Notice that each y_i outputs from the left-most cell during the same cycle as its last input x_i enters there. Thus, the processor array is capable of outputting a y_i every two cycle times, and with *constant-time* response, that is, with a delay independent of m. In addition to the high throughput and low latency properties, the processor array enjoys other desirable properties such as the simple and regular design, and involves no global communications like broadcasting. This is demonstrated by the block diagram given in Figure 3-2. To handle polynomials of large degrees we can either add cells of the same type to the processor array or perform the computation in multiple passes

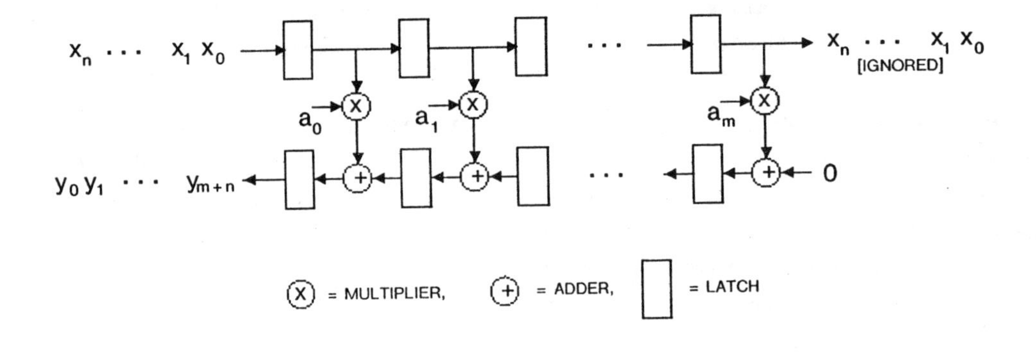

Figure 3-2: Block diagram of the processor array for polynomial multiplication.

3.2. Polynomial Division

Polynomial division is defined as follows. Given polynomials $Y(t) = \sum_{i=0}^{m+n} y_i t^i$ and $A(t) = \sum_{i=0}^{m} a_i t^i$, $a_m \neq 0$, we want to compute the quotient $X(t) = \sum_{i=0}^{n} x_i t^i$ and the remaider $B(t) = \sum_{i=0}^{r} b_i t^i$ such that

$$Y(t) = A(t) \cdot X(t) + B(t), \quad \text{where } r < m.$$

Polynomial division is the inverse of polynomial multiplication. This suggests that by reversing the directions of data streams the processor array for polynomial multiplication can be used to perform polynomial division. The idea indeed works out, after the rights-most cell is changed into a division cell. To facilitate the computation of the remaider, the x input of the cell to the left of the division cell employs a multiplexer. Figure 3-3 depicts how the processor array performs polynomial division when m = 2 and n = 4. In the beginning of the computation the multiplexer is set such that the middle cell inputs results from the division cell. Note that during cycle one, x_2 becomes y_4/a_2, which is its correct, final value. During cycle two, y_3 becomes $y_3 - a_1x_2$ at the middle cell, and the new value of y_3 is divided by a_2 during the next cycle at the right-most cell. The result of the division also becomes the final value for x_1 during cycle three. Similarly, x_0 is computed during cycle five. Immediately after x_0 is passed to the middle cell, the multiplexer switches to accept zero as its input. For the subsequent cycles, coefficients in the remainder are output from the middle cell in the order of b_1 followed by b_0. A block diagram of the processor array for polynomial division is given in Figure 3-4. For the special case that A(t) is monic, that is, $a_m = 1$, the divider can simply be replaced with a straight data path.

In summary, the processor array for polynomial multiplication can perform polynomial division in the same efficiency in terms of throughput and latency, by making a slight modification to its right-most cell.

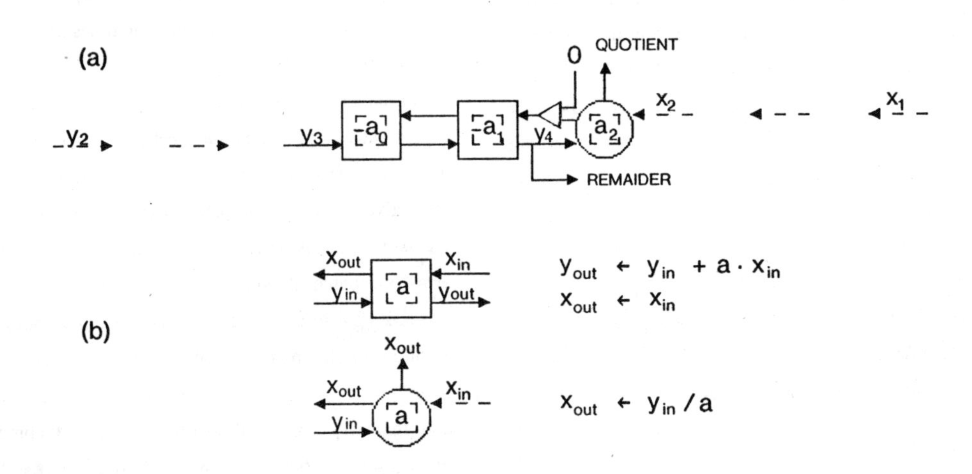

Figure 3-3: (a) Processor array for polynomial division, and (b) definitions of basic cells.

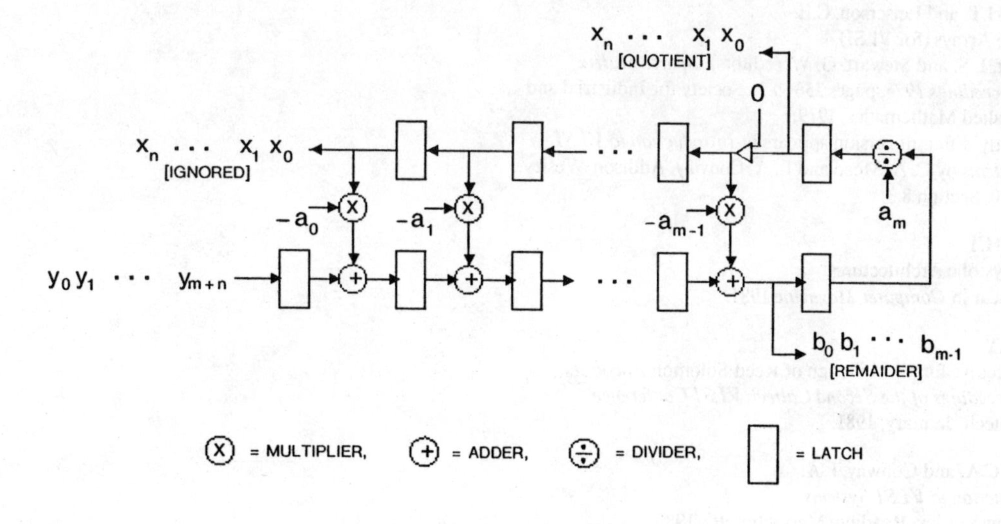

Figure 3-4: Block diagram of the processor array for polynomial division.

4. Concluding Remarks

Similar processor arrays to perform polynomial multiplication and division were previously considered by D. Cohen [1]. In his schemes, an input data item is broadcast to all the cells during every cycle. This requires the use of a bus or some sort of tree network. As the number of cells increases, it will become difficult (especially in VLSI implementations) to keep expanding these *non-local* communication paths to meet the increasing load. Designs in the present paper do not rely on broadcasting; inputs are pulsed through the array *systolically*, while maintaining constant time response and high throughput. This is a characteristic of the *systolic architecture* concept that has been applied to special-pupose designs in many application areas (see, for example, [5, 6]). A VLSI design for polynomial division was also proposed by K. L. Liu [7] to implement the Reed-Solomon encoder. His design is basically the same as Cohen's, and thus requires broadcasting of inputs as well.

Polynomial multiplication and division, and their variants, are basic functional modules in algebraic computation and multi-precision arithmetic. Designs presented in this paper suggest that cost-effective, special-purpose devices for performing various forms of polynomial multiplication and division, or their combinations such as operations involving rational expressions, could conceivably be built using VLSI technology. This is mainly due to the fact that these designs enjoy very regular structure and localized interconnections, and use basically only one type of simple processors. Designs possessing similar advantages are available to some other functional modules useful to symbolic and algebraic computation such as pattern and language recognizers (see, for example, [2, 3]).

Realizing that some routines can probably be carried out cost-effectively with VLSI is only the very first step towards the use of VLSI to resolve some of the computational bottenecks in this area. Much more evaluations and studies, especially from an integrated systems point of view, have to be performed before a clear picture on how to make best use of VLSI can emerge.

References

[1] Cohen, D.
Mathematical Approach to Computational Networks.
Technical Report ISI/RR-78-73, University of Southern California, Information Sciences Institute, November, 1978.

[2] Foster, M. J. and Kung, H.T.
The Design of Special-Purpose VLSI Chips.
Computer Magazine 13(1):26-40, January, 1980.
A preliminary version of the paper, entitled "Design of Special-Purpose VLSI Chips: Example and Opinions", appears in *Proceedings of the 7th International Symposium on Computer Architecture*, pp. 300-307, La Baule, France, May 1980.

[3] Foster, M.J. and Kung, H.T.
Recognize Regular Languages With Programmable Building-Blocks.
In *Proceedings of International VLSI Conference.* August, 1981.

[4] Hong, J-W and Kung, H.T.
I/O Complexity: The Red-Blue Pebble Game.
Technical Report, Carnegie-Mellon University, Department of Computer Science, March, 1981.
Also available in *Proceedings of the Thirteenth Annual ACM Symposium on Theory of Computing*, May 1981.

[5] Kung, H.T. and Leiserson, C.E.
Systolic Arrays (for VLSI).
In Duff, I. S. and Stewart, G. W. (editors), *Sparse Matrix Proceedings 1978*, pages 256-282. Society for Industrial and Applied Mathematics, 1979.
A slightly different version appears in *Introduction to VLSI Systems* by C. A. Mead and L. A. Conway, Addison-Wesley, 1980, Section 8.3.

[6] Kung, H.T.
Why Systolic Architectures.
To appear in *Computer Magazine* 1981.

[7] Liu, K.Y.
Architecture for VLSI Design of Reed-Solomon Encoders.
In *Proceedings of the Second Caltech VLSI Conference.* Caltech, January, 1981.

[8] Mead, C.A. and Conway, L.A.
Introduction to VLSI Systems.
Addison-Wesley, Reading, Massachusetts, 1980.

[9] Nishitani, T., Y. Kawakami, R. Maruta and A. Sawai.
LSI Signal Processing Development for Communications Equipment.
In *Proceedings of ICASSP80*, pages 386-389. IEEE Acoustics, Speech and Signal Processing Society, April, 1980.

[10] Robert N. Noyce.
Hardware Prospects and Limitations.
In Dertouzos, M.L. and Moses, J. (editor), *The Computer Age: A Twenty-Year View*, pages 321-337. IEEE, 1979.

[11] Schirm IV, L.
Multiplier-Accumulator Application Notes.
Technical Report, *TRW* LSI PRODUCTS, January, 1980.

An Algebraic Front-end for the Production and Use of Numeric Programs

Douglas H. Lanam[1]
Computer Science Division
and
Center for Pure and Applied Mathematics
University of California
Berkeley, CA 94720

ABSTRACT

We describe a programming environment which combines the Macsyma algebraic manipulation system with convenient and direct access to numeric Fortran run-time libraries. With this system it is also convenient to generate, compile, load, and invoke totally new Fortran programs which may have been produced by combining algebraically derived formulas and program "templates". These facilities, available on VAX-11 computers, provide an environment for the generation and testing of advanced scientific software. Enhancements of Fortran for high-precision calculations, interval arithmetic, and other purposes are also supported.

1. Introduction

Algebraic Manipulation systems such as Macsyma [Mathlab77] are designed to express concepts from applied mathematics in a form making it possible to compute with appropriate notations algorithmically and symbolically. This differs from the more common numerical approach which would rely on providing insight via tables of numbers or perhaps graphs. Combinations of these two approaches are possibly even more valuable than either alone, yet it has generally been inconvenient to provide both types of facilities in a single programming environment. This paper describes efforts to provide such an environment.

Problems are generated in part because the languages Lisp and Fortran are usually implemented so as to be incompatible. We have to a large extent removed this barrier. In the next two sections we discuss the rationale for for uniting these two languages.

---

1. Work reported herein was supported in part by the U. S. Department of Energy, Contract DE-AT03-76SF00034, Project Agreement DE-AS03-79ER10358.

1.1. Algebraic Programming and Lisp

A number of modern Algebraic Manipulation systems, including Macsyma, use Lisp as their implementation language; this is generally advantageous. There are a number of sophisticated programs which historically have never been implemented in other languages. These include powerful symbolic integration programs, simplifiers, and display packages. Furthermore, Lisp provides an environment which is syntactically and semantically natural to use for extensions of traditional data types and control structures. It provides simple tools for dealing with prefix-tree representations, sets, lists, arrays, etc. The variety of control structures and function disciplines is another convenience. Lisp provides automatic bookkeeping for storage allocation and garbage collection. Lisp provides recursion in data construction and programs, is usually implemented so as to be interactive, with powerful tracing and debugging packages. A major convenience is the fact that the underlying Lisp systems for Macsyma (Maclisp on the PDP-10 and Franz on the VAX-11) provide a true (arbitrary size) "integer" data type as a primitive, out of which further exact arithmetic types such as rational numbers, may be constructed.

By contrast languages like Fortran are primarily oriented toward numeric computation, and primarily floating-point arithmetic, at that, since integers are fixed-size and prone to overflow. The structure of the language inhibits extensibility of data or control structures: If polynomials are represented in a Fortran-based system, they cannot be added with in-fix operators like plus (+); rather, a subroutine-call syntax must be used to deal with constructed objects[2]. As far as the typical Fortran compiler is concerned, the types of each object must be known at compile-time, making an arbitrary and unnecessary restriction on the use of created data types. Lisp, by contrast, can use its run-time interpretation facility (in combination with compilation when appropriate) to provide much greater flexibility.

1.2. Numeric Programming and Fortran

The greatest single advantage Fortran presents over Lisp is the large number of well-designed numerical routines that have been written in Fortran over the last twenty years. Because there are optimizing Fortran compilers in existence for most computers, it is a convenient language for the production and transportation of efficient code within the domain of numeric computation.

---

2. Actually, we overcome some of these barriers by means of the Augment preprocessor, described later.

2. Combining Algebraic and Numeric Programming Systems

Many mathematicians and programmers have expressed interest in using Algebraic Manipulation systems, typically to solve a class of problems to achieve a symbolic solution, and then after some further manipulation, create a numeric (usually Fortran-language) program which will calculate associated numeric quantities efficiently. In some cases, such Fortran programs were intended to be run on computers other than the one performing the algebra. Some programs had to interface with sophisticated graphics systems, or large interactive user-oriented packages. Applications to computer-aided design have been of this type, for example.

One approach to providing a complete set of tools to the user is to make Algebraic Manipulation systems encompass the total spectrum of computation: provide a universal numerical system too. Unfortunately, providing such a system within a Lisp system is costly in implementation, and is not always as effective as alternatives. The most successful melding of numerical computation in a Lisp system is probably Maclisp on the PDP-10, which has a large and complex compiler that attempts to overcome the lack of a compatible Fortran system by allowing numerical constructs to be compiled efficiently. The compiler is several times larger than it might otherwise be, and would have to be larger yet to entirely avoid the occasional production of code inferior to a good Fortran compiler for the PDP-10.

The view we espouse is that the Algebraic Manipulation system should allow the user to call numeric library functions without prejudice as to their original language of implementation. This means

a. The Lisp compiler can be more compact, simpler to write, debug, and transport to new hardware.

b. Advances in the Fortran compiler for the same hardware will automatically be taken into account merely by recompilation.

c. Standard libraries (e.g. (International Mathematical and Statistics Library)) need not be specially converted to run under a Lisp system.

In the remaining sections of this paper we describe an implementation of an algebraic/numeric package using the VAX-11 hardware, the UNIX operating system, and a Lisp level system which is, for the most part, identical to the Macsyma system on the PDP-10. It interacts with the Fortran 77 compiler which is standard under the host operating system.

3. Vaxtran - An Algebraic/Numeric Programming Interface

Vaxtran is a algebraic/numeric package running on the Macsyma system under the VAX/VM/UNIX operating system, which allows users to create numeric Fortran subroutines combining equations which have been created and manipulated in the Macsyma system, templates of Fortran code, and Fortran library routines. The interactive user supplies to the Vaxtran system sufficient information to write a Fortran program, and the system carries out the details. The user must provide, in a format described by examples subsequently, the following data:

a. The name of the function and the type of its return value.

b. The names and types of the function's formal parameters.

c. The code, equations, comments, etc., which form the body of the created function, including, perhaps, sections inserted as the consequence of algebraic computations.

d. The names of libraries which are to be used to resolve subroutine references.

e. (possibly) Descriptions of user-defined "augmented" data types. This is explained briefly in the section on enhanced Fortran.

With the above information supplied, the Vaxtran package will

a. create a Fortran subroutine for the given calculations and parameters, including using default declarations and various templates of its own when necessary.

b. compile the Fortran program,

c. link and dynamically load it with the necessary library functions into the current Macsyma system which the user is running, and

d. create all necessary interfaces needed to call the created Fortran function from Macsyma's top-level evaluator, including input and output when appropriate.

3.1. Common Data Types

The Vaxtran system provides communication between Lisp programs and Fortran by passing (fixed-length) integers, reals, and Macsyma arrays of these into the equivalent Fortran data types. The Vaxtran system also easily allows users to define extended types to the Macsyma and Fortran system. Once the user has supplied Vaxtran with the description of an extended data type, and the conversion routines for the data type, the system treats it like the built-in supplied data types. Currently the system has been extended to understand Kahan-Coonen-Stone IEEE proposed standard arithmetic, a version of interval arithmetic, and Richard Brent's Multiple Precision Arithmetic. Details are available in [Lanam80].

3.2. Common calling sequences/ customized Lisp front-ends

As indicated earlier, the Vaxtran system will automatically write an interface in Lisp for the newly created function so that the user will be able to access the function as though it had been written entirely within Macsyma. This interface routine correctly handles all necessary conversion of parameters and return values, so that, for example, intervals can be treated as arrays of length two in Fortran, but tagged lists in Macsyma. This interface is saved in a convenient file for future re-use along with appropriate library routines and the generated Fortran load module.

3.3. Various other implementation considerations

The tricky problem of identifying Lisp "symbols" with symbols in external program libraries (i.e. program entry points) has been solved by the use of a modified loading program which is used in Franz Lisp. The simplicity of the UNIX operating system contributed significantly toward making this possible. Details can be found in [Lanam80].

Similarly, the UNIX operating system was found to provide convenient facilities in its file system and job control, so that various pieces of text could be passed to the Fortran compiler or its pre-processor, producing object files and associated data bases.

3.4. Enhanced Fortran

The Vaxtran system aids users in interfacing to an enhanced Fortran with additional data types, by means of the Augment Precompiler [Crary74], [Crary75]. This allows users to define new types to Macsyma and Fortran and to treat these new types like standard Fortran data types. With the use of the Augment Precompiler, the user is able to write Fortran code using the new data types in an approximation of the syntax used for standard Fortran data types. An example using KCS/IEEE arithmetic for intervals, follows (for details, see [Lanam80])

4. Example 1.

The following example is a simple Newton iteration routine to find the zeros of a polynomial. We KCS/IEEE Interval arithmetic as implemented by J. Coonen and H. Rabinowitz, [Stevenson81], [Rabinowitz79], and interfaced to Fortran 77 via the Augment precompiler.

In this example, the user first sets up two files to define the function of interest. (The data could be typed directly in to Macsyma, but is then harder to explain). The file below is the main routine which will do the iteration. Into this routine Vaxtran will insert a formula for the function, the derivative of the function, and the initial guess by the user. This file, "batchitr", has the following Fortran/Macsyma fragment in it :

```
      interval g,g2
      interval i
{i=guess;}
10    continue
{g=z}
      if ( (g .lt. 1.0e-12) .and. (-1.0e-12 .lt. g)) goto 20
{g2=zprime;
i=i-g/g2;}
      goto 10
20    continue
{itr=i;}
      return
      end
```

The intention is that two functions, z and zprime will be defined in Macsyma by using the Vaxtran system.

A second file is used to create a function to calculate the function z using interval arithmetic. This file 'batchf' is contains the following text:

```
{ f = z; }
      end
```

Here is a script which uses them to create functions with Vaxtran: The text typed by the user is in bold face.

```
 Vaxima 1.00
 Thu Apr  9 17:42:37 1981

(c1) /* Load in auxiliary routines for using kcs
      * arithmetic with the vaxtran package
      */

/* note: this will cause the vaxtran package to be
  * autoloaded in if it has not been already
  * loaded into vaxima.
  */

loadfile ( "/vb/mac/dhl/kcs" );
/vb/mac/dhl/kcs being loaded.
[fasl /vb/mac/dhl/kcs.o]
[autoload /vb/mac/dhl/vaxtran]
[fasl /vb/mac/dhl/vaxtran.o]
Time= 1666 msec.
```

done (d1)

```
(c2) z:i^3+2*i^2-4;
Time= 16 msec.
```

$$i^3+2i^2-4 \qquad (d2)$$

```
(c3) zprime:diff(z,i);
Time= 16 msec.
```

$$3i^2+4i \qquad (d3)$$

```
(c4) vaxtran( itr( guess: interval ): interval, batchitr);
creating itr
*begin
      subroutine itr(guess,retval)
      interval guess
      interval retval
      interval g,g2
      interval i
      i = guess

10    continue
      g = i**3+2*i**2-4

      if ( (g .lt. 1.0e-12) .and. (-1.0e-12 .lt. g)) goto 20
      g2 = 3*i**2+4*i
      i = i-g/g2

      goto 10
20    continue
      retval = i

      return
      end
*end
[augment & f77 itr]
[ffasl itr.o]
Time= 866 msec.
```

itr created (d4)

```
(c5) vaxtran( f( i: interval ):interval, batchf);
creating f
*begin
      subroutine f(i,retval)
      interval i
      interval retval
      retval = i**3+2*i**2-4

      end
*end
[augment & f77 f]
[ffasl f.o]
/* To see what Augment did with this see appendix A */

Time= 383 msec.
```

f created (d5)

(c6) **itr (interval (3, 3));**
Time= 1233 msec.

$$interval(1.130395434767308x0, 1.130395434767334x0) \quad (d6)$$

(c7) **f (%);**
Time= 83 msec.

$$interval(2.389199948993337x{-}13, 4.600764214046649x{-}13) \quad (d7)$$

(c8) **exit();**
3:14.0 real 37.2 user 19.1 sys

5. Example 2

A simple array example: compute the inverse of a Hilbert matrix two different ways, one way using the IMSL library.

Vaxima 0.35
Mon Jan 19 22:15:03 1981

(c1) **/* set vaxtran to use imsl library**
*****
*** note: this will cause the vaxtran package to be**
*** autoloaded in if it has not been already**
*** loaded into vaxima.**
***/**

newlibrary("-limsld");

/vb/mac/dhl/vaxtran being loaded.
[fasl /vb/mac/dhl/vaxtran.o]

Time= 683 msec.

$$-limsld\ -li77\ -lf77 \quad (d1)$$

(c2) **/* lcm computes the least common multiple of**
*** a list of numbers 1..lcmtop.**
***/**

```
lcm(lcmtop) := block([lcmans],
    lcmans : 1,
    for ilcm thru lcmtop do
        (lcmans : lcmans*ilcm / gcd(lcmans, ilcm)),
    return(lcmans))$
```

Time= 16 msec.

(c3) **m[i,j]:=1/(i+j-1)$**
Time= 16 msec.

(c4) **hilbert[n]:=lcm(2*n-1)*genmatrix(m,n,n);**
Time= 16 msec.

$$hilbert_n := lcm(2n-1)genmatrix(m,n,n) \quad (d4)$$

(c5) **h3 : hilbert[3];**
Time= 850 msec.

$$\begin{bmatrix} 60 & 30 & 20 \\ 30 & 20 & 15 \\ 20 & 15 & 12 \end{bmatrix} \quad (d5)$$

(c6) **/* define a function which calls the imsl library**
*** routine linv2f**
*** - inversion of matrix - full storage mode -**
*** high accuracy solution.**
***/**

```
vaxtran(invert[n,n](matrix[n,n]:double,n:integer,
                    accur:integer):double,batchinvert);
creating invert
*begin
      subroutine invert(matrix,n,accur,retval)
      double precision matrix(n,n)
      integer n
      integer accur
      double precision retval(n,n)
      double precision wkarea(1000)
      integer ier
      integer j
      j = accur
      call linv2f(matrix,n,n,retval,accur,wkarea,ier)
      if ( ier .eq. 129 ) write ( 6, 30 )
30    format("error: matrix is algorithmically singular")
      if ( ier .eq. 131 ) write ( 6,40 )
40    format("error: matrix is too ill-conditioned, for",
     1/,"iterative improvement to be effective ")
      if ( j .ge. 0 ) write ( 6,50 ) accur
50    format(i3," digits in answer were unchanged after ",
     1"improvement")

      end
*end
[f77 invert]
[ffasl invert.o]
Time= 1433 msec.
```

invert created (d6)

(c7) **h3^^-1;**
Time= 566 msec.

$$\begin{bmatrix} \frac{3}{20} & -\frac{3}{5} & \frac{1}{2} \\ -\frac{3}{5} & \frac{16}{5} & -3 \\ \frac{1}{2} & -3 & 3 \end{bmatrix} \quad (d7)$$

(c8) **20*%;**
Time= 133 msec.

$$\begin{bmatrix} 3 & -12 & 10 \\ -12 & 64 & -60 \\ 10 & -60 & 60 \end{bmatrix} \quad (d8)$$

(c9) **invert(h3,3,0);**
15 digits in answer were unchanged after improvement
Time= 116 msec.

$$\begin{bmatrix} 0.15 & -0.6 & 0.5 \\ -0.6 & 3.2 & -3.0 \\ 0.5 & -3.0 & 3.0 \end{bmatrix} \quad (d9)$$

(c10) %*20;
Time= 250 msec.

$$\begin{bmatrix} 3.0 & -12.0 & 10.0 \\ -12.0 & 64.0 & -60.0 \\ 10.0 & -60.0 & 60.0 \end{bmatrix} \quad (d10)$$

(c11) invert(h3,3,0).h3;
15 digits in answer were unchanged after improvement
Time= 266 msec.

$$\begin{bmatrix} 1.0 & -2.220446049250313e-16 & 0.0 \\ 0.0 & 1.0 & 0.0 \\ 0.0 & 0.0 & 1.0 \end{bmatrix} \quad (d11)$$

(c12) exit();
4:33.0 real 19.0 user 14.0 sys

6. Other Work

The package described in this paper is similar to, and based upon a system called Mactran written by Michael Wirth [Wirth80]. His system runs in Macsyma on the PDP-10 at MIT, produces Fortran code, but cannot load such programs into Macsyma and thus is much more limited.

We have added the Brent Arithmetic [Brent76], [Brent79] package into the Vaxtran system, using the Augment interface written by R. Brent. [Brent80]. More details of this can be found in [Lanam80].

We are continuing work in this area with the objective being the development of specialized systems with some understanding of particular problem domains and the capability to maintain user-oriented algebraic and numeric libraries.

7. Acknowledgments

I would like to thank Richard J. Fateman for his help and comments in writing and rewriting this paper. I also would like to thank John Foderaro and Keith Sklower for their help in using Franz Lisp and Vaxima.

8. References

[Brent76] Richard Brent, "A Fortran Multiple-Precision Arithmetic Package". Computer Centre, Australian National University, Canberra, Australia, May 1976.

[Brent79] Richard P. Brent, "MP User's Guide (Third Edition)",Department of Computer Science, Australian National University, Canberra, Australia, December 1979.

[Brent80] Richard P. Brent, Judith A. Hooper, and J.M. Yohe, "An Augment Interface for Brent's Multiple Precision Arithmetic Package", ACM Transactions on Mathematical Software, Vol 6. No 2, pp. 146-149, June 1980.

[Crary74] F.D. Crary, "The Augment Precompiler: I. User's Manual" Mathematics Research Center, University of Wisconsin, Report #1469, December 1974.

[Crary75] F.D. Crary, "The Augment Precompiler: II. Technical Documentation", Mathematics Research Center, University of Wisconsin, October 1975.

[Lanam80] D.H. Lanam, "A Package for Generating and Executing Fortran Programs with Macsyma", Master's Project Report, U.C.Berkeley, December 1980.

[Mathlab77] Mathlab Group, "Macsyma Users' Manual", Laboratory for Computer Science, M.I.T., Version 9, December 1977.

[Rabinowitz79] Henry Rabinowitz, "Implementation of a More Complete Interval Arithmetic", Computer Science Division, E.E.C.S. Department, University of California, Berkeley, 1979.

[Stevenson81] David Stevenson, "A Proposed Standard for Floating-Point Arithmetic", Computer Magazine, Vol. 14, No. 3, March 1981.

[Wirth80] Michael C. Wirth, "On the Automation of Computational Physics", Phd. dissertation, University of California, Davis School of Applied Science, Lawerence Livermore Laboratories. September 1980.

Appendix A

The following text was input to vaxtran for the function *f* in the first example:

```
{ f = z; }
  end
```

With this Vaxtran produced the following Augmented Fortran code:

```
*begin
    subroutine f(i,retval)
    interval i
    interval retval
    retval = i**3+2*i**2-4

    end
*end
```

Augment when given the above program transformed it into the following standard fortran program:

```
      SUBROUTINE F(I,RETVAL)
C           ===== PROCESSED BY AUGMENT
C           --- TEMPORARY STORAGE LOCATIONS ---
C                INTERVAL
      INTEGER INTTMP(4,3)
C           --- GLOBAL VARIABLES ---
C                INTERVAL
      INTEGER I(4), RETVAL(4)
C           ===== TRANSLATED PROGRAM =====
C===== MIXED MODE OPERANDS ACCEPTED =====
C===== MIXED MODE OPERANDS ACCEPTED =====
   CALL INTXBI (I,3,INTTMP(1,1))
   CALL INTXBI (I,2,INTTMP(1,2))
   CALL INTCII (2,INTTMP(1,3))
   CALL INTMUL (INTTMP(1,3),INTTMP(1,2),INTTMP(1,2))
   CALL INTADD (INTTMP(1,1),INTTMP(1,2),INTTMP(1,2))
   CALL INTCII (4,INTTMP(1,1))
   CALL INTSUB (INTTMP(1,2),INTTMP(1,1),RETVAL)

   END
```

Computer Algebra and Numerical Integration

Richard J. Fateman[1]

Computer Science Division
and
Center for Pure and Applied Mathematics
University of California
Berkeley, CA 94720

ABSTRACT

Algebraic manipulation systems such as MACSYMA include algorithms and heuristic procedures for indefinite and definite integration, yet these system facilities are not as generally useful as might be thought. Most isolated definite integration problems are more efficiently tackled with numerical programs. Unfortunately, the answers obtained are sometimes incorrect, in spite of assurances of accuracy; furthermore, large classes of problems can sometimes be solved more rapidly by preliminary algebraic transformations.

In this paper we indicate various directions for improving the usefulness of integration programs given closed form integrands, via algebraic manipulation techniques. These include expansions in partial fractions or Taylor series, detection and removal of singularities and symmetries, and various approximation techniques for troublesome problems.

1. Introduction

For almost twenty years, now, algebraic manipulation programs have been able to compute the answers to typical problems posed in "freshman calculus" texts. In fact, the algorithms and occasional heuristic procedures surpass most freshmen in specific areas. Programs have the potential to be better than professional mathematicians at least in speed and accuracy on many (admittedly not the most esoteric) problems. A highly readable account of the techniques in these programs is given by Moses in [Moses, 71]

The kinds of tasks solved by current programs include both indefinite integration (technically speaking, anti-differentiation) and definite integration. That is, given some $f(x)$, find (some) $F(x)$ where

$$\int f(x)\,dx = F(x) + C$$

1. Work reported herein was supported in part by the U. S. Department of Energy, Contract DE-AT03-76SF00034, Project Agreement DE-AS03-79ER10358.

or given $f(x)$, a, and b, find C where

$$\int_a^b f(x)\,dx = C(a,b)$$

where $F(x)$ and $C(a,b)$ are in general, symbolic expressions. In the cases where a and b are constants (0, π, ∞, $\sqrt{3}$, etc.), then $C(a,b)$ is naturally a constant.

Much has been made of the fact that certain indefinite integrals can be proved to be inexpressible in terms of "elementary functions"(see, for example, [Davenport, 79] for a discussion of the Risch algorithm). In point of fact, the most complete implementation known to this author is that on the MACSYMA system [Mathlab, 77], and that is incomplete. Furthermore, certain classes of integrals can be expressed in closed form by rather straight-forward manipulations, but the answers are so ugly that the MACSYMA program has been constructed to (in essence) refuse to comply with the request. This leads to some confusion as to the significance of an "unintegrated" answer: is the integral inexpressible according to the Risch algorithm; is it that MACSYMA can't find the representation, or is the answer known to be so ugly that the system won't answer. A simple example MACSYMA refuses to integrate (with respect to x) is $1/(x^3+x+1)$.

In any case, computer scientists seem to be pleased with these capabilities; indeed, they seem particularly impressive to some artificial intelligence researchers who cite MACSYMA's integration program as an example of an *Expert System*.

Most applied mathematicians faced with this integration facility view it with interest only briefly, perceiving it more as a parlor trick than a useful tool: the typical applied mathematician is interested in closed form solutions (usually to definite integrations) when they are available, but may be perfectly content with the frequently much easier to compute "numerical value" of the integral.

Some applied mathematicians realize such a facility has an enormous potential. The symbolic nature of the calculation is especially significant when f contains parameters other than x, or the limits of integration are not finite constants, or (much less frequently these days) there is reason to believe that conventional numerical routines provide an insufficiently accurate answer. Then the symbolic result may provide a formula for the integral, and may save a great deal of time or worry. Retaining parameters is especially useful when the parameters are in fact variables of integration in a subsequent (i.e. iterated) integral. Because of this, integration programs are often used in the evaluation of Feynman diagrams which amount to performing iterated integration [Fox and Hearn, 74]. Integration programs are also vital tools in the pursuit of programs which solve more complex problems, such as symbolic solution of differential equations: an improved integration program has a synergistic effect on the layers of user programs relying upon it.

In this paper we address the issue of how to solve definite integration problems using a combination of tools. That is, we assume there is a requirement for calculation of a definite integral, approximately or exactly. We will generally assume that the answer does *not* exist in closed form in terms of elementary functions, or at least the program being used is unable to determine that form.

2. Numeric Integration Techniques

Some of the elaborate and popular programs devised for numerical integration (see, for example, Davis and Rabinowitz, 75], must tread a thin line between taking too much computation time and providing an answer which is more or less often outside acceptable tolerances. In fact, it is easy to show that application of any fixed integration formula (such as Simpson's Rule) or any error-estimating technique (such as the Romberg technique) or any adaptive scheme can result in arbitrarily bad results on some integrands, if the only information available to the integration program is $\{f(a_i)\}_{1\le i\le n}$ for some finite set of rational constants $\{a_i\}_{i\le i\le n}$. Knowing the nature of the canned program, one can construct a function which assumes the value 0 at every evaluation point, but is elsewhere unity. See, for example, the discussion of the integration program in the HP34C hand-held calculator in [Kahan, 80]. Less frivolous functions can be exhibited which feign convergence to the wrong value.

3. Problems with Numeric Methods

Our present concern is the solution of integration problems where the integrand is explicitly presented as an algebraic formula: this by no means eliminates the possibility of "false converge" but does provide another set of problems as indicated in the following sections.

3.1. Infinite or "Large" Intervals

Computational schemes to subdivide an interval for purposes of approximation generally fail for an infinite interval, or an integral, which, though finite, has most of its contribution in a small region. If the integral converges, (something to wonder about too, of course), the integrand must have most of its contributions at finite values: e.g. a shape like a mouse with an infinitely long and decreasingly wide tail, or alternatively the integrand must oscillate or exhibit symmetries which cancel the "infinite" contributions.

An integration of the mouse variety is the value of the well-known error function at ∞.

$$erf(\infty) = \frac{1}{\sqrt{2\pi}}\int_0^{\infty} e^{x^2}dx = 1.$$

This type of integral can be evaluated by determining where the main contributions lie, and how to bound the remaining parts.

As an example of an oscillatory integral, consider the limit as $\alpha\to\infty$ of a symmetric integral:

$$\int_{-\alpha}^{\alpha}\sin x\, dx = \int_0^{\alpha}(\sin x+\sin(-x))\, dx = 0.$$

Recognition of periodic behavior can sometimes be detected by algebraic manipulation programs; symmetries such as that indicated above are often simple matters for a simplification program to prove, although figuring out where the symmetries should be sought is a challenge.

3.2. Expressions Which are Hard to Evaluate Numerically

These are even more difficult to detect: Some expressions cannot be correctly evaluated without a limiting process. One contributing factor to this is that numerical functions generally depend on the accuracy of arithmetic and elementary functions of which they are composed. It is therefore rather important to have a workable model of the arithmetic and accuracy used in the computation. The easiest model is for the most part, to assume that functions and elementary arithmetic operations are entirely accurate, but this is patently impossible. An example from [Kahan 80] illustrates this. Consider the integrand

$$f(u) := \frac{\sqrt{-2\log(\cos(u^2))}}{u^2}$$

as u approaches 0. A calculator would almost inevitably provide the value 0 for small non-zero values of u, yet a careful computation of the Taylor series around $u = 0$ shows the value can be approximated by

$$1+\frac{u^4}{12}+\cdots.$$

Such expansions can be computed by an algebraic manipulation program such as MACSYMA by using Taylor series facilities. Using this series as a start, it becomes apparent that one gets a better answer by integrating, instead of $f(u)$,

$$g(u) := \text{if } (\cos u^2) = 1 \text{ then } 1 \text{ else } f(u)$$

Another example of a dangerous loss in numerical accuracy occurs in a subsequent section, when $x \log x$ needs to be evaluated at $x=0$.

3.3. Singularities on or near the interval of integration

Poles of $f(z)$ can often be found by conventional zero-finding, assuming the function $1/f(z)$ is analytic. It is of course possible to construct arbitrarily ill-behaved functions unless we are willing to restrict the domain of acceptable inputs; this seems to be an endless task, but partial measures are clearly useful.

As an example, it does seem feasible to use rather powerful tools to compute subdivisions of the real line so that a function can be written without absolute-value functions (this requires a program to determine the sign of an expression which can be shown, in general, to be recursively undecidable -- a useful excuse for a failing program [Richardson, 68]) but not a barrier to useful computational procedures.

4. Algebraic Assistance

Algebraic manipulation systems can provide a variety of approaches to obtaining integrals. We will outline a few which promise to be generally useful.

4.1. Qualitative Analysis

Our first approach is to examine the integrand algebraically so as to increase the likelihood of finding a numerical value which is accurate. For example, integration by parts or other methods can sometimes break the integral up into an algebraic formula plus a simpler numeric problem. In the case of integrals with singularities at the endpoints, it is sometimes possible to make algebraic substitutions to remove the undesirable behavior. In the case of "almost-singularities" similar techniques may be used. Oscillatory integrands may yield to special techniques based on averaging. It is desirable to bound the truncation error produced by replacing an infinite interval by a finite one.

Some examples

There are a multitude of related techniques which have been proposed in one form or another for easing the burden of the numerical integration process. These are used when there are singularities in or near the region of integration, reducing an infinite integral to a finite one, or to "bunch up" integrals with long tails so that less time is wasted when a uniform subdivision is used.

In spite of our hope that algorithmic approaches can be found, it appears that heuristics based on pattern matching are about the closest we can come to finding useful transformations. Some examples of such transformations are given in the anecdotal fashion of the usual texts.

Consider the integral I_1, which has as its value, Euler's constant, $\gamma = 0.5772156649...$:

$$I_1 = \gamma = \int_0^1 -\ln(-\ln x)dx$$

This integrand is singular at 0 and at 1, being $-\infty$ and ∞ respectively. The zero-crossing occurs when x is e^{-1}. Judicious substitutions indicated below provide some help.

$$I_1 = \int_0^1 -\ln(\frac{(1-x)\ln x}{x-1})dx$$

$$= \int_0^1 -\ln(1-x)dx + \int_0^1 -\ln(\frac{\ln x}{x-1})dx = 1 - \int_0^1 \ln(\frac{\ln x}{x-1})dx$$

where now the singularity at $x=1$ has disappeared and the integrand starts at ∞ but remains positive until it drops to zero at x=1.

The substitution $x=t^2$ transforms the problem to

$$1 - 2\int_0^1 \ln(\frac{2\ln t}{t^2-1})t\ dt$$

which in turn can be transformed to

$$1 - 2\int_0^1 [\ln 2 + \ln(\frac{\ln t}{t-1}) - \ln(t+1)]t\ dt$$

and in this case the first two terms of the remaining integrand can be treated exactly:

$$I_1 = \frac{3}{2} - 2\ln 2 - 2\int_0^1 t\ \ln(\frac{\ln t}{t-1})\ dt.$$

This, finally has a seductively simple profile, although evaluation at $t=0$ must be done with some foresight. (I am indebted to W. Kahan for the neat resolution of this final form.) The several substitutions seem particularly appropriate given their contexts in producing factors inside the logarithm function.

An easily applied, and somewhat general technique which is quoted in several texts (e.g. Greenspan and Benney, 73; Dahlquist and Bjorck, 74] is applicable to integrals of the form

$$\int_0^1 \frac{f(x)}{x^\alpha}dx$$

where $0<\alpha<1$ and f(x) is non-zero at 0. The example integrands given in the two cited texts are in fact $x^{-1/2}\cos x$ and $x^{-1/2}e^x$.

Taking the latter example, the "Taylor Series" (as generated by MACSYMA) is actually a Laurent series:

$$\frac{1}{\sqrt{x}} + \sqrt{x} + \frac{x^{3/2}}{2} + \cdots$$

and by truncating the series at the non-negative exponent, and integrating, we identify an appropriate substitution in a relatively automatic fashion. That is, by setting

$$t = \int \frac{1}{\sqrt{x}}dx = \sqrt{x} + C$$

and substituting $x=t^2$ we find the integral reduces to

$$2\int_0^1 e^{t^2}dt.$$

To aid in such substitutions, there is a function "changevar" in MACSYMA which accomplishes the simultaneous substitution of t for x and $2t\ dt$ for dx. [In fact it took some prompting to prevent it from producing

$$\int_0^1 \frac{te^{t^2}}{|t|}dt$$

which required further substitution for simplification namely: $t/|t| = 1$ when $0 \le t \le 1$.]

In general, if f(x) can be expanded in series about an endpoint, then the order of the pole can be determined, the finiteness of the integral can be judged, and a substitution found.

Similar goal-oriented substitutions for other patterns can be devised, although a simple structure for sure-fire winning transformations is elusive.

4.2. Taylor Series Expansions

We have already indicated a potential role for Taylor series in deriving endpoint conditions for integrals; it is of course trivial to integrate a Taylor series, and if there is sufficient reason to believe the series is a good approximation for the integrand over the range of integration, this route is easily followed.

4.3. Rational Function Integration

It would seem that rational function integration would be simple to provide, since there are well-known techniques taught in freshman calculus for solving "any" problem of this class. Unfortunately, freshmen are easily and often deceived.

The standard technique expands the function given as a sum of a polynomial and a "partial fraction expansion", the integral of which can be written down as a polynomial plus a sum of (perhaps complex) logarithms.

The deception practiced on freshmen is that all the textbook examples have denominators which factor neatly into quadratic and linear factors. This makes it possible to perform the necessary partial fraction expansion. MACSYMA is quite good at factoring polynomials as computer programs go, but not all polynomials factor so easily. Even those polynomials which do have finitely representable algebraic factors may appear so complicated that few people would choose to see them. Hence MACSYMA's refusal tc integrate $1/(x^3+x+1)$. There are also those polynomials whose factors cannot be represented by radicals. These are necessarily of degree 5 or more, although not all polynomials of degree five or more are irreducible. Using various algebraic number techniques it is possible but expensive to write such integrals down.

An alternative approach is to factor the denominator over the "floating-point complex numbers," which has been well-studied from a numerical standpoint. The partial fraction decomposition and integration steps then result in a formula which can be evaluated at upper and lower bound to find a specific definite integral. The formula can also be examined for its behavior at various limiting conditions (near a pole or infinity).

The question arises as to how this compares to straightforward quadrature of the original function: Why should we trade a zero-finding program for a quadrature program? In fact, we expect that one integration of a rational function on a single interval which is situated well away from any zeros of the denominator would be done better by any of a number of well-known quadrature programs (e.g. see [Davis and Rabinowitz, 75]. Near a zero of the denominator, or in particular when zeroes of the numerator and denominator are close, this algebraic/numerical approach should pay off with higher accuracy.

An example of a simple program for partial fraction integration using floating point numbers is given in the Appendix.

4.4. Reduction of Infinite Ranges

As one step in the evaluation of a definite integral, one sometimes has to reduce an infinite interval to a finite one. This may allow the use of a conventional quadrature formula, if one is fortunate enough as to have introduced only mild singularities in the integrand. Thus for real $a, b > 0$

$$\int_0^\infty \frac{t^{a-1}}{(1+t)^{a+b}}dt$$

can be reduced (via $s' = t/(t+1)$) to

$$\int_0^1 s^{a-1}(1-s)^{b-1}ds.$$

Either of these integrals defines the Beta function:

$$B(a,b) = \frac{\Gamma(a)\Gamma(b)}{\Gamma(a+b)}.$$

Assuming we were unable to compute these integrals in closed form, we would probably find the second more tractable in spite of some trouble at $t=1$ for some values of a and b.

Sometimes these transformations do not come readily to hand. For example, using the same transformation on the following integral produces an uncomfortable integrand:

$$\int_2^\infty (\ln x)^{-\ln x}dx \ = \ \int_{2/3}^1 (\frac{x}{1-x})^{-\ln(\ln x - \ln(1-x))}dx.$$

A better transformation (suggested by W. Kahan) is $x = e^{1/u}$ which produces (on a hand-held calculator, using 255 samples!)

$$\int_0^{1/\ln 2} e^{u^{-1}-2\ln u + u^{-1}\ln u}du \ = \ 5.037176561 \pm 7\times 10^{-9}.$$

Nevertheless, there will remain cases in which we would like to break up an integral into two (or more) parts.

$$I_0 = \int_a^\infty f(x)dx = \int_a^b f(x)dx + \int_b^\infty f(x)dx = I_1 + I_2.$$

The integral I_1 is to be evaluated numerically. The integral I_2 represents a truncation error in taking I_1 to be the value of I_0. We wish to bound this error in magnitude by choosing a dominating function $g(x)$ such that $g(x) > |f(x)|$ on (b,∞) and that $\int_b^\infty g(x)dx$ is easily evaluated (for example, by closed form integration).

Example: suppose we wish to evaluate

$$I_0 = \int_0^\infty e^{-x^2}dx.$$

Choose $b=3$. We can dominate

$$I_2 = \int_3^\infty e^{-x^2}dx \ \text{ by } \ \int_3^\infty e^{-3x}dx = \frac{e^{-9}}{5} = 0.00002...$$

The contribution to truncation error by choosing the cutoff 3, is bounded in magnitude, and the difference between I_0 and I_1 calculated accurately, should be no more than that given by I_2.

In fact if we have been asked to find a finite integral such as

$$\int_0^5 e^{-x^2}dx$$

the same technique could be used: the contribution between 5 and ∞ is so small that for purposes of this integration, $5 = \infty$.

Details on the construction of integrable dominating functions are deferred to a full version of this paper.

4.5. Other techniques

Space considerations dictate omission of discussion of various ad hoc tricks involved with integration by parts, and asymptotic expansions.

5. Error Bounding Techniques

One justification for dealing with more elaborate systems including algebraic manipulation is the possibility of obtaining better accuracy estimates. Two such techniques are indicated briefly below.

5.1. Interval Arithmetic

An extreme approach to guaranteeing bounds on truncation error is the use of interval arithmetic. For example, one can compute the Riemann sum corresponding to an integral on a finite domain, but using interval arithmetic. In interval arithmetic, each real number is represented by a pair of representable numbers which bound from above and below, the real number. Following, for example, [Moore, 1966], what one can do is compute $f(X)$ for X an interval, $[a_i, a_{i+1}]$, yielding an interval $[\min(f(X)), \max(f(X))]$, and again by using interval arithmetic, compute the Riemann sum as an upper and lower bound on the integral:

$$\int_{a_0}^{a_n} f(x)dx \ \text{ is contained in } \sum_{i=0}^{n-1} f([a_i, a_{i+1}])(a_{i+1}-a_i)$$

These error bounds in terms of interval arithmetic are generally quite pessimistic. One could use other quadrature formulas with similar results. Elaborations of interval arithmetic to allow for parameters have been constructed, but produce extremely complex arithmetical formulas.

5.2. Taylor Series

If we are concerned with a function which is analytic near the interval of integration, and preferably analytic in some larger region, one approach is to use Taylor series to approximate f, integrate the Taylor series term by term, and provide a simple formula (a polynomial!) for the result. The round-off error involved in the evaluation of a polynomial is simple to calculate, (we have programmed such a function in MACSYMA), and [Fateman, 77] or [Adams, 67] explains the technique).

The truncation error consists of the remainder term integrated over the interval. Since the remainder term is not (by hypothesis) easy to compute, we seek an easily computed, and reasonably tight, rigorous bound on this. Algebraic techniques provide a method for computing a high derivative, although for a possibly complicated function this can be costly; then finding its maximum on the interval of integration can be difficult. The use of interval analysis, initially appealing is actually a rather poor idea here: one can usually obtain some number, but it is likely to be grossly pessimistic. One may be better off estimating the integral by drawing a rectangle about the maximum excursions of the integrand.

The same approach of evaluating some other quadrature formula and its associated error formula is clearly another possibility, if the details of rounding error and trucation can be dealt with.

6. Summary

We have described in broad outline, but with a few suggestive details, how one might use algebraic manipulation systems in conjunction with the arsenal of numerical quadrature programs, to provide more accurate and complete answers to definite integration problems.

7. References

[Adams, 67] D. A. Adams, "A Stopping Criterion for Polynomial Root Finding," *Comm. ACM 10:10*, Oct. 1967, pp. 15-25.

[Dahlquist and Bjorck, 74] G. Dahlquist and A. Bjorck, *Numerical Methods*, (trans: Ned Anderson), Prentice-Hall, Englewood Cliffs, N.J.

[Davenport, 79] J. Davenport, "Integration of Algebraic Functions," *Symbolic and Algebraic Computation*, Lecture Notes in Computer Science 72, Springer-Verlag, Berlin, pp. 415-425.

[Davis and Rabinowitz, 75] P.J. Davis and P. Rabinowitz, *Methods of Numerical Integration*, Academic Press, New York.

[Fateman, 77] R. J. Fateman "An Improved Algorithm for the Isolation of Polynomial Real Zeros," *Proc. 1977 MACSYMA Users' Conf.* NASA CP-2012

[Fox and Hearn, 74] J. A. Fox and A. C. Hearn, "Analytic Computation of Some Integrals in Fourth Order Quantum Electrodynamics," *J. Comp. Physics*, 14, pp. 301-317

[Greenspan and Benney, 73] H. P. Greenspan and D. J. Benney, *Calculus: an Introduction to Applied Mathematics*, McGraw-Hill, New York.

[Kahan, 80] William M. Kahan, "Handheld Calculator Evaluates Integrals," *Hewlett-Packard Journal*, August, 1980 (23-32).

[Mathlab, 77] The Mathlab Group, *MACSYMA Reference Manual* , version 9, Laboratory for Computer Science, MIT.

[Moore, 66] R. E. Moore, *Interval Arithmetic*, Prentice-Hall, N.Y. 1966.

[Moses, 71] Joel Moses, "Symbolic integration, the stormy decade," *Comm. ACM* 14:8 (August, 1971) (548-560) [Richardson, 68] Daniel Richardson,"Some unsolvable problems involving elemntary functions of a real variable," *J. Symbolic Logic* 3 (1968) (511-520).

Appendix: A Macsyma Floating Point Partial Fraction Integration Program

```
/*Floating point rational integration program.
 P is arbitrary rational function in x */

fpint(p,x):=
block ([num,den,div,res,keepfloat,lroots,
     distinctroots,qq,i,c],
     local(r,q),
/* simplify the input */
     p:rat(p), num:ratnumer(p), den:ratdenom(p),
/* separate polynomial and rational parts */
     div:divide(num,den),
/* integrate polynomial if any, and initialize result */
     res: integrate(ratdisrep(div[1]),x),
/* set num to numerator/leading coefficient of denom */
     num: ratdisrep(div[2]/ratcoef(den,x^hipow(den,x))),
     keepfloat:true,
     r[i]:=0,
/* call the IMSL polynomial zero-finder */
     lroots:zrpoly(den,x),
/* The number of roots at c, a point in the complex plane
  is stored in r[c] */
     for c in lroots do  r[c]:r[c]+1,
/* distinctroots is set to a list of the roots [c1,c2, ... ] */
     distinctroots: map(first,rest(arrayinfo(r),2)),
/* qq is a reconstruction of the denominator, factored */
     qq:1,
     for c in distinctroots do qq:qq*(x-c)^r[c],
/* qq[c] is qq with the r[c] roots at c taken out */
     for c in distinctroots do q[c]:qq/(x-c)^r[c],
/* The formula below gives the (constant) residue
needed in the numerators in the partial fraction expansion.
Integration of expressions of the form K/(x-c)^i is easy */
     for c in distinctroots do for i:1 thru r[c] do
     res: res+1/(r[c]-i)!
        *ratsimp(subst(c,x,diff(num/q[c],x,r[c]-i)))
           *(if i=1 then log(x-c) else (x-c)^(1-i)/(1-i)),
     return(res));
```

```
(c2) prob:1/(x^3+x+1);
Time= 16 msec.
```

$$\frac{1}{x^3+x+1} \tag{d2}$$

```
(c3) fpint(prob,x) ;
Time= 2666 msec.
```

$$-\frac{0.430462 i \log(x+1.16154 i-0.341164)}{1.16154 i-1.023491} \tag{d3}$$

$$-\frac{0.430462 i \log(x-1.16154 i-0.341164)}{1.16154 i+1.02349}$$

$$+\,0.41723 \log(x+0.682328)$$

Tracing Occurrences of Patterns in Symbolic Computations

F.Gardin & J.A. Campbell

Department of Computer Science, University of Exeter,
Exeter EX4 4QL, England

ABSTRACT: A report is made on the present state of development of a project to construct a tracing aid for users of symbolic computing systems that are written in LISP (or, in principle, any similar high-level language). The traces in question are intended to provide information which is primarily in terms that are natural for a user, e.g. on patterns of actions performed on his data, or patterns occurring in the data themselves during the operation of his program. Patterns are described in a syntax which is inspired by SNOBOL.

1. INTRODUCTION

Most tracing facilities in symbolic computing systems are oriented towards the base which supports the computations (e.g. LISP, and/or provision of lists showing sequences of function-calls), rather than the structure intrinsic to any one particular user's computation. Thus, if there is a system error, the designer of the system can expect to learn the cause of the problem from a backtrace of some kind, and if the user makes a mistake (e.g. improper use of types), one usually hopes that some kind of error-message facility, or semi-amateur interpretation of a backtrace, will put him right. However, there is relatively little diagnostic support for certain classes of error-free computation in which a user, amateur or specialist, may need to learn about what is going on beneath the surface. In this paper, we report on the current state of work on the design and construction of a package to provide some of the relevant information.

There seem to be two main contexts in which beneath-the surface information is useful. The first is that of the search for the optimal way to program the computation of a quantity whose mathematical properties and defining relations are known, but which may be computed by a variety of paths, some very inefficient. (As a simple example, the Y_{2n} functions (1) have no denominators apart from factors $1/2^{2n-1}$, yet a redefinition of their computation in terms of $X_{2n} = 2^{2n-1}Y_{2n}$ can decrease the time for the computation by about 30% in a standard system (2). The second is one in which some or all of a large computed result may be available in a form which allows the exercise of no real intuition about its meaning, so that it is desirable to re-state and shorten the result by identification and naming of sub-expressions. (Y. Sundblad has given an example of a symbolic computation in which the quantity computed was seen eventually to be $(r_{31}r_{42}+r_{32}r_{41})/(r_{31}r_{42}-r_{32}r_{41})$, where $r_{ij}^2 = \frac{1}{2}(a_i-a_j)$, but only after considerable manipulation of an expression which involved the a_i symbols and which, when computed by naive programming, occupied about 30 pages of output).

In both contexts, the information that a user needs can be described as information about patterns - patterns of bindings and calls to functions, with results, in the first case, and patterns of occurrence and recurrence of sub-expressions inside larger expressions, in the second. Therefore we have decided to treat the appropriate trace computations as manipulation of patterns. Fortunately there is already one familiar syntax and informal semantics in existence for a widely-used pattern-matching langiage: SNOBOL (3). We have introduced into our syntax conditional syntax, namely

IF x MATCHES y THEN .... ,

with MATCHES being an infix operator. The match is undefined if neither x nor y is of PATTERN type; otherwise, the standard interpretation is that an instance of y is sought in x, with any bindings mentioned in y being available for later use if the matching operation succeeds.

The SNOBOL examples tabulated above refer to strings of characters, which are helpful in the unusual cases where (as in our practice with the first example of Section 4) a user may wish to manipulate pieces of S-expressions as strings. More commonly, however, the user deals with 1D mathematical expressions, for which it may indeed be most convenient to make reference to sub-expressions as strings including infix mathematical operators, and function and array names. It is simple to interpret the mathematical meanings of such quantities in our scheme, and to translate them into the corresponding S-expression forms used by particular symbolic computing systems.

As an alternative to this approach, we recognize that many examinations of patterns within expressions require isolation of terms, or pieces of terms, which cannot be found, except in lucky cases, by combinations of operations like substitution and differentiation which exist already in (e.g.) REDUCE, and which are awkward to write in the string notation. Therefore we also admit operators COEFF, NUMERATOR, DENOMINATOR, TRIGPART, POLYPART etc. (the list is not completed at the time of writing) which select the appropriate parts (as a product of factors, if this is the correct form) of one term of an expression, INDEPENDENT(x,y), which gives ("matches") the parts of the term x that are independent of y, and FIRST and REST, which separate the first term from the rest of an expression. Of course, some systems to which the package may be applied eventually will have some or all of these operations, perhaps under different names, directly available to the user. As users, we would like to encourage such a trend! Implementation of the operators here, for any one system, demands the same knowledge of the internal representation of mathematical expressions for that system which is used for the translations of the strings mentioned above, but in principle this information is always easily accessible (though unfortunately not always in the system's manual for users).

In addition to the ability to describe patterns of data, our notation permits reference to system functions or groups of functions in connection with requests for traces. These requests, to search for structure in the actual operation of a system in carrying out a computation, are given in terms of an object which we call a "field", which is the concatenation of names of functions or procedures, variables (included in the syntax although we are not yet happy with our ways of reaching local system-variables inside procedures of a system), operators, and strings, in the order in which we expect them to occur during execution. A field is delimited by square brackets; examples occur below. Fields are also patterns.

3. A GRAPHICAL FEATURE

One of the original motivations for the present project comes from the work of Fitch and Garnett (5) on plotting occupied storage (in CAMAL, which administered its storage by a reference-count method) against time. They presented an example of the use of the plot to deduce from closely-spaced repetitions of one distinctive type of visible pattern that a particular computation was being repeated unnecessarily - perhaps in a circumstance like a := f(x), b := f(x)+c, where a greatly improved alternative would have been a := f(x), b := a+c. In practice, one would not expect the optimization to be so obvious, because the form of the appropriate pattern need not be at all obvious a priori. Therefore it is advisable to give a user the option of seeing some picture of his computation, to allow him to focus on subjectively interesting shapes in the picture, and to act on the information even without any detailed knowledge of the internal behaviour of the system that is carrying out the computation.

There is a very simple but very useful graphical treatment of this type which works even when reference-count measurements of storage are not possible. This is the automatic assignment of a unique serial number to each system function that is called, followed by the plotting of a histogram in which the function-names appear on the horizontal axis and the serial numbers, suitably scaled, on the vertical basic elements which are in close correspondence with the basic pattern elements in SNOBOL. Apart from the establishment of patterns, it is possible for a user to express commands for their control and manipulation in a language of essentially the form of PASCAL or ALGOL 60. Because of the wide diffusion of REDUCE for symbolic computing, we have adopted the ALGOL-like notation of user-defined REDUCE procedures for this purpose.

Although we argue primarily from the standpoint of convenience to the user, above, it is also likely that a specialist concerned to improve the use of a particular symbolic computing system will be able to investigate its internal behaviour by formulating requests for tracing patterns.

At present, we are using LISP as a base for our work, because we assume that the widest applicability of a package for pattern-tracing in symbolic computing will be in connection with systems written in LISP. An interpreter for the high-level ALGOL-like language, in LISP, has the job of modifying the code in LISP S-expression form for a symbolic computing system to produce the desired additional pattern-handling behaviour. The test-bed for its performance at present is a collection of LISP functions for the standard operations (simplification, differentiation etc.) of symbolic computing, with data in ordinary prefix notation.

Below, we consider first the syntax and the possibilities for the description of small patterns, and then some practical examples.

2. SYNTACTIC DETAILS

Firstly, the standard syntax for REDUCE procedures or instructions is assumed. The types of variables named in ref. 4 are INTEGER, REAL and SCALAR, with the latter (essentially a general symbolic mathematical expression) being the default type. In practice, since values of all of these types are representable as LISP S-expressions, programmers intermix those types and can dispense with their type-declarations. Because most descriptions of patterns do not translate directly into S-expressions, we introduce the new type PATTERN, which is distinguished by variable-names beginning with $. Values of patterns are attached to pattern variables through the normal binding operation :=, after which they may be referenced by name in REDUCE-like procedures.

The simplest pattern in SNOBOL is a string. Patterns at the next level of complexity are built up from strings by concatenation (x y to indicate x followed by y; we emphasise that each should match something in turn during a pattern-match by writing the concatenation as x AND y), and any position in a pattern may be occupied by two or more alternatives (e.g. (x | y) in SNOBOL, x OR y in our scheme).

The power of a pattern-matching language comes from its admission of generalized patterns in place of any specific x. In SNOBOL, for example, LEN(n) is a general instance of any pattern formed from n successive characters in a string. Generalized patterns can be defined either through small programs which test for specific strings which fit the pattern or, equivalently, through statement of the test(s) which a string must pass in order to be a specific instance of the pattern. The following, which are some examples of SNOBOL patterns together with our equivalents, demonstrate this idea and introduce our notation.

```
        LEN(5)     =  5 (char)
  SPAN('ABCD')     =  $OURSPAN := (A OR B OR C OR D)
                      (num) $OURSPAN
           ARB     =  £
   ANY('ABCD')     =  $OURSPAN
    BREAK('A')     =  (num) ((char) AND (NOT(A)))
NOTANY('ABCD')     =  (char) AND (NOT(A OR B OR C OR D))
            @X     =  @(num)
      DUPL(X,N)    =  N $X
       (X . K)     =  ($X k)
       (X $ K)     =  ($X k)
```

We allow certain single symbols, like £ for arbitrary expressions, to stand for commonly-needed patterns which require no further annotation, while bracketed lower-case names are used for other primitive patterns. "Other" covers the use of binding: if a pattern made up by concatenation of smaller patterns is matched successfully against some structure, it is often desirable to know which parts of the structure have matched various components of the pattern. When a component is primitive, e.g. (num), we use (num n) if the part of the structure matching (num) is to be bound to a local variable n.

Pattern-matching is the fundamental "semantic" operation, and actions follow according to whether or not a match succeeds. We incorporate both the match and the control of actions by a slight extension of

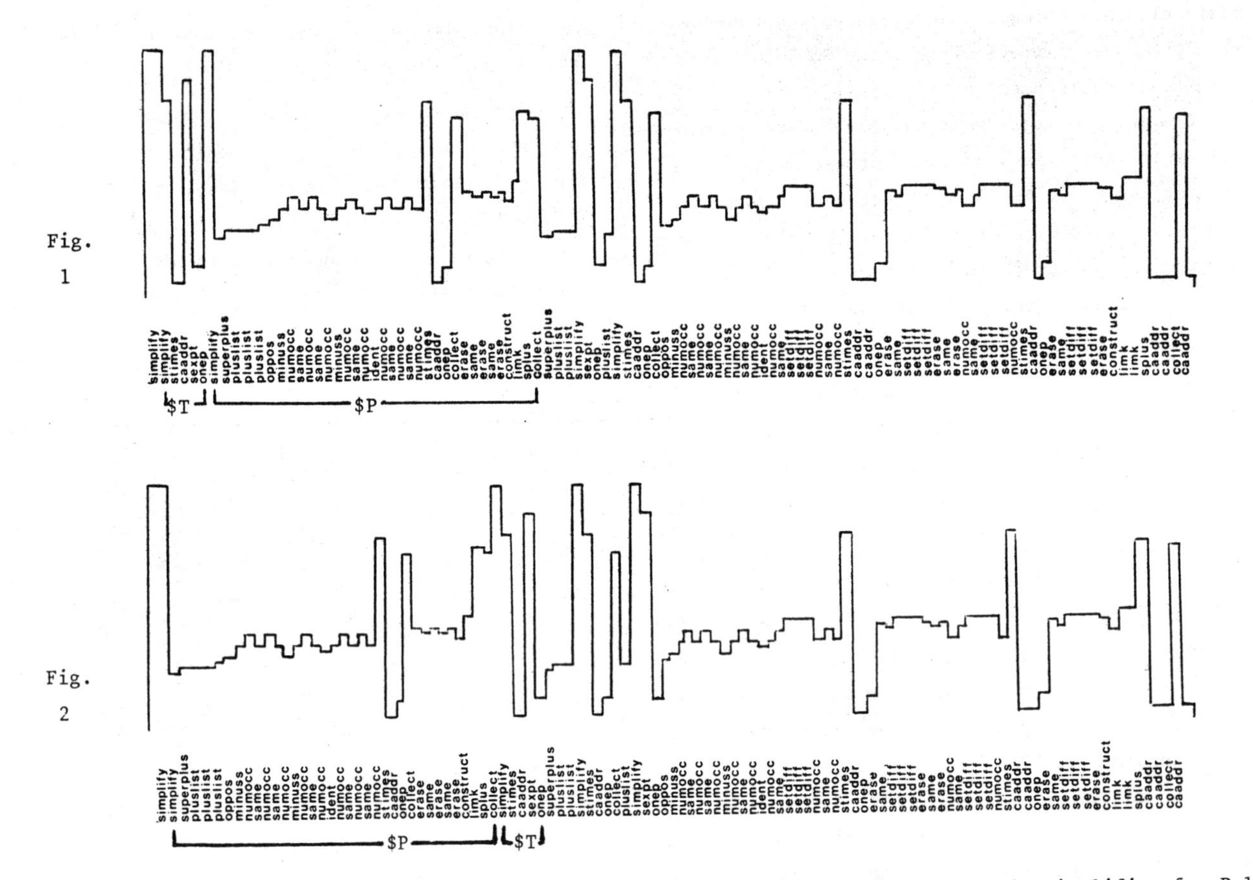

Fig. 1

Fig. 2

axis, as in Figs. 1 and 2. Although a user may know nothing of the details of these functions, he can detect repetitions of blocks of calls by their distinctive repeated shapes, and ask further "tracing" questions about them by giving names to such blocks. For example, the instruction

GRAPH([(proc e) , (proc f)])

produces a histogram for the entire computation, in which e and f are bound to the names of the first and last system functions called. (, does for arbitrary runs of procedures what £ does for arbitrary patterns among strings). Interesting blocks of calls, such as those marked $T and $P in Figs. 1 & 2, may be assigned PATTERN names for later reference in other requests for traces. A non-specialist user only has to infer, bound by name, and then describe his trace requests in terms of blocks of calls of whose detailed behaviour he does not need to be aware. A situation of this type is described in the next Section.

4. SOME SIMPLE EXAMPLES

The earliest useful exercise which we have treated was the identification of an inadvertent typing error in a standard LISP algebraic simplifier for Polish-prefix expressions. It produced the correct result for the input '(PLUS (TIMES A A)(PLUS A A)), but the result '(TIMES 2 A) for '(PLUS (PLUS A A) (TIMES A A)). We first monitored the operation of the simplifier on '(TIMES A A), requesting that the sequence of names of functions thereby called should be bound to $T, and then did the same for '(PLUS A A), binding the resulting sequence (field) to $P. To test that the simplifier had no fault causing asymmetric treatment of the two arguments of PLUS, we re-ran the simplification of '(PLUS (TIMES A A) (PLUS A A)), monitoring for a successful match to the pattern of calls defined in our syntax by [, $P , $T ,], and the alternative form in searching for a match to [, $T , $P ,].

Both matches succeeded, indicating absence of any problems of asymmetry. We then searched for the most general pattern of calls occurring in a simplification of expressions of the form (PLUS q q), where q = (PLUS _ _), by imposing the field-pattern [, $A:=[((num)(proc))] , $A , (NOT($A)) ,], where ((num)(proc)) here matches any sequence of function-calls relevant to simplifications of expressions in PLUS, and where just 3 occurrences of the sequence are specified. The most general sequence here, SUPERPLUS-PLUSLIST-PLUSLIST, was then found to occur in both correct (see Figs.) and incorrect simplifications of expressions involving only one PLUS monitored with the field-pattern [, $A , NOT($A) ,]. After exhaustive searches for evidence of anomalous sequences of calls distinguishing the erroneous example from correct computations, with the same syntactic descriptions of patterns, it was possible to ask for a detailed trace of the behaviour of the 5 function-calls before and after the point at which the anomaly (named $U after a successful local search for a match in the previous run) arose. (The relevant command is TRACE(-5,5)([, $U])). We should stress that, up to this point in the tracing, a user would have needed no special knowledge of the overall structure of the simplifier, but only some general intuition about what to expect in computations, e.g. that a piece of data containing three PLUS operators should invoke the same PLUS-simplifier three times. With the indicated output of the final pattern-based trace in hand, it was possible to inspect the definitions of only a few short LISP functions against a known correct source, and to discover that a 1 had been typed erroneously as -1.

The example above is designed to illustrate a little of the syntax of pattern-description, and to demonstrate an incidental debugging use of the pattern-tracing facility which removes the need for a user inexperienced in LISP to become familiar with LISP tracing facilities in order to detect mistakes in a system of programs. However, the main intended use of pattern-tracing is to examine behaviour of an error-free program, and to break computed symbolic mathematical expressions into parts, by essentially non-analytic operations, to allow significant regularities to be detected. This facility makes explicit use of variables to name instances of patterns that arise during computations. The patterns for 2 recent examples are as follows:

1) In an expression which is a sum of terms whose common form is $n\Pi_i a_i^{p_i}$ where n, i and p_i are integers and the a_i are distinct symbols, some terms which are bilinear in the a_i have the same coefficient n (e.g. $8448a_1a_6 + 8448a_2a_5 + 8448a_3a_4$). The general pattern used in monitoring the overall expression is [£ $A £ ((num)[£ $A]) £] , with

$L ;= [(name)(num)] and $A := [(num 1)(2['*' $L])].

Here, (num 1) acts as a global variable whose value (the value of the coefficient) is available outside the pattern-match if needed; the final part of the definition of $A represents 2 occurrences of the sequence matching a * followed by $L.

2) In the same problem (6) (in which, incidentally, the behaviour monitored successfully in example 1 led to an attempt (also successful) to find a proof by hand that that behaviour would always occur), it is possible to find patterns of the form $n_1 a_i \prod_{j\neq i} a_j^{p_j} + n_2 a_i^{q_i} \prod_{j\neq i} a_j^{q_j}$, where n_1/n_2 is a "small" positive integer. Assuming that ↑ denotes exponentiation, the relevant pattern is

([£ $B £ $C £] OR [£ $C £ $B £])

with

```
$B := [((?k OR (num k)) '*' $F (NOT('↑'))]
$F := [A (num 4)]
$C := [((num 3) AND (((?k OR (num k))/(num 3)) =
                    $SMALL)) '*' $F '↑' ((num 2) > 1)]
$SMALL := 1 OR 2 OR 3 OR 4 OR 5 OR 6 OR 7 OR 8
                                            OR 9    ,
```

where (num k) binds an integer value to the local variable k, and ?k means "if k is already defined in this local context, then the value of k, else fail".

If this match succeeds, e.g. in detecting $27432a_2^2a_4 + 54864a_2a_3^2$, then the information (num 2) = 2 for the power of a_2, (num 3) = 27432 and (num 4) = 2 for the distinguished subscript is all global and available for later use

5. DISCUSSION

To date, we have been able to express a variety of patterns of function-calls mixed (by means of AND and OR connectives) with references to patterns (written as strings of the characters which the user would see in 1D input and output) in the syntax, whose generality is roughly the generality of SNOBOL. Therefore it can be expected that a user may request monitoring for quite general patterns appearing either in the behaviour of a symbolic computing system during execution of his computation or, probably more usefully, in the evolution of his data in that computation. Our pattern scheme is still under development, and examples of requests for complex traces of these types, for particular computations, will be welcomed.

The effect of the processing of a pattern here is essentially to insert appropriate modifications in the interpreted LISP code of a symbolic computing system. The price to be paid for pattern-tracing is thus a loss of efficiency by comparison with the normal use of any system which is presented in compiled form. However, if a certain function needs modification, it seems quite practicable to override its compiled form by input of a copy of its high-level definition (e.g. from secondary storage), modification of the latter by the tracing package, and subsequent compilation of the modified definition.

We are not yet sure of the practical limitations of an implementation of automatic searches for "sufficiently interesting" patterns of data where multiple alternatives may be equally interesting in principle, at least until the user has had the opportunity to give his verdict (e.g., say, searches for possible factors of the last symbolic expression given in Section 4, in which it may be desired to find resolutions of each numerical coefficient into a product of two integers). Our experience with PROLOG (7), which supports and even unduly encourages this kind of computational behaviour, suggests that the only safe automatic searches may need such tight limitations that they become trivial, and therefore of trivial value to the user. We intend, nevertheless, to try to escape from that impasse by implementing more selective forms of automatic backtracking (8) than have been used in general for PROLOG and similar systems for work in artificial intelligence.

ACKNOWLEDGEMENT

This project has been supported by the Science Research Council, U.K., under grant GR/B/0151.7.

REFERENCES

1: N. Fröman, Arkiv för Fysik 32, 541 (1966); J.A. Campbell, J. Comp. Phys. 10, 308 (1972)

2: Y. Sundblad, SIGSAM Bull. A.C.M., nr. 24, 18 (1972)

3: R.E. Griswold, J.F. Poage and I.P. Polonsky, "The SNOBOL4 Programming Language" (Prentice-Hall Inc., Englewood Cliffs, New Jersey, 1971)

4: A.C. Hearn, "REDUCE 2 User's Manual", UCP-19, Computational Physics Group, University of Utah, Salt Lake City, Utah; March 1973

5: J.P. Fitch and D.J. Garnett, Measurements on the Cambridge Algebra System, in "Proceedings of the A.C.M. International Symposium, Venice", pp. 139-147 (ACM Headquarters, New York, 1972)

6: P.O. Fröman and E. Walles, "Systematization of the Evaluation of Integrals by the Method of Steepest Descents", report nr. 31, Institute of Theoretical Physics, University of Uppsala, Sweden, August 1980

7: D.H.D. Warren, L.M. Pereira and F.C.N. Pereira, SIGART Newsletter of the A.C.M., nr. 64, pp. 109-115 (1977); H. Coelho, J.C. Cotta and L.M. Pereira, "How to solve it with PROLOG" (Laboratório Nacional de Engenharia Civil, Lisbon, 1979)

8: L.M. Pereira and A. Porto, in "5th. Conference on Automated Deduction", eds. W. Bibel and R. Kowarski, pp. 306-317 (Springer-Verlag Notes on Computer Science, vol. 87, 1980)

The Automatic Derivation of Periodic Solutions to a Class of Weakly Nonlinear Differential Equations

JOHN FITCH
SCHOOL OF MATHEMATICS, UNIVERSITY OF BATH, ENGLAND
ARTHUR NORMAN AND MARY ANN MOORE
UNIVERSITY OF CAMBRIDGE COMPUTER LABORATORY, CAMBRIDGE, ENGLAND

Summary

There is a variety of standard mathematical techniques for finding Poisson series expansions of periodic solutions to perturbed variants of the differential equation y" + y = 0. The calculations involved in applying these techniques are often substantial, and several algebra systems were initially designed specifically for use on such problems. The mathematical theory of oscillations is, however, sufficiently rich that to date, even when special purpose algebra systems are used, each new problem requires fresh analysis and the design of a new (user level) program. In this paper we describe a hierarchy of packages we are developing that analyse members of a class of perturbation problems, apply transformations to avoid difficulties detected and deal with the technical problems of communicating the knowledge so gained to an algebra system.

Introduction

This paper is concerned with the application of symbolic algebra to the solution of nonlinear differential equations. The problems we consider here all concern mildly nonlinear oscillations, either representing the response of some system to a periodic forcing function, or modelling a self-sustaining vibration. The nonlinearities are mild in the sense that the differential equations must contain an explicit small parameter, and we seek solutions expressed as series in this parameter. In the limiting case where the small parameter vanishes we expect the equations to reduce to the form y" + y = f(t) where f is some (possible zero) periodic driving term.

Equations of this form are well known, and there is a substantial mathematical literature on techniques for solving them. From the point of view of computer algebra the major application area where problems of this form have arisen is Celestial Mechanics. As is clearly shown by Delaunay's mammoth work (1), when high order expansions have to be produced the formal algebra involved can get very hard indeed. A number of algebra systems have been designed specifically to do this work: typical applications of them are described in (2), (3) and (4). Although these systems are somewhat specialised, in that they work conveniently and efficiently with Poisson series, they still have to be given explicit instructions showing how each new problem should be solved.

It would clearly be convenient if a standard set of subroutines, capable of solving all oscillation problems, could be implemented as part of one of these algebra systems. As this is not easy, we propose instead to cope with the mathematical quirks and computational curiosities by the use of a preprocessor that accepts the algebraic form of one of our problems, and synthesises an algorithm for solving it. This package can call on a library of standard techniques (such as the method of strained co-ordinates), and can perform small amounts of symbolic algebra in analysing and preconditioning the problems that are presented to it.

This paper describes our preprocessor and traces its early development, showing how a desire to combine generality, flexibility and efficiency has led to a gradual growth in the sophistication of our package. Our discussion here concerns the very simplest types of nonlinear oscillations, where the vibrations studied are strictly periodic. This case covers both autonomous systems and forced ones, but in the latter case requires the response of the system to lock onto the forcing term.

The Basic Problem

The sort of series expansion methods with which we are concerned can be applied to any weakly nonlinear equation of the form

$$Ly = \varepsilon g(y,t) + f(t)$$

where L is a linear operator, and f and g are arbitrary (smooth) functions. The parameter ε is assumed to be small and we will want to obtain expansions for y in terms of it. We will be concerned solely with the particular case where y is a simple variable, and where the equation is of the form

$$y'' + y = \varepsilon g(y,t) + f(t) \qquad (*)$$

where y'' represents d^2y/dt^2. Although apparently simple, this problem includes versions of the Duffing and Mathieu equations (see 5) as well as forms of the Liénard equation (6).

Our concern is not with the theory of existence and stability, but with the practical problems of deriving explicit solutions, expressed as series in ε. These series give global solutions to the equation, and evaluating them numerically may be an attractive alternative to the use of step-by-step numeric integration, or hybrid methods such as pas de gèant (7).

The solution to (*) is going to be expressed as a series in the form

$$y = y_0 + \varepsilon y_1 + \varepsilon^2 y_2 + \ldots$$

where each y_i is a (finite) sum of terms:

$$y_i = \sum_j (a_{ij}\cos(jt) + b_{ij}\sin(jt))$$

It will sometimes be convenient to collect harmonic terms together (rather than powers of ε), and so we have also

$$y = \sum_j (a_j\cos(jt) + b_j\sin(jt))$$

where a_j and b_j are power series in ε.

In practice the situation described above is somewhat complicated for autonomous systems, where we have to write $\cos(j\omega t)$ and $\sin(j\omega t)$ in place of $\cos(jt)$ and $\sin(jt)$ in the above, and allow the frequency ω to be a power series in ε, typically with leading term of unity. We will use the symbol τ to represent stretched time, defined by $\tau=\omega t$.

The technique used to find the functions y_i is due to Lindstedt and Poincaré (8,9). In the context of computer algebra it is a variation of the repeated approximation theme presented by Barton and Fitch (10). We start by assuming that the zero-order solution y_0 is known. This is then substituted into the right hand side of (*) to give a residue r_0 (of order ε) as a series in sines and cosines. The equation

$$y'' + y = r_0$$

can now be solved to find y correct to the first order in ε. By repeated use of this idea we develop an iterative method which obtains the i+1st order solution for y from the residue r_i. If this iteration was all that had to be implemented our task would be trivial. There are, however, four additional points that require attention:

(1) The general solution to $y'' + y = r_i$ will involve a complementary function containing two arbitrary constants. Some way of determining the values of these constants must be found.

(2) If the residue r_i were to involve $\sin(t)$ or $\cos(t)$ the related solution for y would contain a term $t\sin(t)$ or $t\cos(t)$, which would correspond to an unwanted non-periodic (or secular) component in the full solution.

(3) For autonomous systems we will need to develop an approximation for the period of the oscillation.

(4) In an autonomous system the transformation $t \to t + c$, for any constant c, leaves the differential equation unchanged. This fact should be reflected in the form of solution we obtain.

It is clear that the first and third introduce degrees of freedom which the second and fourth must eliminate. For different equations the correspondence between the place where freedom is introduced and the place where it is removed can vary; it is not, for instance, the case that the coefficient of the cos(t) term in the complementary function can always be determined by examination of the sin(t) resonance term.

An Interpretive Program

The first version of our system was a CAMAL (11) program, Alkahest I. The major feature of this program was the way in which it handled the secularity constraints. Various algebraic symbols were reserved to represent constants introduced into the solution but not yet given a value. A set of subroutines and tables kept track of which symbols were in use, and made it possible to free and then re-allocate them. Using these extra variables, the developing solution for y would normally be of the form:

$$y = \langle\text{known terms}\rangle + \alpha\cos(t) + \beta\sin(t)$$

with α and β represented explicitly as indeterminates.

When solving (at the next order in ε) the equation

$$y'' + y = r_i$$

resonance terms are found. These must be linear in some of the previously undetermined parameters. Since there is a table showing which free parameters exist, this can be checked. When a search has found which symbol can be eliminated, the newly discovered value must be substituted back into all expressions that might depend on it.

This program can be thought of as mimicking the action of a human mathematician in solving the equations. It uses place-holder variables to keep track of quantities that await evaluation, and in effect re-analyses the problem at every step of the work.

The main problem with the initial system is, of course, that it is clumsy and not very efficient. The place-holder variables and substitutions in terms of which it works represent good technique in hand calculation but are a needless expense when the algebra is to be done by machine. Experience has shown that when high order perturbation expansions are required even specialised algebra systems may need substantial amounts of time and store, and so it is worthwhile trying to produce good programs for them.

Transition to a Compiler

To gain the flexibility we needed, we replaced the stand-alone CAMAL program by a package, written in LISP, that could generate CAMAL programs. The very first version of this always printed out, line for line, the CAMAL form of Alkahest I. Not surprisingly this version was working very rapidly. It was then easy to arrange for some of the more static tests in the code to be superceded by the use of compile time tests and selective code generation.

We felt, however, that at this stage it was important to break away from the low level details of CAMAL programming; Alkahest II had to be concerned with synthesising an algorithm rather than transcribing a program. To move in that direction the code to print CAMAL statements was replaced by more abstract constructions, reflecting a high level view of the algebraic operations that have to be performed. These call on a set of detailed translation routines that write out a program to perform the operations. Separate sets of translation routines are provided so that the generated program can be in a variety of languages. We have implemented three such sets, one for CAMAL, one for REDUCE (12), and one for a private system that exists just as a set of FORTRAN callable subroutines.

Automatic Problem Analysis

The package described above had two substantial drawbacks. The first of these involved the amount of detailed guidance it needed from its user: even for the very restrictive class of problems we consider it needed half a dozen flags or variables set before it could get started. These included information as to whether time-stretching should be used, and what sort of boundary conditions (eg. phase and amplitude constraints) could be applied, as well as the more obvious details concerning initial approximations to the frequency and form of the solution. When considering a new problem we often found ourselves finding out how to set these flags by experimentation, and wondering how certain we could be that the program would fail gracefully when they were wrong.

The other problem concerns efficiency directly, and reliability indirectly: it is the use of place-holder variables, and in particular the way in which we introduce them indiscriminately and later organise a search to see which ones can be eliminated. Quite apart from any cost there might be in an algebra system organising its data structures to allow for extra indeterminates, searching expressions for them and then back substituting for their values is clumsy and time consuming.

The solution to both of these problems involved making our package do some analysis before starting to generate code. Here we will concentrate on showing how the analysis can lead to the removal of all place-holder variables, all run time searching and all run time substitutions. Implicit in all this is a study of each problem that is, at first, detailed enough to provide a good check that the user's specification of flags is correct, and which can later take over full responsibility for much detailed control.

The form our analysis takes at present is adequate for the major families of equation we consider. Its current limitations seem to be closely bound up with the problems of working with coupled oscillators and with multiple time scales, and so will be noted but not explored further at this stage. To generate an algorithm for solving some particular equation, we will prove, by construction, the inductive step that shows a solution to order n+1 can be derived from one to order n. Since an initial solution will have been provided by the user, we will then be ready to generate solutions to any order. The important part of the inductive step involves showing that there will always be a way of avoiding resonance in the equation, and the consequent introduction of secular terms in the solution. We are thus led to suppose we have a solution to order n of the form

$$y = \hat{y} + \alpha\varepsilon^n\cos(\tau) + \beta\varepsilon^n\sin(\tau)$$

$$\omega = \hat{\omega} + \gamma\varepsilon^{n+1}$$

where τ is ωt, and α,β and γ are still to be determined. In certain cases ω remains 1, that is all values of γ are zero, and there is no difference between t and τ. In general these values of y and ω can be substituted into the right hand side of the differential equations, and will give a residue r_n. This residue will usually involve terms in $\sin(\tau)$ and $\cos(\tau)$ and we will have to choose α, β and γ to make these vanish. The heart of a proof, and the key to generating efficient algorithm involves tracing how the $\sin(\tau)$ and $\cos(\tau)$ terms depend on α, β and γ. This must be done without detailed knowledge of the rest of the developing solution, but can (and will) depend on the initial approximation y_0.

Our program does this using what is in effect a special purpose algebra subsystem. It treats all variables as power series in ε, and so considers y as

$$y = y_0 + \varepsilon y_1 + \dots + \varepsilon^n y_n$$

The coefficients in these series are then each split up to show their dependencies on the parameters: an item with a hat represents one not depending on any of α, β or γ. Thus we have, for instance

$$y_0 = \hat{y}_0$$

$$y_n = \hat{y}_n + \alpha\cos(\tau) + \beta\sin(\tau)$$

$$\omega_{n+1} = \gamma$$

We now use this algebraic subsystem to evaluate the coefficients of ε^{n+1} in the right hand side of the equation. Polynomial manipulation is done using the special rules

$$(a \pm b)_n = a_n \pm b_n$$

$$(a\ .\ b)_n = a_0 b_n + a_n b_0$$

$$(\varepsilon a)_n = a_{n-1}$$

where in the product rule we have been able to abandon almost all of the cross terms because they can never contribute to any dependence on α or β.

As an example we will work through the application of these rules to the expression $\varepsilon y'(1-y^2)$ which appears in the van der Pol equation. We obtain the sequence of transformations:

$$(\varepsilon y'(1-y^2))_{n+1}$$

$$\to \quad (y'(1-y^2))_n$$

$$\to \quad y'_0(1-y^2)_n + y'_n(1-y^2)_0$$

$$\to \quad -2y'_0 y_0 y_n + y'_n(1-y^2_0)$$

$$\to \quad -2y'_0 y_0(\alpha\cos(\tau) + \beta\sin(\tau)) + (-\alpha\sin(\tau) + \beta\cos(\tau))(1-y^2_0)$$

where all terms independent of α or β have been discarded. In this case we would then substitute in the known value for y_0 which is 2 cos(τ) and then linearise the products of sines and cosines.

The resulting expression then shows explicitly how the free parameters introduced at one step in the solution affect the results obtained at the next. For the van der Pol equation we find that when the variable frequency is allowed for the coefficient of the sine resonance term will have the forms $\hat{s} + 2$ and the cosine one will be $\hat{c} + 4$. Thus it will always be possible to give values to α and γ in order to cancel the unpredictable values $\hat{s}$ and $\hat{c}$. Furthermore the precise form of the equations that have to be solved have been found, in that the coefficients of α and γ are constant expressions. The parameter β, which is not determined by secularity considerations, will have to be fixed by applying an arbitrary phase constraint to the solution: $y'(0) = 0$ is standard and convenient.

This has shown how α and γ can be found, given values for $\hat{s}$ and $\hat{c}$ in our example. It is however easy to find these since they are just the coefficients of sin(τ) and cos(τ) that appear if we substitute $\alpha = \gamma = 0$ into the right hand side of the equation. Rather than introduce symbols α and γ and later substituting zero values for them we can of course just leave them out of the computation. If we do this, it will be necessary to correct the calculations when α and γ are finally given their true values. Fortunately the analysis that shows how resonance terms are affected can also be used to show how other quantities depend on the indeterminates.

From this example it is easily seen that other equations can be treated in a similar fashion, with different coefficients for the constants and different distribution of constants in the equations.

The Generated Program

Figures 1 and 2 show fragments of the CAMAL and REDUCE programs that our system generates when asked to solve the van der Pol equation. Its analysis of the problem is as described above, and the constants 2 and 4 we found can be seen on the lines marked (1) and (2). The code to adjust the calculated value of y to allow for a non-zero α is marked (3).

It can be seen that CAMAL and REDUCE call for substantially different code generation. In CAMAL the equation $y'' + y = r$ is solved by using operators HEXPAND, HPARSE and PCOEFFT to split r into terms each of which can be integrated easily. In REDUCE it is slightly more complicated to separate out the resonance terms, but most of the work of solving the equation can be done using pattern matching, expressed as LET statements in the program. In each case the program involves features that could confuse straightforward users of an algebra system. In CAMAL full use has been made of the editing routines. For instance if X is a variable holding a harmonic term p(a,b)sin(nt), after the call HPARSE(X,I[0]) the array element I[1+'t'] will contain the number n. In REDUCE we have to set, and later restore, the flag RESUBS to make the pattern work as required.

Since we generate the programs automatically, we only have to consider these technical points once in the compiler rather than once for each problem to be solved. We are thus in a position to seek the assistance of the designers of particular algebra systems, and to try to incorporate into our package their ideas about how their system should be used.

Problems Investigated

The algebra that our package has to perform when analysing problems is generally orders of magnitude smaller than that involved when running the generated programs. Long and complicated-looking expressions in an equation are therefore no problem to us. Given this, the current capabilities and limitations of our system can be illustrated by giving simple examples of the classes of equations it can handle. In these examples it is appropriate to note how apparently minor changes in the form of algebraic expressions can change the category of a problem. and thus result in the need for a noticeably different solution method.

1. $y'' + y = \varepsilon y^3$ This autonomous version of the Duffing equation is conservative, and so has a periodic solution of any given amplitude a. The frequency of this solution will depend on a and ε. This equation is discussed at length elsewhere (13).

2. $y'' + y = \varepsilon(y')^3$ Our system can determine that this equation does not have any periodic solutions. The $\varepsilon(y')^3$ term is a damping one, and there is nothing to counteract it.

3. $y'' + y = \varepsilon y'(1-y^2)$ This is the van der Pol equation, the classic example of an equation with a limit cycle. By using more complicated functions on the right hand side it is easy to construct systems which have several distinct limit cycles: the initial approximation y_0 can be used to define which one is to be found. The analysis of this problem is as given above.

4. $y'' + y = \varepsilon y^3 + \cos(2t)$ If a forcing term is added to any of the above equations the response

of the system may lock into phase with it. If this is possible our system will detect this and solve the equation.

5. $y'' + y = \varepsilon y^3 + \varepsilon^{50} y'(1-y^2)$ The system described here fails to cope with this example, which was constructed to show that certain problems can need indefinite amounts of analysis before they can be solved completely. The difficulty here is that the εy^3 term (which represents a conservative system) masks the ε^{50} term, even though the damping term eventually dominates. If the ε and ε^{50} in this example change places our system copes with it quite happily.

6. $y'' + y = \varepsilon(y^2 + \cos(t))$ At first sight the cos(t) forcing term in this example leads to a resonance. However, because the forcing function is small this resonance interferes with the non-linear behaviour of the equation. To model the resulting behaviour it is necessary to work in terms of multiple time scales, which is beyond the scope of the present phase of our work.

The user interface for our system is a mini language consisting of keywords followed by flag values, numbers or algebraic expressions. Thus the complete input to our package for the first example above is

```
TITLE AUTONOMOUS DUFFING VERSION 1
NONLINEARPART E*Y**3
ZEROAPPROXN A*COS(T)
```

together with an indication of what language should be generated and how high an order of solution is required. Of these three lines of data only the second is strictly necessary. All additional information needed to complete the solution can be discovered automatically. Problems of the above scale take a second or so to process on an IBM370/165: our package is coded in LISP and runs in 350K bytes.

Conclusions

The problem we have considered is a classical one, and has a substantial literature, both in application areas and as a branch of mathematics. Our concern has been to build a tool that can take reasonably well behaved problems and apply standard techniques to solve them. This has led us into the field of problem analysis and classification: textbooks tend to describe methods and classes of problem to which they apply, but often leave the inverse problem of choosing a method given a problem to some combination of intuition and trial and error.

We believe that our work has relevance in computer algebra as well as in the field of differential equations. It demonstrates that relatively small scale algebra in a preprocessor can enhance the efficiency of later large scale algebraic computations. In this way it points to the possibility of optimising compilers for algebra systems; the classes of transformation relevant in these would be very different from those used in optimising compilers for, for example, BLISS or FORTRAN. In that we can generate programs in several languages we claim that our package constructs algorithms rather than just code. This seems useful given the current gross differences between the languages accepted by existing algebra systems.

Even though the analysis we do seems small when measured against the calculations done by some algebra systems, it is more than elaborate enough to give a human opportunity to lose a term or drop a sign. Thus our preprocessor, by accepting a compact and explicit description of a problem, should increase the reliability of the solution finally obtained. This point is strengthened by the fact that our system, both at compile and run time, repeatedly checks for anomalies and difficulties. We have already mentioned that we can tune the code that we generate to the system on which it will run. With typical algebra systems this can make dramatic difference to space and time utilisation.

The indications from mathematically biased papers is that we should never expect to solve all perturbation problems mechanically. As each new class is brought under control, a new set of unpleasant cases is uncovered. In this first step we have dealt with a range of problems which, while at first sight severely restricted, are of substantial physical relevance. We have also put ourselves in a position to attack the next class up, where mechanised analysis seems yet more important. While doing this we have tried to move beyond the stage where a machine acts purely as an assistant performing straightforward manipulation of symbols; we want to let the computer loose on some mathematics.

References

1. Delaunay, C (1860) "Théorie du Mouvement de la Lune" Gauthier-Villars, Paris. Vol.1.

2. Deprit, A; Henrard, J and Rom, A (1970) "Lunar Ephemeris: Delaunay's Theory" Astronomical Journal 75 p747-50

3. Deprit, A and Rom, A(1970) "The Main Problem of Artificial Satellite Theory for Small and Moderate Eccentricities" Celestial Mechanics 2 p166-206

4. Jeffreys, W.H (1970) "An Automated List Processor for Poisson Series" Celestial Mechanics 2 p474-80

5. Nayfeh, A.A (1973) "Perturbation Methods" John Wiley & Sons

6. Liénard, A(1928) "Etude des Oscillations Entretenue" Revue Générale de l'Electricité p901

7. Nadeau, A; Guyard, J and Feix, M R (1974) "Algebraic-Numerical Method for the Slightly-perturbed Harmonic Oscillator" Math Comp 28 p1057 - 66

8. Lindstedt, A (1882) "Ueber die Integration einer für die Strorungstheorie wichtigen Differentialgleichung" Astron Nach 103 col 211-20

9. Poincare, H (1893) "Les Méthodes nouvelles de la mécanique céleste" Vol 2 Gauthier-Villars, Paris (English Translation: NASA TTF-450 1967)

10. Barton, D and Fitch J.P (1972) "Applications of Algebraic Manipulation Programs in Physics" Report on Progress in Physics <u>35</u> p235-314

11. Fitch, J.P (1974) "CAMAL User's Manual" University of Cambridge Computer Laboratory

12. Hearn, A.C (1973) "REDUCE User's Manual" University of Utah UCP-19

13. Fitch, J.P (1975) "Course Notes" SIGSAM Bulletin <u>35</u> p4-8


```
         |***** INVERTOP *****
          | Now invert operator. In this case D^2+1
         F=0
         G=0
         Y=0
         1:
         ->2 IF Z=0
         HEXPAND(Z,Z[1])
         Z=Z[1]
         X=Z[0]
          | X is leading term
         HPARSE(X,I[0])
         ->3 IF 1-I[1+'t'].2~0
         ->4 IF I[0]~0
         F=PCOEFFT(X)
         ->1

         4:
         G=PCOEFFT(X)
         ->1

         3:
          | Now integrate
         Y=(X+Y-YI[1+'t'].2/(1-I[1+'t'].2)
         ->1

         2:
         TEXT:Secular sin term elimination - :
         PRINT(F)
         NEWLINE
(1)      V=(-F)/2
         PRINT(V)
(3)      Y=(-(3Vsin[3t]-8Y))/8
         B=(eB+Vcos[t])/e
         TEXT:Secular cos term elimination - :
         PRINT(G)
         NEWLINE
(2)      V=(-G)/4
         PRINT(V)
         C=C+V
```

Figure 1: Part of CAMAL program for van der Pol equation

```
         %***** INVERTOP *****;
          % Now invert operator. In this case D^2+1;
         SS:=0;
         SC:=0;
         Y:=SUB(SIN(T)=0,ZZ);
         SS:=SUB(SIN(T)=1,(-(Y-ZZ)));
         Y1:=Y;
         Y:=SUB(COS(T)=0,Y1);
         SC:=SUB(COS(T)=1,Y1-Y);
         OFF RESUBS;
         FOR ALL NN LET COS(NN*T)=COS(NN*T)/(1-NN**2);
         FOR ALL NN LET SIN(NN*T)=SIN(NN*T)/(1-NN**2);
         Y:=Y;
         OFF RESUBS;
         FOR ALL NN CLEAR COS(NN*T),SIN(NN*T);
         WRITE "Secular sin term noticed; Solve ";
         WRITE SS;
(1)      V:=(-SS)/2;
         WRITE V;
(3)      Y:=(-(3*V*SIN(3*T)-8*Y))/8;
         B:=(V*COS(T)+E*B)/E;
         WRITE "Secular cos term noticed; Solve ";
         WRITE SC;
(2)      V:=(-SC)/4;
         WRITE V;
         CC:=CC+V;
```

Figure 2: Part of REDUCE program for van der Pol equation

User-based Integration Software

JOHN FITCH, SCHOOL OF MATHEMATICS
UNIVERSITY OF BATH, ENGLAND

Introduction

Algorithms are not the same as user software; which is like saying that pure mathematics is not the same as its application. In earlier algebra conferences Risch and Norman (1976) and Norman and Davenport (1979) have described a new method for symbolic indefinite integration. This paper is based on the work described there, but considers the process of turning this important area of algorithm research into a module for the REDUCE algebra system, in a form that can be used by the regular REDUCE user in the same way that he would differentiate or perform substitutions. The software package that is described is the Utah version (REDUCE INT) of the original Norman and Moore (1976) implementation. Rather than describe the method again, the paper concentrates on the software engineering and human engineering aspects of the package, and changes to the procedure that may not be mathematically justified but provide a powerful tool to the end user.

Origins of the Package

The starting point of the REDUCE INT package was the pilot implementation of the new integration method of Risch, undertaken by Moore under the direction of Norman in the symbolic mode of REDUCE. This system was also the ancestor of the ISIS package (Norman, Moore and Davenport,1978), and the thesis system of Davenport (1981).

The first attempt in Utah to produce a system with user characteristics was by Harrington (1979), where a series of heuristic and pattern matching techniques were used to complement the inadequacies of the earliest version of the Cambridge code. The system described in this paper is not related to Harrington's system, except in ancestry. It was developed separately and in parallel, using the other system as a measure of success. The REDUCE package as presently distributed has no heuristic searching, and is only non algorithmic in the senses outlined below, under the heading of degree bound. The current package began as an attempt to modify the ISIS package for Standard LISP (Marti et al, 1978) from the initial Cambridge LISP (Fitch and Norman, 1977). In the course of this simple task a number of modifications were suggested, and so yet another symbolic integrator was born.

The Basic Method

The REDUCE INT package is derived from the new integration method presented by Risch at the SYMSAC 76 conference. In order to make the modifications and departures from the original method more clear a shortened version of the algorithm is presented here with a simple example. Further details can be found in Norman, Moore and Davenport (1978) or Fitch (1981). The method begins by identifying a set of functions which are independent of each other, written as exponential or logarithm functions. Thus for example the integration of

$$\frac{x^2+x+1}{x^3+2x^2+x} - \frac{1}{(x+1)\log^2(x+1)}$$

identifies log(x+1) and x as being independent. We call these z_1 and z_0 respectively, and placing the integrand over a common denominator we get

$$\frac{z_1{}^2 z_0{}^2+z_1{}^2 z_0+z_1{}^2-z_0{}^2-z_0}{z_1{}^2 z_0{}^3+2z_1{}^2 z_0{}^2+z_1{}^2 z_0}$$

The Risch generalisation of the Liouville theorem tells us that if we factorize the denominator to linear factors, then we can obtain the denominator of the integral by subtracting one from all powers except exponential factors which are left alone. In the case of our example the denominator is $z_0 z_1{}^2(z_0+1)^2$, and so the denominator of the integral is $z_1(z_0+1)$. The solution of the integration now has the form of a polynomial in the z variables divided by the denominator, plus a linear sum of logarithms of the linear factors. Our example thus has the solution

$$\frac{\sum u_{ij} z_0{}^i z_1{}^j}{z_1(z_0+1)} + a \log z_1 + b \log (z_0+1) + c \log z_0$$

The method proceeds by formally differentiating this form, and equating coefficients. In our example if we do this, and cancel the denominators we get

$$(1-b-c)z_1^2z_0^2+(1-b-2c)z_1^2z_0+(1-c)z_1^2-az_1z_0^2-az_1z_0$$

$$-z_0^2-z_0$$

$$=\sum u_{ij}\{(i-1)z_1^{j+1}z_0^{i+1}+(j-1)z_1^{j}z_0^{i+1}+iz_1^{j+1}z_0^{i}\} \quad (*)$$

We have to solve these equations for the constants a, b and c, and coefficients u_{ij}. The method of solution proposed by Norman is related to the repeated approximation method much used in series expansion, and turns out to be a generalization of integration by parts. Particular u_{ij} are chosen to explain the leading terms in the right hand side. This may add other terms, which are of a lower order. One can view this method as an interleaving of the generation of the equations with their solution. The constants a, b and c can be chosen to give a solution. If the equations cannot be solved then there is no closed solution.

We will return to this example later.

Adaption of Messages to the User

When building a research package it is usual practice to build in diagnostic aids. In many cases this takes the form of additional print statements that are only activated if a special switch is on. Such a system was in the original integration package. Of course the messages that it produced were related to the mathematical and algorithmic behaviour of the integral. For example in the algorithm outlined in the previous section, it is necessary to identify the linear factors of the denominator, and to construct the denominator of the answer. It is apparent to anyone well versed in computer algebra that this may involve a square free factorization, and a subsequent complete factorization. However, to the user, who just wishes to do an integral, these terms are of little or no interest. The ability to suppress all these messages is already available in the research tool, and for many purposes that is all that is requfred. However, if the integral does not succeed in some way, or the user is curious about the manner of production of the integral, a similar system of tracing is desirable; but it must be modified to relate to the terms of the integral.

This change was made to the INT package at the request of some users, who were interested in a general sort of way, as well as some whose interest lay deeper, and in particular in teaching some of the principles of the integrator. A side effect of this change was to demythologize the package, and has led in part to the large quantity of advice and suggestions that have marked this project. It was possible to watch the behaviour of recalcitrant integrals, and report them together with suggestions, and eventually in some cases with code for the improvement. It has been my experience in undertaking this work that this apparently unimportant change has had significant effect on the overall level of attainment of the package.

Changes in Algorithm

The message changes described in the previous section are largely cosmetic. Of more significance is the change that has been made in the fundamental method.

The mathematical background of the integration method is to be found in the theory of differential fields, and the definitions of the exponential and logarithm functions. The foundation papers on integration proceed by considering the integrations of expressions expressed in this way. The Norman-Moore implementation did depart from this by substituting arctangents for the logarithms of complex quantities, and by the subsequent allowance of the tangent function. To a large extent this is a renaming exercise. However, the normal form of expression that the user wishes to integrate involves a larger class of functions, including the sine and cosine functions. The normal solution to this is to transform the integrand into the equivalent exponential form, or in the case of the Norman-Moore code to the tangent of half angle form, and this transformation was the responsibility of the user. In the INT package we took a different approach.

If the decidability aspects of the Risch integration algorithms are ignored, then one can view it as a method. We can express the integrand in terms of any function, as long as we know how to differentiate them, and the z variables in which the numerator is expressed contain the differential closure of the set of functions. By this it is meant that if the integrand contains a sine function then we should consider that the answer may contain both sine and cosine. With this the basic method as outlined above can be applied. If this succeeds in obtaining an integral, then the problem has been solved; if this integration fails by finding a term that cannot be explained, then one cannot conclude that the integral is not elementary. This approach allows sin(x) to be integrated very quickly to cos(x) without the introduction of tan(x/2) or exp(ix), or by using pattern matching. Further, if REDUCE is told that

```
FOR ALL X DF(F(X),X) = 1 + F(X)**3
```

then the program can integrate $f(x)(1+f^3(x))$ to $f^2(x)$. This does not help for a number of problems that cannot be done by this trick. In this case the INT package makes the transformation of sine and cosine functions to the half angle form, and other similar transformations, and then calls the whole integration package recursively. This recursive call is of an expression that fits in the class treated by the mathematical theory, and so the program should provide a decision procedure.

In the event of the recursive integration giving an elementary integral, it is required to attempt to express the answer in an acceptable form. If a half angle transformation has been made, then in many cases it should be transformed back. But this is not always the case. Consider for example the integration of 1/sin(x). The first attempt to integrate will fail, and so the transformation system will give the integrator

$$\frac{1 + \tan^2(x/2)}{2 \tan(x/2)}$$

which can be integrated to log(tan(x/2)). While it is possible to express this as log(sin(x)/(1+cos(x))), this is more complex. Following the advice of Hearn such an adaptive transformation and untransformation system was written by Harrington, and incorporated in the package.

A similar problem is what to do when the integral is not elementary. From a theoretical point of view it is sufficient to determine this point. From the stand point of a user orientated package it is important to give an answer that is both formally correct and useable in subsequent calculations. This was done by many packages for symbolic integration, including ISIS and the SIN system from Moses (1971). However, a further stage is necessary. The integral must be expressed in a form that is as simple as possible. As the Risch-Norman process is a generalisation of integration by parts, it is to be hoped that the part of the integrand that cannot be integrated is smaller (in the sense of being of lower degree) than the original input. But when this simple principle is applied to the INT program with its recursive call, it is possible that what is simpler in the inner integral is not mapped back to a simple form in the answer class of functions. The package must be willing to deal with this eventuality, and check to see after the transformation back whether any residual part is more complex than the original problem. This problem is exacerbated by the modifications that were made to the degree bound, as described later.

There is one later stage to be applied to the part of an integral that cannot be done. The integration operator is linear, and so forms such as INT(2*SIN(X)/X,X) can be written as 2*INT(SIN(X)/X,X). This linearization allows for a representation that is nearer a canonical form, and so it will be possible to apply a pattern matcher afterwards to recognise at least simple occurrences of higher transcendental functions. This is done in INT, and is the only concession to heuristic methods.

The Degree Bound Problem

At times in this paper the Risch-Norman method has been referred to as an algorithm. This is only true if there is a degree bound on the numerator of the integral in the formal form of the solution. This bound has proved to be a point of difficulty with the method. In the early presentation (Risch & Norman, 1976) it was assumed that the integrand gave rise to a natural degree bound, and this was not explicit as the Norman solution method does not need it. It was subsequently discovered that there are integrals where this bound was incorrect. Considerable mathematical effort has been put into determining the correct form of the bound; Norman and Davenport (1979) treat this subject, and show that the problem arises with unforeseen and chance cancellations. The ISIS program with the phantom term machanism is capable of detecting the possibility of this cancellation, and hence in the case of a failed solution stating whether it is a decision procedure.

In a user environment all this mathematics can only confuse. Except for the cases where the required answer is only whether the integral exists in finite terms, all the user requires is an integrator that surpasses him by an order of magnitude, and is frequently happy with a formal integral if it differentiates correctly. For the REDUCE INT package it was decided as a matter of policy to provide a program that was <u>not</u> a decision procedure, but was a successful integration program. We were willing to do a number of things to reduce the likelihood of a misleading result even if the mathemtical basis of the method were not sound. One of the most important features of the Risch-Norman method is that it is based on the answer differentiating to the integrand, and so as was mentioned above with respect to the inclusion of non-elementary functions, any integral that is produced will be correct.

While we await a solution to the full degree bound problem, and will incorporate the concluding ideas as soon as we are able, in the mean time we have instituted a method that has no rigorous foundation, and no justification other than the fine performance. This is by two stages; a degree bound determined by a heuristic, and a variation on the method of solution of the linear equations for the coefficients of the numerator.

The degree bound used is in two parts. For each of the z variables take one higher than the largest power that occurs in the integrand. This is the natural degree bound, (3,3) in our example, which can be apparently justified, although it is sometimes too low. Then the right hand side of the equation (*) above is inspected for factors of the form (j-n) with n being an integer. If the n is larger than the bound for that variable (determined by j) obtained so far, then n is used. (In the example above (i-1) and (j-1) occur, but the (3,3) already obtained dominates). This value was chosen by observation of the integrals that give trouble, and is related to the phantom term method that Norman has been investigating in a more rigorous fashion. Rang (1979) has claimed to prove this bound is correct, but the proof is as yet non rigorous.

With this bound the Norman solution is relaxed to allow for the integration by parts to go on the 'wrong' direction, until the bound is reached. This does, of course, on occasion give rise to an excessive search for the non zero coefficients of the numerator. One can characterize the Norman phantom term method as looking for precisely those coefficients that are non zero beyond this natural bound, while the method used in INT is to put a larger box round the space of $u_{ij..}$ outside which we believe the coefficient to be all zero. This is sketched in figure 1.

With such a pragmatic and ad hoc method the only justification can be the performance. This has proved to be good. There are a number of integrals that this program can do, which fail on both ISIS and SIN. The main problem is that on occasions the computing time used is expensive, and further development is needed to limit this. We look forward to the time when the mathematical background will be sufficiently developed for these pragmatic methods to be abandoned.

Documentation

It is a truism that software without manuals and user documentation is of no value or use. The integration program has been provided with a short description of its use, with references to the open literature, and descriptions of the various flags that can be set. In addition there is a file of test examples that demonstrate the range of capabilities of the program.

However, this information is insufficient to generate confidence in a package that purports to be an answer to a problem that every user knows is difficult, and from which many users have been actively discouraged for a number of years. To answer this a set of four tutorial articles have been prepared, describing the background of the various integration methods, and giving enough information for a user to watch the trace output from the program and feel confident that he understands what is happening. We have already seen that when users understand this output they are capable of providing constructive criticism of the program, with suggestions for improvement. This is an area of help that we ignore at our peril.

The Algebraic Case and Higher Transcendental Functions

The program as currently working does not pretend to be capable of integrating algebraic extensions, although of course with the relaxed view taken of the allowed functions, provided that extra logarithm terms are not needed, it is likely that the program will be able to do some of these integrals. Work is in progress to incorporate parts of the Davenport integrator (Davenport, 1979) where at least the most difficult cases are not encountered. It is hoped that this will be done in the same pragmatic style as the rest of the package.

The other continuing problem in integration is the treatment of the higher transcendental functions. While we await a full mathematical treatment of this area, the INT program can deal with a limited number of these functions through the normal mechanisms. For example, as long as REDUCE is informed how to differentiate the dilogarithm function, INT is capable of integrating functions like x dilog(x). What it cannot do is recognise dilogarithms when they arise anew, and so far a variety of attempts to include this capability automatically have failed. The best that this package can offer at present is to use pattern matching after the integrator.

Conclusion

This paper has described the modifications made to a Risch-Norman integrator in the pursuit of a user package that can be distributed as widely as REDUCE itself. By a number of ways, relaxing the expression class, a fake degree bound and others, it has sought to provide the user with a package that can be used to perform integrations as part of other calculations. Already reports from diverse sites have indicated that user reaction is satisfactory, and while it is capable of improvement and the authors must never rest satisfied, it is fulfilling a need.

Apart from the original work of A C Norman and P M A Moore, without which this system could not have been begun, I would like to acknowledge the part played by A C Hearn, S J Harrington, J H Davenport, D Dahm and D Morrison, who have variously provided ideas, code and encouragement.

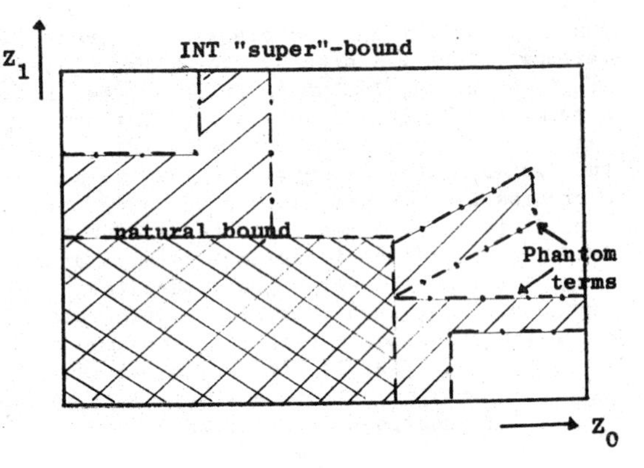

Figure 1: Diagram of Natural Bound (— — — — —) Phantom Bound(— · — · —) and Super Bound (————————)

References

J H Davenport, On the Integration of Algebraic Functions, Springer-Verlag Lecture Notes in Computer Science 102, 1981

J P Fitch, Integration Notes, REDUCE Newsletter, to appear

J P Fitch & A C Norman, Implementing LISP in a High Level Language, Software Practice & Experience, 7 713, 1977

S J Harrington, A New Symbolic Integration System in REDUCE, Computer Journal 22, 1979

J B Marti et al, Standard LISP Report, University of Utah, UCP-60, 1978

J Moses, Symbolic Integration, The Stormy Decade, Comm ACM 14 584, 1971

A C Norman & P M A Moore, Implementing the New Risch Integration Algorithm, Proc.4th Int.Coll on Advanced Computing Methods in Theoretical Physics, 1977

A C Norman, P M A Moore & J H Davenport, ISIS Method, Internal Report, 1978

A C Norman & J H Davenport, Symbolic Integration - The Dust Settles?, EUROSAM 79, Springer-Verlag Lecture Notes in Computer Science, 72 398 1979

H Rang, Private Communication, 1979

R Risch & A C Norman, A New Integration Algorithm, SYMSAC 2, Verbal Presentation, 1976

AUTHOR INDEX